Instructor's Solutions Manual with Transparency Masters *to Accompany*

DIGITAL DESIGNING WITH PROGRAMMABLE LOGIC DEVICES

JOHN W. CARTER
University of North Carolina, Charlotte

Prentice Hall
Upper Saddle River, New Jersey   Columbus, Ohio

Printed in the United States of America

10 9 8 7 6 5 4 3 2 1

ISBN: 0-13-378365-0

CONTENTS

Notes to the Instructor

The text for which this manual has been prepared was written to support an upper-division (Junior/Senior level) collegiate course in digital electronics. The course was designed to be part of a program in Electrical Engineering Technology, Computer Engineering Technology, Electrical Engineering, Computer Engineering, or Computer Science. This course would have a typical academic prerequisite of a course in digital electronic devices which includes the subjects of binary systems, combinational and sequential logic circuits. An introductory knowledge of microprocessors or microcontrollers would also be advantageous. The instructor should have a solid background in the use of programmable logic devices (ROMs, PALs, FPLAs) and the use of the MC68HC11 microcontroller.

With this background in mind, the intent of this course is to introduce the application of programmable logic devices into the design complexity spectrum between discrete logic circuits and microcontrollers. It is the thesis of the author that such programmable logic devices provide a cost-effective solution for the implementation of digital circuitry ranging from circuits with a few discrete devices, through relatively complex state machines. This latter circuit can be used in a wide variety of process control and monitoring applications. If the designer has a solid background in the application of programmable logic devices, the use of the more expensive and complex microcontroller or microprocessor may be avoided, saving a manufacture in development and production costs.

This course is designed to support a three or four semester-hour course with or without a laboratory. If a laboratory environment is to be applied, the practical Exercises in the text may be used as the laboratory assignments. The lab facilities will need to include prototyping and test equipment necessary to build discrete logic circuits, a variety of programmable logic devices (particularly useful are EPROMs and GALs), and the hardware/software system to program the devices. This text uses the CUPL programming language. The compiler for this language is available from Logical Devices, Inc. 692 S. Military Trail, Deerfield Beach, FL. 33442. A free starter kit is available in a student version. It contains all of the significant features of the CUPL language and can be used to implement most of the applications in this text.

Course Outline

The following course outline may be used in the preparation of lesson plans. It is designed for a 15-week semester with either 45 lectures (three 50-minute lectures per week) or 30 lectures (two 75-minute lectures per week.) Lecture numbers for each are indicated in the first two columns. The Practical Exercises are designed to be used as laboratory assignments. When there is no lab component to the course, these can be used as homework assignments.

50 75 Tentative Course Outline

50 min	75 min		Reference
1	1	**Background Review**	**Earlier Dig Texts**
2		**An Introduction to Programmable Logic Devices**	**Chapter 1**
		Sum of Products Expressions	
3	2	Implementation with Combinational Logic	
4		Implementation with Programmable Logic	
5	3	PAL Output Macrocells, Example PLD Architectures, PLD Programming	
		[Assign LAB 1: Practical Exercises 1.1, 1.2, 1.3, 1.4, 1.5]	
6		**Synchronous Binary Counter Design**	**Chapter 2**
		The Synchronous Counter, Excitation Logic, State Diagrams, State Tables	
7	4	Extracting SOP Expressions, Counter Construction	
8	5	Hazard Protection	
		[Assign LAB 2: Practical Exercise 2.1, 2.2, 2.3, 2.4]	
9	6	**Decision Making with Binary Counters**	**Chapter 3**
		State Machine Programming, Decision Making in State Tables	
10		Karnaugh Mapping Input Arguments, Design Method Summary	
11		Input Forming Logic Simplification	
		[Assign LAB 3: Practical Exercise 3.1, 3.2, 3.3]	
12	7	**Output Forming Logic Design**	**Chapter 4**
		Simple Output Forming Design, State Transition Definitions	
13		A Complete State Machine Design	
14	8	Timing Diagrams, Handling Multiple Inputs	
		[Assign LAB 4: Practical Exercise 4.1, 4.2, 4.3]	
15	9	Unconditional Output Forming Logic, Conditional Output Forming Logic	
		[Assign LAB 5: , Practical Exercise 4.4, , 4.6 or 4.7]	
16	10	**EXAM 1**	
17	11	**Programming PLD Circuits with CUPL**	**Chapter 5**
		Development Sequence, CUPL Program Structure and Elements	
18		Combinational Circuit Design, Intermediate Variables	
19	12	Sequential Circuit Design	
		[Assign LAB 6: Practical Exercise 5.1, 5.2, 5.3]	
20		CUPL Programming Environment	

min	min		Reference
21	13	**State Machine Design Examples**	**Chapter 6**
		Pattern Recognition, Sorting Objects in an Assembly Line	
22		Glitch-Free Three-Bit Up-Down Counter, PM Type Stepper-Motor Controller	
		Wait State Generation in an Imbedded Processor	
		[Assign LAB 7: Practical Exercise 5.4, 5.5 or 5.6]	
23	14	Priority Interrupt Encoder with Interrupt Mask	
		Programmable Monostable Multivibrator, Combinational Shifter	
24	15	Four-Bit Combinational Barrel Shifter, Eight-Bit Combinational Barrel Shifter	
		Digital Waveform Synthesizer/Generator	
		[Assign LAB 8: Practical Exercise 7.1, 7.2, 7.3 or 7.4 or 7.5]	
25		**ROM-Based Controller Architecture**	**Chapter 7**
26	16	Microprogrammed Design	
27		**Microcontroller Architecture**	**Chapter 8**
		The Microprocessor as a State Machine, Programming the Processor	
		[Assign LAB 9: Practical Exercise 7.6, 7.7, 7.8, 7.9]	
28-30	17-19	Microprogramming Primer, Microcontrollers	
31	20	**EXAM 2**	
		[Assign LAB 10: Practical Exercise 8.1, 8.2, 8.3, 8.4, 8.5, 8.6, 8.7, 8.8, 8.9]	
32	21	**The M68HC11 Microcontroller**	**Chapter 9**
		General Description	
33	22	Instruction Formats, Addressing Modes	
34		Instruction Set	
35	23	Port A: Digital I/O, Pulse Accumulator and Timer Functions	
36	24	Port B: Digital Output Port	
37		Port C: Digital I/O	
38	25	Port D: Digital I/O and Serial Communications	
39	26	Port E: Digital Input and Analog Output	
40-44	27-29	**Address Bus Decoding and Logic Design**	**Chapter 10**
		Address Bus Decoding	
		Circuit Design Methodology	
		[Assign LAB 11: Practical Exercise 10.1, 10.2]	
		[Assign LAB 12: Practical Exercise 10.3, 10.4]	
45	30	**EXAM 3**	

ERRATA (Errors in published text, first edition.)

Page 109 Figure 4.11 Low-assertion circuit uses a **NAND** logic gate at its output.

Page 300 Figure 9.2. **SCI Labels** Reversed

Page 304 "If manual RESET is desired, the **MC34164** may be used ..."

Page 305 Figure 9.6 Label MISI should be **MOSI**

Page 308 Figure 9.8 Label POIC should be **PIOC**

Page 308 "In this mode, the **Port B** and **Port C** input/output functions are lost."

Page 311 Figure 9.10 Label MISI should be **MOSI**

Page 312 "Table 9.4 illustrates the instruction **set** for the MC68HC11 microcontroller."

Page 317 Program Line 20: Instruction should be **ABA**.

Page 319 Program Line 32: **BCLR** TMSK2,X,**$20** ;PAOVI=0; Disable overflow int.

Page 319 Program Line 44: **BCLR** TMSK2,X,**$20** ;PAOVI=0; Disable overflow int.

Page 326 Figure 9.16 Label POIC should be **PIOC**

Page 327 Program Line 11: "... sets bit **2** of Port B ..."

Page 327 Program Line 13: "... ;Port **B[2]** = 1"

Page 328 Figure 9.17 Label POIC should be **PIOC**

Page 329 Program Line 14: "... sets bit **2** of Port **C** to a logical 1."

Page 329 Program Line 17: "... ;Port **C[2]** = 1"

Page 329 Program Line 22: "... ;Get Port **C** data"

Page 332 Table 9.8: 4.912 MHz should be **4.9152 MHz**.

Page 332 Table 9.8: 3.864 MHz should be **3.6864 MHz**.

Page 336 Program Line 15: LDAA **#$23**

Page 336 Program Line 16, 17, 19, 36, 38: Use **Indexed X** addressing.

Page 337 Program Line 19, 20, 22, 43: Use **Indexed X** addressing.

Page 357 "ACIA enable = ... **!A2**" (Add operator to summation.)

Page 366 Figure 10.22 and FIGURE 10.23: **Remove second PIA**.

Page 367 CUPL Program Line 5: "RAM: **2Kx8**"

Practical Exercise 1.1: SOP Expression Conversions

Consider the following SOP expressions:

$$W = f(A,B,C) = \overline{A}\,\overline{B}\,\overline{C} + \overline{A}\,\overline{B}C + A\overline{B}\,\overline{C} + A\overline{B}C$$

$$X = f(A,B,C,D) = \overline{A}B + A\overline{B}C\overline{D} + \overline{B}CD$$

$$Y = f(A,B,C) = A\overline{B} + \overline{B}C$$

1. Convert each expression to its equivalent typographic form.

$$W = f(A,B,C) = (!A\&!B\&!C)\#(!A\&!B\&C)\#(A\&!B\&!C)\#(A\&!B\&C)$$

$$X = f(A,B,C,D) = (!A\&B)\#(A\&!B\&C\&!D)\#(!B\&C\&D)$$

$$Y = f(A,B,C) = (A\&!B)\&(!B\&C)$$

2. Use Boolean algebra to convert each expression to its equivalent summation form.

$$W = \sum(0,1,4,5)$$

$$X = \sum(3,4,5,6,7,10,11)$$

$$Y = \sum(1,4,5)$$

3. Use Karnaugh Maps to convert each expression to its equivalent summation form.

$$W = \sum(0,1,4,5)$$

$$X = \sum(3,4,5,6,7,10,11)$$

$$Y = \sum(1,4,5)$$

Consider the following SOP expressions:

$$T = f(A,B,C) = \sum(0,3,4,5)$$

$$U = f(A,B,C,D) = \sum(0,1,2,8,9,11,14,15)$$

$$V = f(A,B,C) = \sum(1,3,4,6,7)$$

1. Convert each expression to its equivalent conventional form.

$$T = f(A,B,C) = \overline{ABC} + \overline{A}BC + A\overline{BC} + A\overline{B}C$$

$$U = f(A,B,C,D) = \overline{A}\,\overline{B}\,\overline{C}\,\overline{D} + \overline{A}\,\overline{B}\,\overline{C}D + \overline{A}\,\overline{B}C\overline{D} + A\overline{B}\,\overline{C}\,\overline{D} + A\overline{B}\,\overline{C}D + A\overline{B}CD + ABC\overline{D} + ABCD$$

$$V = f(A,B,C,D) = \overline{A}\,\overline{B}C + \overline{A}BC + A\overline{B}\,\overline{C} + AB\overline{C} + ABC$$

2. Reduce each expression to its simplest SOP form.

$$T = f(A,B,C) = \overline{B}\,\overline{C} + \overline{A}BC + A\overline{B}$$

$$U = f(A,B,C,D) = \overline{B}\,\overline{C} + \overline{A}\,\overline{B}\,\overline{D} + ABC + \begin{Bmatrix} ACD \\ A\overline{B}D \end{Bmatrix}$$

$$V = f(A,B,C,D) = \overline{A}C + \overline{A}C + \begin{Bmatrix} AB \\ BC \end{Bmatrix}$$

Practical Exercise 1.2: Multiplexer Implementation of SOP Expressions.

Draw the schematic diagram of a digital circuit which uses multiplexers to realize each of the following:

$$W = f(A,B,C) = \overline{A}\,\overline{B}\,\overline{C} + \overline{A}\,\overline{B}C + A\overline{B}\,\overline{C} + A\overline{B}C$$

$$X = f(A,B,C,D) = \overline{A}B + A\overline{B}C\overline{D} + \overline{B}CD$$

$$Y = f(A,B,C) = A\overline{B} + \overline{B}C$$

First, convert each expression to its equivalent summation form.

$$W = \sum(0,1,4,5)$$

$$X = \sum(3,4,5,6,7,10,11)$$

$$Y = \sum(1,4,5)$$

Then, implement using multiplexers:

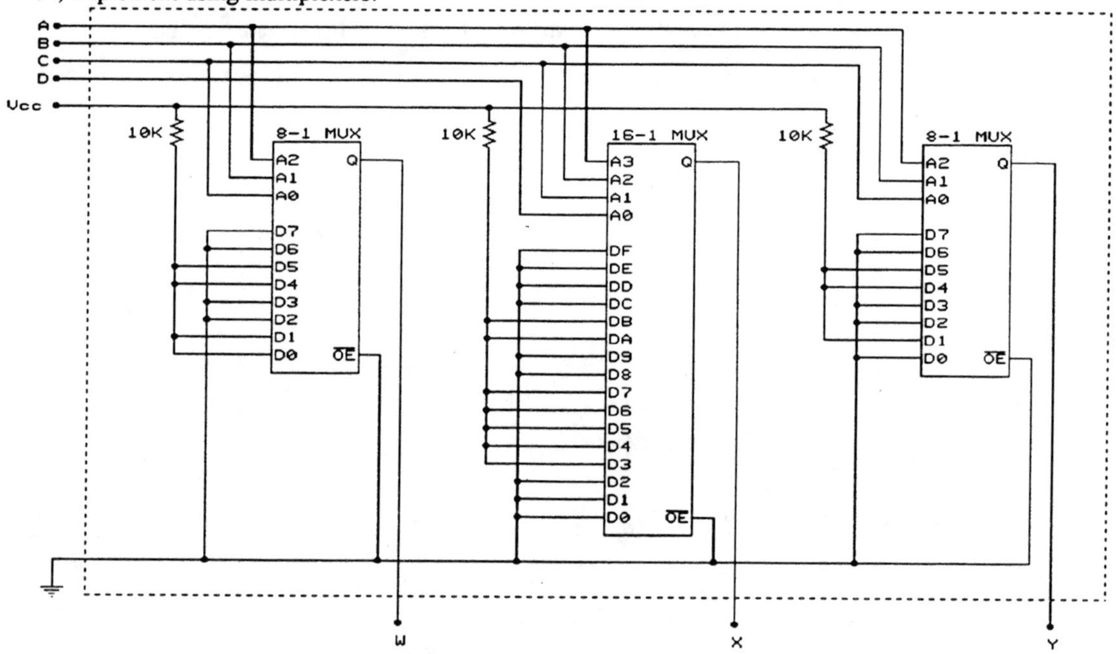

Practical Exercise 1.3: ROM Implementation of SOP Functions.

Draw the schematic diagram of a ROM implementation of the following functions:

$$W = \overline{A}\,\overline{C} + A\overline{C}\,\overline{D} + A\overline{B}D$$

$$X = A\overline{C} + BD + AB + \overline{A}\,\overline{B}C\overline{D}$$

$$Y = \overline{C}\,\overline{D} + \overline{A}B + B\overline{C} + A\overline{B}D$$

$$Z = \overline{C}D + AB\overline{C} + \overline{A}B\overline{D} + A\overline{B}C\overline{D}$$

Convert each expression to summation form:

$$W = f(A,B,C,D) = \sum(0,1,4,5,8,9,11,12)$$

$$X = f(A,B,C,D) = \sum(2,5,7,8,9,12,13,14,15)$$

$$Y = f(A,B,C,D) = \sum(0,4,5,6,7,8,9,11,12,13)$$

$$Z = f(A,B,C,D) = \sum(1,4,5,6,9,10,12,13)$$

Then, combine the true minterms for each ROM output on the schematic:

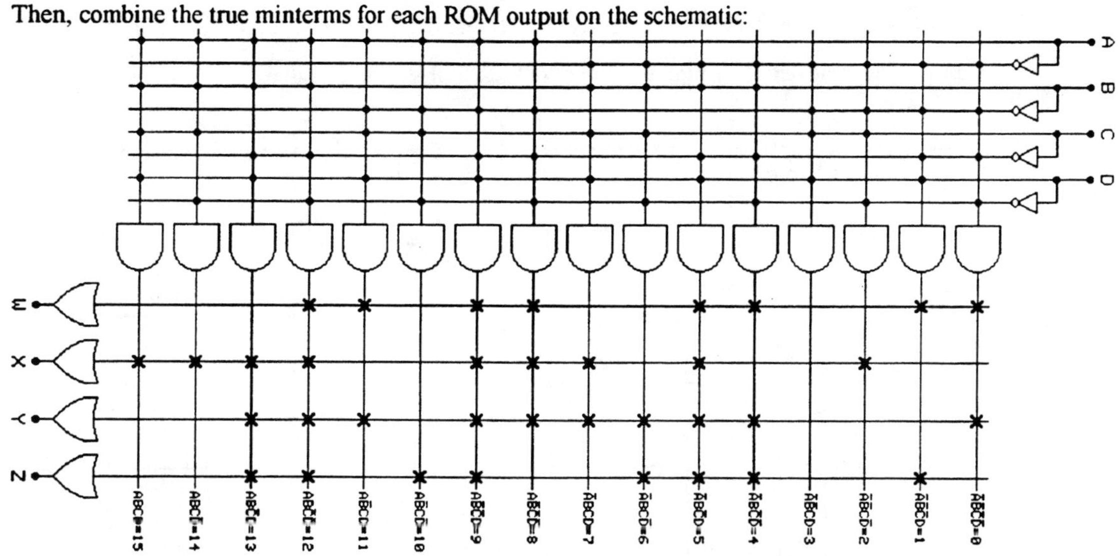

Practical Exercise 1.4: PAL Implementation of SOP Functions.

Consider the following functions:

$$W = \overline{A}\,\overline{C} + A\overline{C}\,\overline{D} + A\overline{B}D$$

$$X = A\overline{C} + BD + AB + \overline{A}\,\overline{B}C\overline{D}$$

$$Y = \overline{C}\,\overline{D} + \overline{A}B + B\overline{C} + A\overline{B}D$$

$$Z = \overline{C}D + AB\overline{C} + \overline{A}B\overline{D} + A\overline{B}C\overline{D}$$

Using a PAL with summates four minterms, we obtain the following schematic diagram:

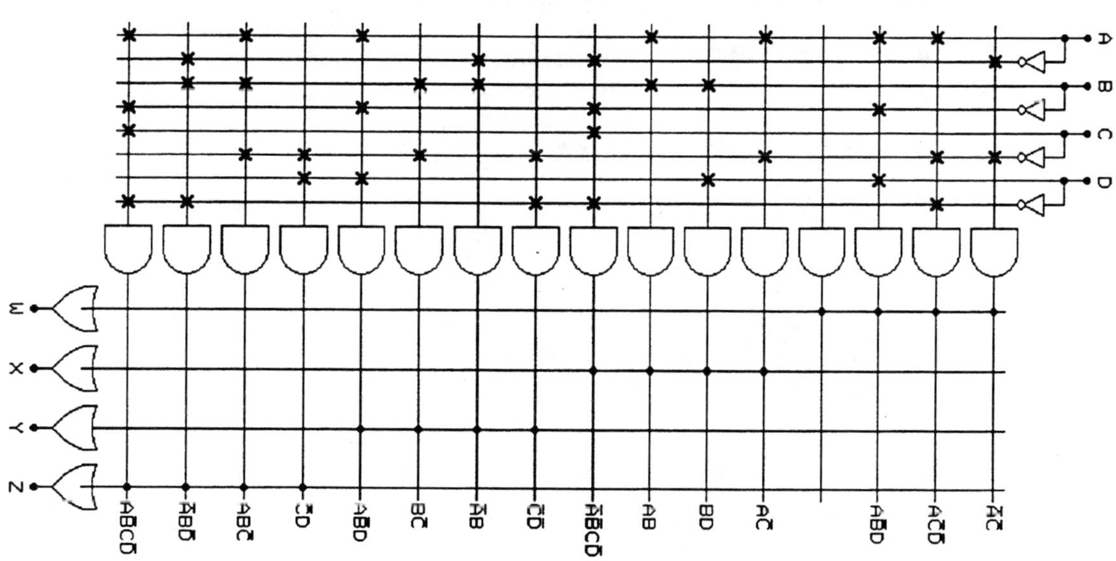

Practical Exercise 1.5: BCD to Seven-Segment Code Converter.

Connect the indicated LED segment output to each four-bit input pattern which is designated to enable it:

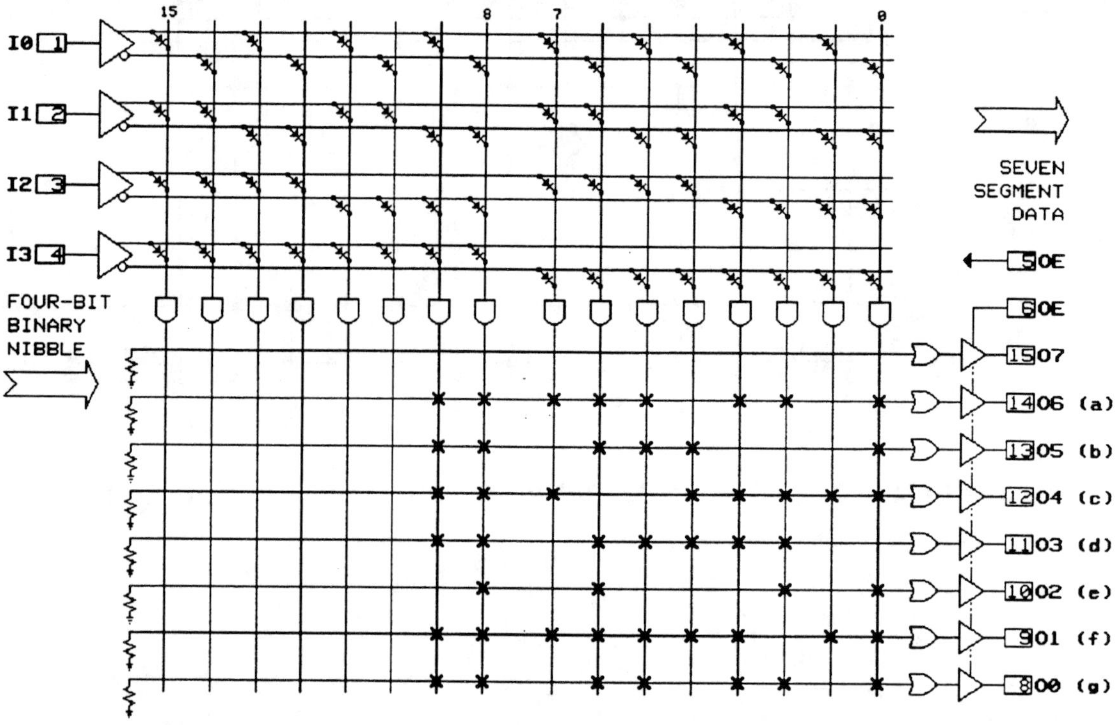

FIGURE 1.32 BCD to seven-segment code definitions and PROM.

Practical Exercise 2.1: JK-type Synchronous Binary Counter

Using the method described in this chapter, redesign the problem in Practical Example 2.2 using JK-type flip flops. The following is the state diagram of the problem.

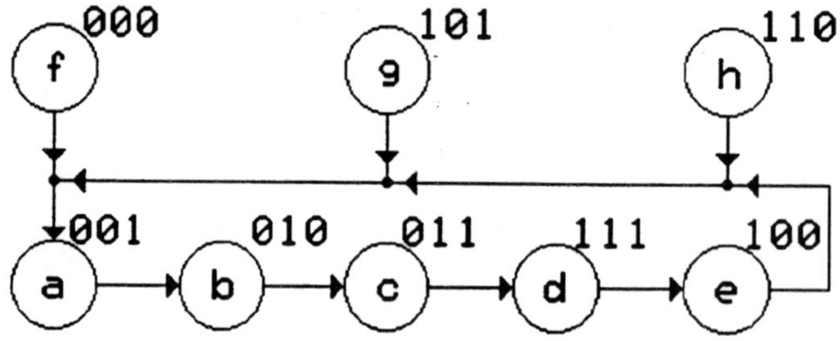

First, the state table is created from the above state diagram. This table defines the logic functions needed for the J and K inputs of each respective flip-flop. The state table and resulting logic functions are indicated below.

PRESENT STATE ABC	NEXT STATE ABC	Input Forming Logic					
		J_A	K_A	J_B	K_B	J_C	K_C
000	001	0	X	0	X	1	X
001	010	0	X	1	X	X	1
010	011	0	X	X	0	1	X
011	111	1	X	X	0	X	0
100	001	X	1	0	X	1	X
101	001	X	1	0	X	X	0
110	001	X	1	X	1	1	X
111	100	X	1	X	1	X	1

$J_A = BC$

$K_A = \bar{B} + \bar{C}$

$J_B = \bar{A}C$

$K_B = A$

$J_C = 1$

$K_C = AB + \bar{A}\bar{B}$

Practical Exercise 2.2: SR-type Synchronous Binary Counter

Using the method described in this chapter, redesign the problem in Practical Example 2.2 using SR-type flip flops. The following is the state diagram of the problem.

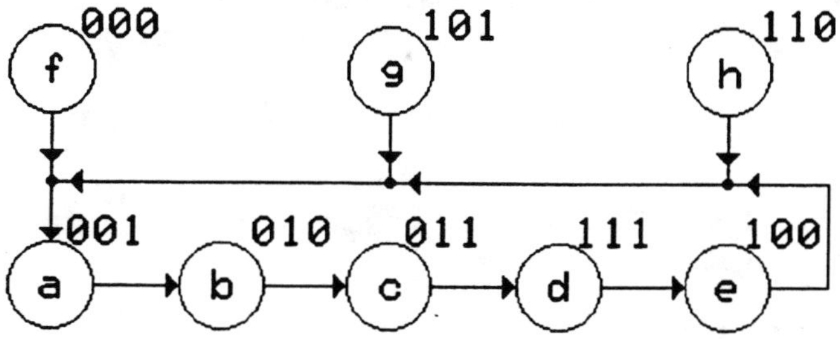

First, the state table is created from the above state diagram. This table defines the logic functions needed for the S and R inputs of each respective flip-flop. The state table and resulting logic functions are indicated below.

PRESENT STATE ABC	NEXT STATE ABC	Input Forming Logic					
		S_A	R_A	S_B	R_B	S_C	R_C
000	001	0	X	0	X	1	0
001	010	0	X	1	0	0	1
010	011	0	X	X	0	1	0
011	111	1	0	X	0	0	0
100	001	0	1	0	X	1	0
101	001	0	1	0	X	X	0
110	001	0	1	0	1	0	X
111	100	X	0	0	1	X	0

$S_A = BC$

$R_A = \bar{B} + \bar{C}$

$S_B = \bar{A}C$

$R_B = A$

$S_C = \bar{C}$

$R_C = ABC + \bar{A}\bar{B}C$

Practical Exercise 2.3: Glitch-Free Counter Design

Considering the solution of Practical Example 2.3, use the same method to produce four other sequences which are glitch free. Start each sequence at *State-0*. Any sequence which, when displayed on a next-state K-map, which consists of a closed loop of six states is a correct solution. There are numerous correct solutions. Below are a few examples:

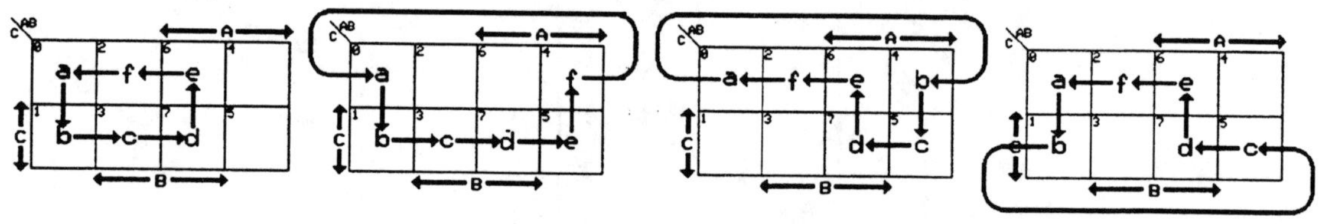

Practical Exercise 2.4: Decision Making in Glitch-Free Counters.

Using the same problem definition as Practical Example 2.4, determine at least two other correct solutions.

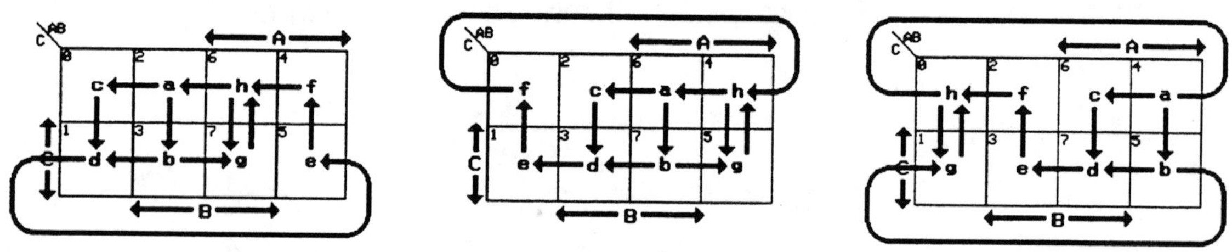

Practical Exercise 3.1: Creating a State Diagram from a State Machine Program

Consider the following state machine program:

```
SEQUENCE [ ] {
          PRESENT a    NEXT b;
          PRESENT b    IF R NEXT e;    DEFAULT NEXT c;
          PRESENT c    NEXT d;
          PRESENT d    NEXT c;
          PRESENT e    IF R NEXT f;    DEFAULT NEXT k;
          PRESENT f    NEXT g;
          PRESENT g    NEXT h;
          PRESENT h    IF R NEXT a;    DEFAULT NEXT i;
          PRESENT i    NEXT j;
          PRESENT j    NEXT c;
          PRESENT k    NEXT g;
          }
```

Task: Use this program to (1) generate a state diagram, (2) assign numeric values to each state in the counter disallowing glitches, (3) rewrite the program and state diagram using the assigned numeric values for the states, and (4) draw a block schematic diagram.

Shown below is the next-state Karnaugh map used to design a glitch-free sequence and the resulting state machine program. Since there are numerous correct glitch-free count sequences, there are numerous correct solutions.

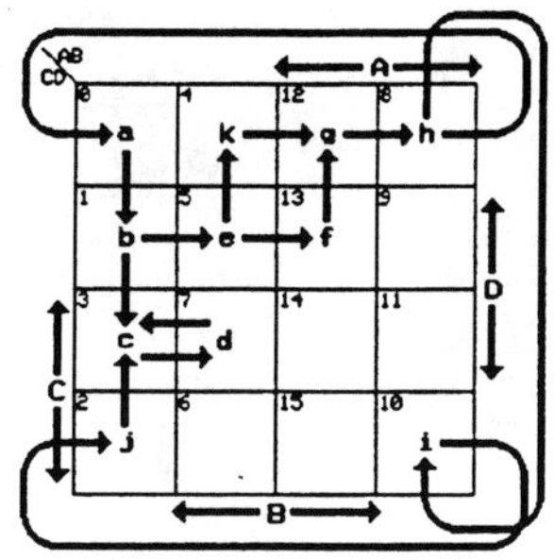

```
SEQUENCE [A,B,C,D] {
        PRESENT 'b'0000      NEXT 'b'0001;
        PRESENT 'b'0001      IF R NEXT 'b'0101;    DEFAULT NEXT 'b'0011;
        PRESENT 'b'0011      NEXT 'b'0111;
        PRESENT 'b'0111      NEXT 'b'0011;
        PRESENT 'b'0101      IF R NEXT 'b'1101;    DEFAULT NEXT 'b'0100;
        PRESENT 'b'1101      NEXT 'b'1100;
        PRESENT 'b'1100      NEXT 'b'1000;
        PRESENT 'b'1000      IF R NEXT 'b'0000;    DEFAULT NEXT 'b'1010;
        PRESENT 'b'1010      NEXT 'b'0010;
        PRESENT 'b'0010      NEXT 'b'0011;
        PRESENT 'b'0100      NEXT 'b'1100;
        }
```

Practical Exercise 3.2: Decision Making in a Binary Counter

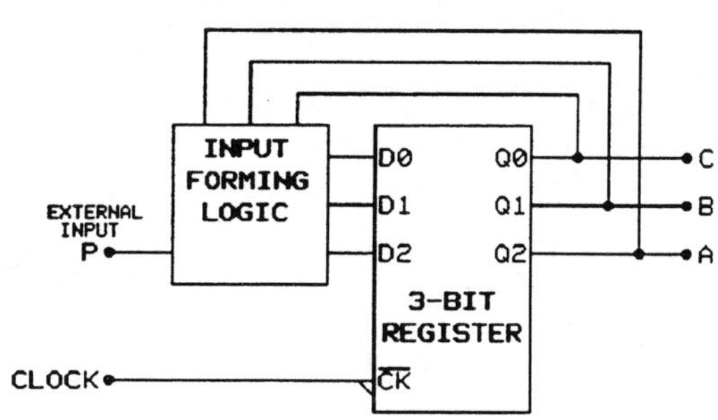

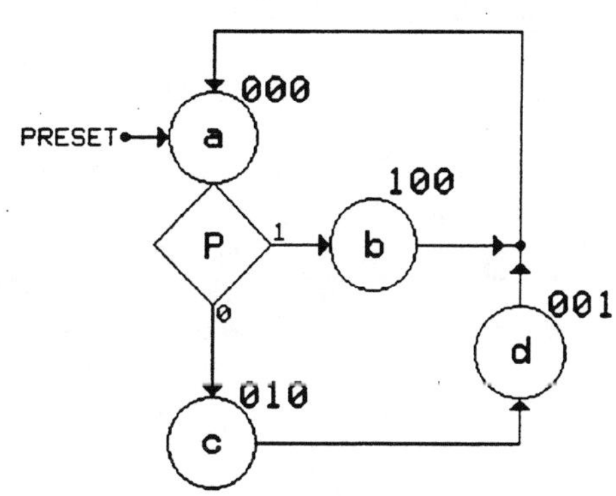

Task: (1) Write the state machine program, (2) create the state table, (3) determine the minimum input forming logic equations, and (4) draw the schematic diagram of the counter which exhibits the count sequence shown. Illustrated below is the state machine program, the state table, and the minimum input forming logic equations:

```
SEQUENCE [A,B,C]   {
        PRESENT 'b'000                           \* State A *\
                IF !P           NEXT 'b'010;
                DEFAULT         NEXT 'b'100;
        PRESENT 'b'100          NEXT 'b'000;     \* State B *\
        PRESENT 'b'010          NEXT 'b'001;     \* State C *\
        PRESENT 'b'001          NEXT 'b'000;     \* State D *\
        }
```

	PRES ABC	INPUT P	NEXT ABC	D_A	D_B	D_C
a	000	0 1	010 100	P	$\bar{P}$	0
d	001	*	000	0	0	0
c	010	*	001	0	0	1
	011	*	***	*	*	*
b	100	*	000	0	0	0
	101	*	***	*	*	*
	110	*	***	*	*	*
	111	*	***	*	*	*

* = Don't Care

$$D_A = \bar{A}\bar{B}\bar{C}P$$

$$D_B = \bar{A}\bar{B}\bar{C}\bar{P}$$

$$D_C = B$$

Practical Exercise 3.3: Circuit Complexity in Controlled Binary Counters

Consider the following controlled counter:

```
SEQUENCE [A,B,C,D] {
    PRESENT a   IF (!R & S) NEXT b;    DEFAULT NEXT d;
    PRESENT b   IF (R # !S) NEXT c;    DEFAULT NEXT a;
    PRESENT c   IF (!R & !S) NEXT d;   DEFAULT NEXT e;
    PRESENT d   IF S NEXT a;           DEFAULT NEXT d;
    PRESENT e   NEXT f;
    PRESENT f   NEXT g;
    PRESENT g   NEXT h;
    PRESENT h   IF !R NEXT e;          DEFAULT NEXT d; }
```

Task One: Select state values, $A=0, B=2, C=4, D=8, E=1, F=3, G=12, H=9$. (minimum 1's)
Task Two: Select state values, $A=0, B=4, C=12, D=8, E=13, F=15, G=11, H=9$. (glitch-free)

For each task draw the (1) State Diagram, (2) Next-State Map, (3) State Table, (4) Karnaugh Maps, (5) Schematic Diagram. How many logic gate inputs were required to implement the circuit? How many logic gate inputs were required to implement the circuit? The resulting state tables and input forming logic expressions are indicated below:

STATE	N	PRES ABCD	INPUT RS	NEXT ABCD	D_A	D_B	D_C	D_D
a	0	0000	00 01 10 11	1000 0010 1000 1000	$R+\bar{S}$	0	$\bar{R}S$	0
e	1	0001	**	0011	0	0	1	1
b	2	0010	00 01 10 11	0100 0000 0100 0100	0	$R+\bar{S}$	0	0
f	3	0011	**	1100	1	1	0	0
c	4	0100	00 01 10 11	1000 0001 0001 0001	$\bar{R}\bar{S}$	0	0	$R+S$
	5	0101	**	****	*	*	*	*
	6	0110	**	****	*	*	*	*
	7	0111	**	****	*	*	*	*
d	8	1000	*0 *1	1000 0000	1	0	0	0
h	9	1001	0* 1*	0001 1000	R	0	0	$\bar{R}$
	10	1010	**	****	*	*	*	*
	11	1011	**	****	*	*	*	*
g	12	1100	**	1001	1	0	0	1
	13	1101	**	****	*	*	*	*
	14	1110	**	****	*	*	*	*
	15	1111	**	****	*	*	*	*

* = Don't Care

$$D_A = \bar{A}\bar{B}\bar{C}\bar{D}R + \bar{A}\bar{B}\bar{C}\bar{D}\bar{S} + CD + \bar{A}\bar{B}\bar{D}\bar{R}\bar{S} + ADR + AB + A\bar{D}$$

$$D_B = CD + C\bar{D}\bar{R} + C\bar{D}\bar{S}$$

$$D_C = \bar{A}\bar{B}\bar{C}\bar{D}\bar{R}S + A\bar{C}\bar{D}$$

$$D_D = \bar{A}\bar{C}D + \bar{A}B\bar{C}\bar{D}R + \bar{A}B\bar{C}\bar{D}S + AD\bar{R} + AB$$

STATE	N	PRES ABCD	INPUT RS	NEXT ABCD	D_A	D_B	D_C	D_D
a	0	0000	00 01 10 11	1000 0100 1000 1000	$R+\bar{S}$	$\bar{R}S$	0	0
	1	0001	**	****	*	*	*	*
	2	0010	**	****	*	*	*	*
	3	0011	**	****	*	*	*	*
b	4	0100	00 01 10 11	1100 0000 1100 1100	$R+\bar{S}$	$R+\bar{S}$	0	0
	5	0101	**	****	*	*	*	*
	6	0110	**	****	*	*	*	*
	7	0111	**	****	*	*	*	*
d	8	1000	*0 *1	1000 0000	$\bar{S}$	0	0	0
h	9	1001	0* 1*	1101 1000	1	$\bar{R}$	0	$\bar{R}$
	10	1010	**	****	*	*	*	*
g	11	1011	**	1001	1	0	0	1
c	12	1100	00 01 10 11	1000 1101 1101 1101	1	R+S	0	R+S
e	13	1101	**	1111	1	1	1	1
	14	1110	**	****	*	*	*	*
f	15	1111	**	1011	1	0	1	1

* = Don't Care

$$D_A = \bar{A}R + \bar{A}S + A\bar{B}\bar{C}\bar{D}\bar{S} + AB + AD$$

$$D_B = \bar{A}\bar{B}\bar{R}S + \bar{A}BR + \bar{A}B\bar{S} + A\bar{B}\bar{C}D\bar{R} + AB\bar{C}\bar{D}R + AB\bar{C}\bar{D}S + B\bar{C}D$$

$$D_C = BD$$

$$D_D = ABR + ABS + BD + A\bar{B}\bar{C}D\bar{R} + AC$$

The second example is glitch-free, and is slightly more complex. The relationship between states 0-7 created a lot of useful "don't care" states, further simplifying this solution. If this pattern were not evident, this design would be far more complex than the first.

Practical Exercise 4.1: Unconditional Output Forming Logic

Consider the following state machine definition.

- Draw the State Table and State Machine Program
- Determine the Input Forming Logic (IFL) by using Karnaugh Maps
- Draw a schematic diagram, including the IFL, the binary register and the OFL.

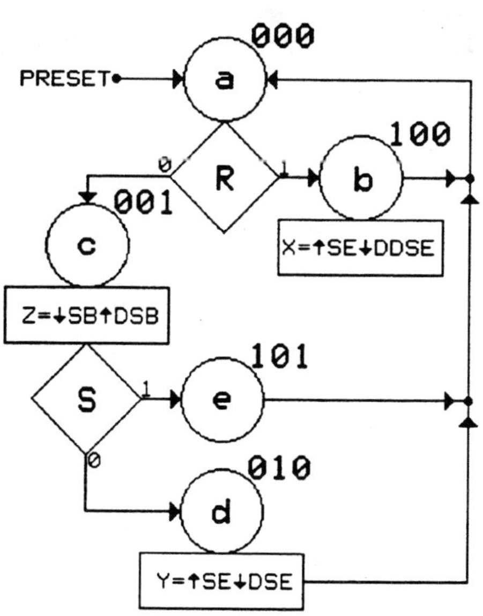

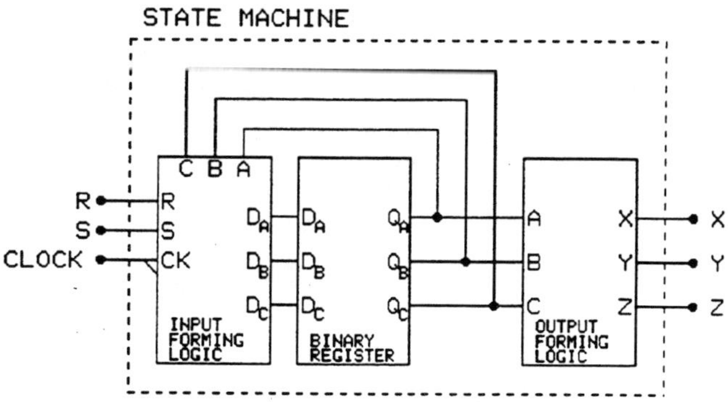

Illustrated below is the state machine program, state diagram, logic expressions and OFL schematics for the design.

```
SEQUENCE [A,B,C]   {
        PRESENT 'b'000                                         \* State A *\
                IF !R           NEXT 'b'001;
                DEFAULT         NEXT 'b'100;
        PRESENT 'b'100          NEXT 'b'000  OUT X;            \* State B *\
        PRESENT 'b'001  OUT Z;                                 \* State C *\
                IF !S           NEXT 'b'010;
                DEFAULT         NEXT 'b'101;
        PRESENT 'b'010          NEXT 'b'000  OUT Y;            \* State D *\
        PRESENT 'b'101          NEXT 'b'000;                   \* State E *\
        }
```

	PRES ABC	INPUT RS	NEXT ABC	D_A	D_B	D_C	X	Y	Z
a	000	0X 1X	001 100	R	0	$\bar{R}$			
c	001	X0 X1	010 101	S	$\bar{S}$	S			↓SB↑DSB
d	010	**	000	0	0	0		↑SE↓DSE	
	011	**	***	*	*	*			
b	100	**	000	0	0	0	↑SE↓DDSE		
e	101	**	000	0	0	0			
	110	**	***	*	*	*			
	111	**	***	*	*	*			

* = Don't Care

$$D_A = \bar{A}\bar{B}\bar{C}R + \bar{A}CS$$

$$D_B = \bar{A}C\bar{S}$$

$$D_C = \bar{A}\bar{B}\bar{C}\bar{R} + \bar{A}CS$$

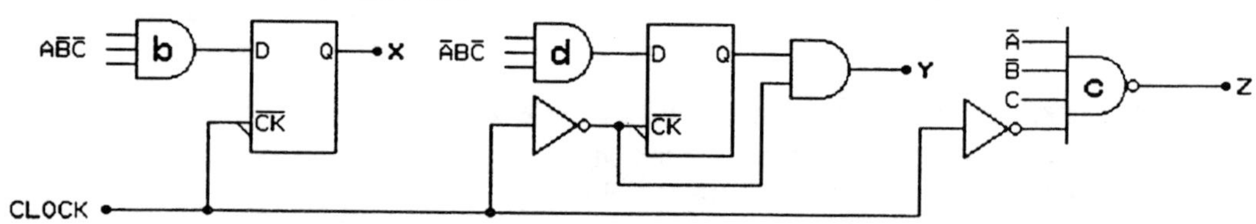

Note that the three-input and logic used to generate the state output signals, X, Y and Z, is available directly from the state machine outputs when the above sequence program is used. That is, if a PLD is programmed with the above sequence statements, the three-input logic gates are included in the PLD, and are not to be added to external circuitry.

Practical Exercise 4.2: State Machine Timing Diagram

Task: construct a timing diagram which describes the logic levels of all of the variables defined by the state diagram of Practical Exercise 4.1. Below is an example timing diagram:

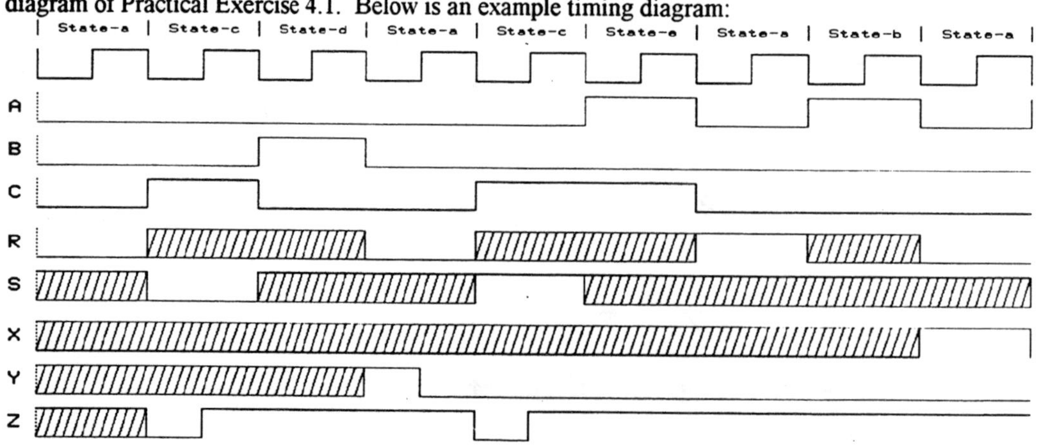

Practical Exercise 4.3: A Complete State Machine Design

Consider the following state diagram. Develop a complete state machine design by completing each of the following tasks: (1) Assign state values for simplest implementation; (2) draw a timing diagram; (3) write a state machine program; (4) develop a state table which includes present state, input stimulus, next state, and output signal information; (5) develop the set of Karnaugh maps in order to (6) draw a detailed schematic diagram.

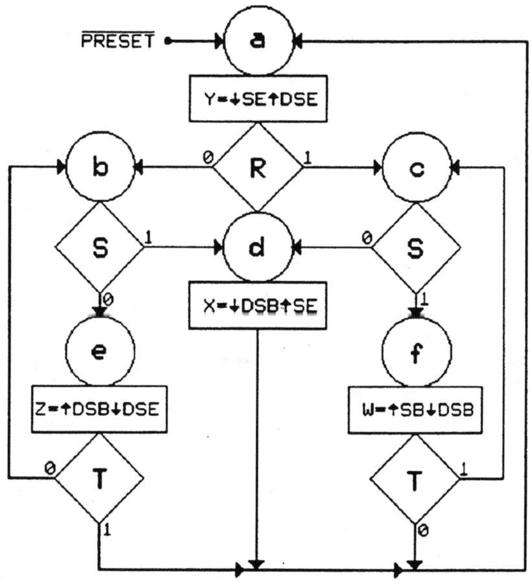


```
SEQUENCE [A,B,C]   {
        PRESENT 'b'000  OUT Y;                              \* State A *\
                IF !R           NEXT 'b'001;
                DEFAULT         NEXT 'b'100;
        PRESENT 'b'001                                      \* State B *\
                IF !S           NEXT 'b'011;
                DEFAULT         NEXT 'b'010;
        PRESENT 'b'100                                      \* State C *\
                IF !S           NEXT 'b'010;
                DEFAULT         NEXT 'b'110;
        PRESENT 'b'010          NEXT 'b'000  OUT X;         \* State D *\
        PRESENT 'b'011  OUT Z;                              \* State E *\
                IF !T           NEXT 'b'001;
                DEFAULT         NEXT 'b'000;
        PRESENT 'b'110  OUT W;                              \* State F *\
                IF !T           NEXT 'b'000;
                DEFAULT         NEXT 'b'100;
        }
```

Note: State numbers were selected to minimize the complexity of the design by (1) assigning numbers with the fewest 1's to the states which are most frequent in the "Next State" column of the state table, and (2) equally distributed those 1's among the three counter bits. Casual observation of the state diagram shows three logical 1's in each counter bit of the "Next State" column. This will generate a maximum of nine minterms, three to each of the three expressions. The D_B expression only required two minterms since two 1's were in State B.

	PRES ABC	INPUT RST	NEXT ABC	D_A	D_B	D_C	W	X	Y	Z
a	000	0** 1**	001 100	R	0	$\bar{R}$			↓SE↑DSE	
b	001	*0* *1*	011 010	0	1	$\bar{S}$				
d	010	***	000	0	0	0		↓DSB↑SE		
e	011	**0 **1	011 000	0	$\bar{T}$	$\bar{T}$				↑DSB↓DSE
c	100	*0* *1*	000 100	S	0	0				
	101	***	000	0	0	0				
f	110	**0 **1	000 100	T	0	0	↑SB↓DSB			
	111	***	***	*	*	*				

* = Don't Care

$$D_A = \bar{A}\bar{B}\bar{C}R + A\bar{B}\bar{C}S + ABT$$
$$D_B = \bar{A}\bar{B}C + BC\bar{\bar{T}}$$
$$D_C = \bar{A}\bar{B}\bar{C}\bar{R} + \bar{A}\bar{B}C\bar{S} + BCT$$

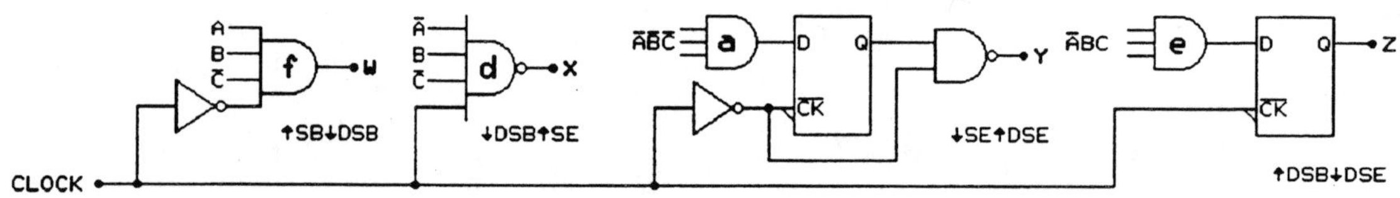

Practical Exercise 4.4: Output Pulses

Task: Given the following state diagram, develop a detailed state machine program, timing diagram, state table, minimum IFL and OFL circuitry.

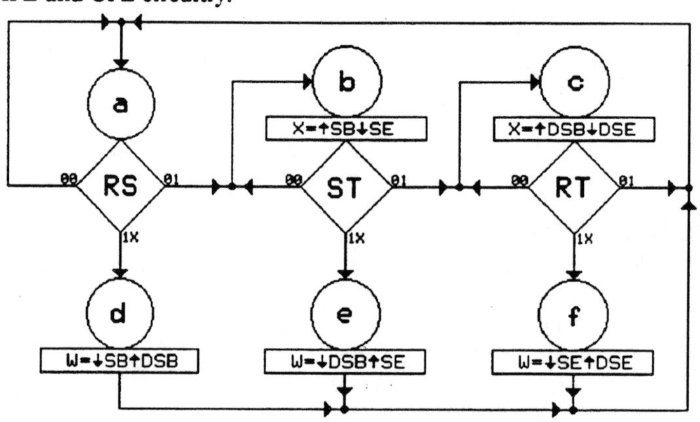

Illustrated below is the state table and resulting logical expressions for this problem:

	PRES ABC	INPUT RST	NEXT ABC	D_A	D_B	D_C	W	X
a	000	00* 01* 10* 11*	000 001 011 011	0	R	R+S		
b	001	*00 *01 *10 *11	001 010 100 100	R	$\bar{R}S$	$\bar{R}\bar{S}$		↑SB↓SE
c	010	0*0 0*1 1*0 1*1	010 000 101 101	R	$\bar{R}\bar{S}$	R		↑DSB↓DSE
d	011	***	000	0	0	0	↓SB↑DSB	
e	100	***	000	0	0	0	↓DSB↑SE	
f	101	***	000	0	0	0	↓SE↑DSE	
	110	***	***	*	*	*		
	111	***	***	*	*	*		

$$D_A = \bar{A}\bar{B}CR + B\bar{C}R$$
$$D_B = \bar{A}\bar{B}\bar{C}R + \bar{A}\bar{B}C\bar{R}S + B\bar{C}\bar{R}\bar{T}$$
$$D_C = \bar{A}\bar{B}\bar{C}R + \bar{A}\bar{B}\bar{C}S + \bar{A}\bar{B}C\bar{R}\bar{S} + B\bar{C}R$$

Illustrated below is the output forming logic for this problem. First note that an AND logic gate is used to summate each of those outputs which form the low-assertion level of the W output. OR logic is used to summate each of those outputs which form the high-assertion level of the X output.

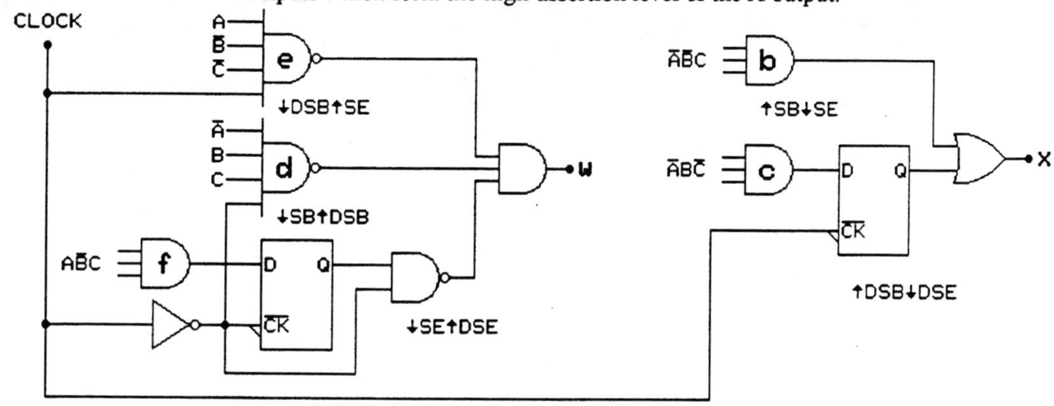

Practical Exercise 4.5: Output Levels

Task: Given the following state diagram, develop a detailed state machine program, timing diagram, state table, minimum IFL and OFL circuitry.

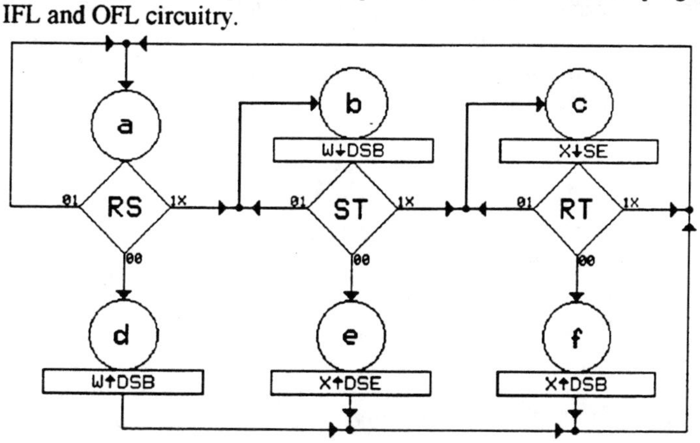

In this exercise, outputs W and X are to be maintained between states by a Set-Reset type flip-flop. The following is the state table and input forming logic expressions for this problem.

	PRES ABC	INPUT RST	NEXT ABC	D_A	D_B	D_C	W	X
a	000	00*	011	0	$\bar{R}\bar{S}$	$R+\bar{S}$		
		01*	000					
		10*	001					
		11*	001					
b	001	*00	100	$\bar{S}\bar{T}$	S	$\bar{S}T$	↓DSB	
		*01	001					
		*10	010					
		*11	010					
c	010	0*0	101	$\bar{R}\bar{T}$	$\bar{R}T$	$\bar{R}\bar{T}$		↓SE
		0*1	010					
		1*0	000					
		1*1	000					
d	011	***	000	0	0	0	↑DSB	
e	100	***	000	0	0	0		↑DSE
f	101	***	000	0	0	0		↑DSB
	110	***	***	*	*	*		
	111	***	***	*	*	*		

$$D_A = \bar{A}\bar{B}C\bar{S}\bar{T} + B\bar{C}\bar{R}\bar{T}$$

$$D_B = \bar{A}\bar{B}\bar{C}\bar{R}\bar{S} + \bar{A}\bar{B}CS + B\bar{C}\bar{R}T$$

$$D_C = \bar{A}\bar{B}\bar{C}R + \bar{A}\bar{B}\bar{C}\bar{S} + \bar{A}\bar{B}\bar{S}T + B\bar{C}\bar{R}T$$

The following is the output forming logic for this exercise. Note how the Set-Reset flip-flops are used to maintain the logic levels between states. The Set and Reset inputs of the flip-flops are enabled at the appropriate time in the appropriate state using a signal which is always asserted high. Note also that the X output is set in two different states, so an OR logic gate is used to combine those two states into an input signal to the flip-flop.

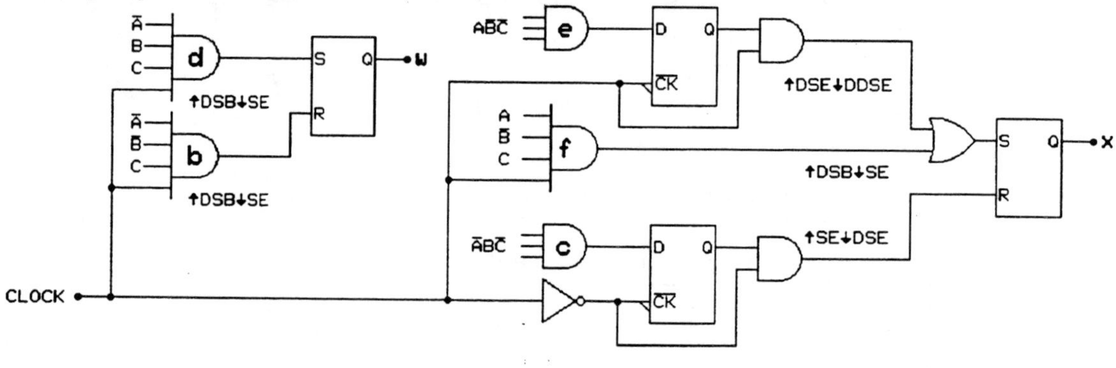

Practical Exercise 4.6: A Simple Candy Machine Controller

Using the same approach as Practical Example 4.5, design a candy machine controller with the following constraints. Show all development steps in detail.

Inputs:	No Activity	
	Press Coin Return	
	Input Quarter	
	Select Chewing Gum	(25¢ Cost)
	Select Potato Chips	(50¢ Cost)
	Select Chocolate Bar	(50¢ Cost)
Outputs:	Dispense Quarter	
	Dispense Chewing Gum	
	Dispense Potato Chips	
	Dispense Chocolate Bar	

The state diagram for the problem is shown to the right. Note that this is similar to Practical Exercise 4.5, with two additions.

- First, a 25¢ product, CG_O is available in State B.
- Second, it is possible to input a quarter in State C. If this is done, the amount of money in the machine is greater than 50¢ so the quarter is dispensed and control returns to the same state.

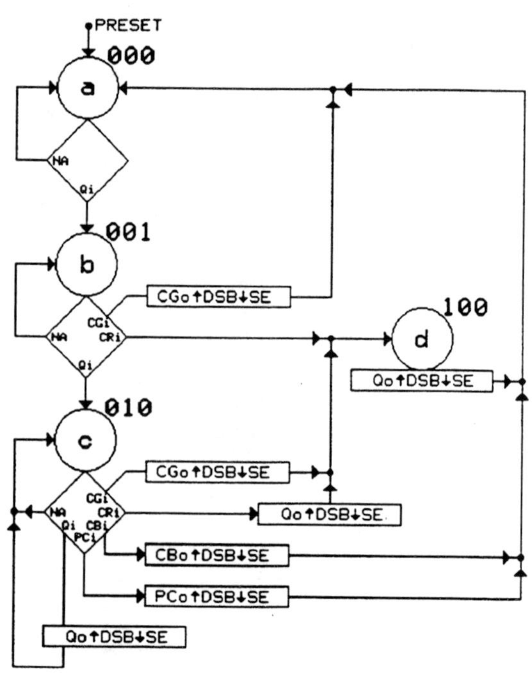

Next, consider the state table shown on the following page. In order to simplify the design, an input, NA is provided to indicate No Activity. It is generated by using NOR logic on all inputs. That is, all inputs to the state machine are NOR'ed together to create the NA signal which is used to indicate no activity.

	PRES ABC	INPUTS NA N	Qi Q	CRi R	CBi C	PCi P	CGi G	NEXT ABC	D_A	D_B	D_C	Qo	CBo	PCo	CGo
a	000	1	0	0	0	0	0	000	0	0	Q	0	0	0	0
		0	1	0	0	0	0	001				0	0	0	0
b	001	1	0	0	0	0	0	001	R	Q	N	0	0	0	0
		0	1	0	0	0	0	010				0	0	0	0
		0	0	1	0	0	0	100				0	0	0	0
		0	0	0	0	0	1	000				0	0	0	↑DSB↓SE
c	010	1	0	0	0	0	0	010	R+G	N+Q	0	0	0	0	0
		0	1	0	0	0	0	010				↑DSB↓SE	0	0	0
		0	0	1	0	0	0	100				↑DSB↓SE	0	0	0
		0	0	0	1	0	0	000				0	↑DSB↓SE	0	0
		0	0	0	0	1	0	000				0	0	↑DSB↓SE	0
		0	0	0	0	0	1	100				0	0	0	↑DSB↓SE
	011	X	X	X	X	X	X	XXX	X	X	X	X	X	X	X
d	100	X	X	X	X	X	X	000	0	0	0	↑DSB↓SE	0	0	0
	101	X	X	X	X	X	X	XXX	X	X	X	X	X	X	X
	110	X	X	X	X	X	X	XXX	X	X	X	X	X	X	X
	111	X	X	X	X	X	X	XXX	X	X	X	X	X	X	X

The resulting input forming logic expressions are:

$$D_A = CR + BR + BG \quad D_B = CQ + BQ + BN \quad D_C = \overline{A}\,\overline{B}\,\overline{C}\,Q + CN$$

Since all of the output forming logic uses the same AST method, its implementation is simpler than that of most previous examples. Consider the following solution. Note how an AND logic gate was placed on the output of each OFL circuit in order to provide the proper DSB-SE pulse.

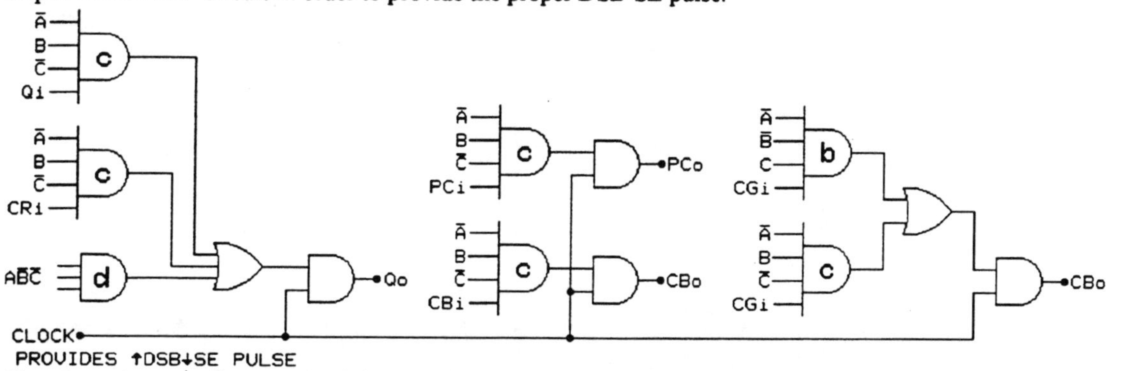

Also, from the state diagram, the following state machine program can be created:

```
SEQUENCE [A,B,C]   {
        PRESENT 'b'000                                    \* State A *\
                IF Qi           NEXT 'b'001;
                DEFAULT         NEXT 'b'000;
        PRESENT 'b'001                                    \* State B *\
                IF Qi           NEXT 'b'010;
                IF CRi          NEXT 'b'100;
                IF CGi          NEXT 'b'000 OUT CGo;
                DEFAULT         NEXT 'b'001;
        PRESENT 'b'010                                    \* State C *\
                IF Qi           NEXT 'b'010 OUT Qo;
                IF CRi          NEXT 'b'100 OUT Qo;
                IF CBi          NEXT 'b'000 OUT CBo;
                IF PCi          NEXT 'b'000 OUT PCo;
                IF CGi          NEXT 'b'100 OUT CGi;
                DEFAULT         NEXT 'b'110;
        PRESENT 'b'100          NEXT 'b'000  OUT Qo;      \* State D *\    }
```

Practical Exercise 4.7: A Moderate Candy Machine Controller

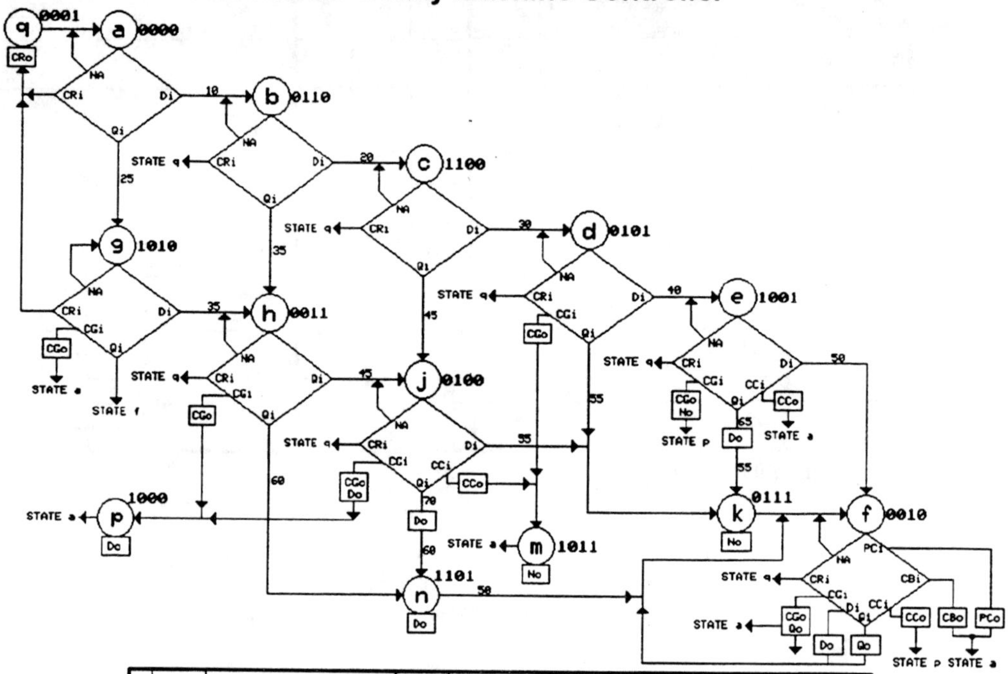

	PRES WXYZ	NA N	Qi Q	Di M	CRi R	CGi G	CCi C	CBi B	PCi P	NEXT WXYZ	Dw	Dx	Dy	Dz	Qo	Do	No	CRo	CGo	CCo	CBo	PCo
a	0000	1	0	0	0	0	0	0	0	0000	Q	M	Q+M	R	0	0	0	0	0	0	0	0
		0	1	0	0	0	0	0	0	1010												
		0	0	1	0	0	0	0	0	0110												
		0	0	0	1	0	0	0	0	0001												
q	0001	*	*	*	*	*	*	*	*	0001	0	0	0	1	0	0	0	1	0	0	0	0
f	0010	1	0	0	0	0	0	0	0	0010	C	0	N+Q +M	R	Q+G	M	0	0	G	C	B	P
		0	1	0	0	0	0	0	0	0010												
		0	0	1	0	0	0	0	0	0010												
		0	0	0	1	0	0	0	0	0001												
		0	0	0	0	1	0	0	0	0000												
		0	0	0	0	0	1	0	0	1000												
		0	0	0	0	0	0	1	0	0000												
		0	0	0	0	0	0	0	1	0000												
h	0011	1	0	0	0	0	0	0	0	0011	Q+G	Q+M	N	N+Q +R	0	0	0	0	G	0	0	0
		0	1	0	0	0	0	0	0	1101												
		0	0	1	0	0	0	0	0	0100												
		0	0	0	1	0	0	0	0	0001												
		0	0	0	0	1	0	0	0	1000												
j	0100	1	0	0	0	0	0	0	0	0100	Q+G +C	N+Q +M	M+C	Q+M +R+C	0	Q+G	0	0	G	C	0	0
		0	1	0	0	0	0	0	0	1101												
		0	0	1	0	0	0	0	0	0111												
		0	0	0	1	0	0	0	0	0001												
		0	0	0	0	1	0	0	0	1000												
		0	0	0	0	0	1	0	0	1011												
d	0101	1	0	0	0	0	0	0	0	0101	M+G	N+Q	Q+G	1	0	0	0	0	G	0	0	0
		0	1	0	0	0	0	0	0	0111												
		0	0	1	0	0	0	0	0	1001												
		0	0	0	1	0	0	0	0	0001												
		0	0	0	0	1	0	0	0	1011												
b	0110	1	0	0	0	0	0	0	0	0110	M	N+M	N+Q	Q+R	0	0	0	0	0	0	0	0
		0	1	0	0	0	0	0	0	0011												
		0	0	1	0	0	0	0	0	1100												
		0	0	0	1	0	0	0	0	0001												
k	0111	*	*	*	*	*	*	*	*	0010	0	0	1	0	0	0	1	0	0	0	0	0
p	1000	*	*	*	*	*	*	*	*	0000	0	0	0	0	0	1	0	0	0	0	0	0
e	1001	1	0	0	0	0	0	0	0	1001	N+G	Q	Q+M	N+Q +R	0	Q	G	0	G	C	0	0
		0	1	0	0	0	0	0	0	0111												
		0	0	1	0	0	0	0	0	0010												
		0	0	0	1	0	0	0	0	0001												
		0	0	0	0	1	0	0	0	1000												
		0	0	0	0	0	1	0	0	0000												
g	1010	1	0	0	0	0	0	0	0	1010	N	0	N+Q +M	M+R	0	0	0	0	G	0	0	0
		0	1	0	0	0	0	0	0	0010												
		0	0	1	0	0	0	0	0	0011												
		0	0	0	1	0	0	0	0	0001												
		0	0	0	0	1	0	0	0	0000												
m	1011	*	*	*	*	*	*	*	*	0000	0	0	0	0	0	0	1	0	0	0	0	0
c	1100	1	0	0	0	0	0	0	0	1100	N	N+Q +M	0	M+R	0	0	0	0	0	0	0	0
		0	1	0	0	0	0	0	0	0100												
		0	0	1	0	0	0	0	0	0101												
		0	0	0	1	0	0	0	0	0001												
n	1101	*	*	*	*	*	*	*	*	0010	0	0	1	0	0	1	0	0	0	0	0	0
	1110	*	*	*	*	*	*	*	*	0010	*	*	*	*	*	*	*	*	*	*	*	*
	1111	*	*	*	*	*	*	*	*	0010	*	*	*	*	*	*	*	*	*	*	*	*

From the state table, logic expressions can be derived using variable-entered maps. For example:

$$D_W = \bar{W}\bar{X}\bar{Z}Q + \bar{W}\bar{X}YZQ + \bar{W}\bar{X}YZG + \bar{W}\bar{X}Y\bar{Z}C + \bar{W}X\bar{Y}G + \bar{W}X\bar{Y}\bar{Z}C + \bar{W}X\bar{Y}ZM + \bar{W}XY\bar{Z}M + W\bar{X}\bar{Y}ZN + W\bar{X}\bar{Y}ZG + WY\bar{Z}N + WX\bar{Z}N$$

$$D_X = \bar{W}\bar{Y}\bar{Z}M + \bar{W}\bar{X}YZQ + \bar{W}\bar{X}YZM + X\bar{Z}M + X\bar{Z}N + X\bar{Y}\bar{Z}C + W\bar{X}\bar{Y}ZQ + \bar{W}X\bar{Y}ZN + \bar{W}X\bar{Y}ZQ$$

$$D_Y = \bar{W}\bar{Y}\bar{Z}M + \bar{W}\bar{X}\bar{Z}Q + \bar{W}\bar{X}YN + \bar{W}\bar{X}\bar{Z}M + Y\bar{Z}N + Y\bar{Z}Q + \bar{W}X\bar{Y}ZQ + \bar{W}X\bar{Y}ZG + \bar{W}X\bar{Y}\bar{Z}C + \bar{W}X\bar{Y}ZQ + XYZ + WXZ + WY\bar{Z}M + W\bar{X}\bar{Y}ZQ + W\bar{X}\bar{Y}ZM$$

$$D_Z = \bar{W}\bar{Z}R + \bar{W}\bar{Y}\bar{Z}Q + \bar{W}X\bar{Y}\bar{Z}C + \bar{W}X\bar{Y}\bar{Z}M + \bar{W}\bar{Y}Z + \bar{W}\bar{X}YR + \bar{W}\bar{X}YZN + \bar{W}\bar{X}YZQ + X\bar{Y}\bar{Z}M + X\bar{Z}R + WY\bar{Z}M + Y\bar{Z}R + W\bar{X}\bar{Y}ZN + W\bar{X}\bar{Y}ZQ + W\bar{X}\bar{Y}ZR$$

$$Q_O = \bar{W}\bar{X}Y\bar{Z}Q + \bar{W}\bar{X}Y\bar{Z}G \qquad N_O = XYZ + WYZ + W\bar{X}\bar{Y}ZG$$

$$D_O = \bar{W}\bar{X}Y\bar{Z}M + \bar{W}X\bar{Y}\bar{Z}Q + \bar{W}X\bar{Y}\bar{Z}G + WXZ + W\bar{X}\bar{Y}\bar{Z} + W\bar{X}\bar{Y}ZQ$$

$$CR_O = \bar{W}\bar{X}\bar{Y}Z \qquad CG_O = \bar{W}X\bar{Y}G + \bar{W}\bar{X}YG + W\bar{X}\bar{Y}ZG + WY\bar{Z}G$$

$$CC_O = \bar{W}X\bar{Y}\bar{Z}C + \bar{W}\bar{X}Y\bar{Z}C + W\bar{X}\bar{Y}ZC \qquad CB_O = \bar{W}\bar{X}Y\bar{Z}B \qquad PC_O = \bar{W}\bar{X}Y\bar{Z}P$$

A more effective solution is illustrated in Chapter 7.

Practical Exercise 5.1 CUPL Program of a Combinational Circuit

Write a CUPL program which will program a 16L8 PAL device to act as a code converter which will accept a three-bit number and complement any numbers which are odd. First, the truth table for the code converter is developed:

INPUTS			OUTPUTS		
A	B	C	X	Y	Z
0	0	0	0	0	0
0	0	1	1	1	0
0	1	0	0	1	0
0	1	1	1	0	0
1	0	0	1	0	0
1	0	1	0	1	0
1	1	0	1	1	0
1	1	1	0	0	0

$X = \bar{A}\bar{B}C + \bar{A}BC + A\bar{B}\bar{C} + AB\bar{C}$

$X = A \oplus C$

$Y = \bar{A}\bar{B}C + \bar{A}B\bar{C} + A\bar{B}C + AB\bar{C}$

$Y = B \oplus C$

$Z = 0$

```
Name            P05-01;
Revision        01;
Date            05/21/96;
Designer        J. Carter;
Company         University of North Carolina, Charlotte;
Location        None;
Assembly        None;
Device          P16L8;
Format          j;
/*************************************************************************/
/*   Practical Example 5.1                                               */
/*   Three-Bit Code Converter                                            */
/*************************************************************************/
/* INPUTS */          PIN 2 = A;
                      PIN 3 = B;
                      PIN 4 = C;
/* OUTPUTS */         PIN 19 = x;
                      PIN 18 = Y;
                      PIN 17 = Z;
/* EXPRESSIONS */     X = A $ C;
                      Y = B & C;
                      Z = 'b'0;
```

Practical Exercise 5.2 CUPL Program of a Sequential Circuit

Consider a synchronous binary counter which will count in the repeating sequence: 0 - 2 - 4 - 5 - 7 - 3 - ... If the counter is in a state outside the sequence, cause it to go to count 0.

1. Write a CUPL program which uses discrete logic expressions to implement the counter design on a 16R4.
2. Write a CUPL program which uses SEQUENCE statements to implement the counter design on a 16R4.

The following is the state table, input forming logic functions and CUPL program of this problem:

PRESENT			NEXT			INPUT LOGIC		
A	B	C	A	B	C	D_A	D_B	D_C
0	0	0	0	1	0	0	1	0
0	0	1	0	0	0	0	0	0
0	1	0	1	0	0	1	0	0
0	1	1	0	0	0	0	0	0
1	0	0	1	0	1	1	0	1
1	0	1	1	1	1	1	1	1
1	1	0	0	0	0	0	0	0
1	1	1	0	1	1	0	1	1

$DA = \bar{A}B\bar{C} + A\bar{B}$

$DB = \bar{A}\bar{B}\bar{C} + AC$

$DC = A\bar{B} + AC$

```
Name          P05-02A;
Revision      01;
Date          05/21/96;
Designer      J. Carter;
Company       University of North Carolina, Charlotte;
Location      None;
Assembly      None;
Device        P16R4;
Format        j;
/************************************************************************/
/*   Practical Example 5.2a                                             */
/*   Three-Bit Counter using Logic Expressions for IFL                  */
/************************************************************************/
/* COUNTER BITS */ PIN 17 = A;
                   PIN 16 = B;
                   PIN 15 = C;
/* EXPRESSIONS */  A.D = (!A &  B& !C) # (A & !B);
                   B.D = (!A & !B& !C) # (A &  C);
                   C.D = ( A & !B)     # (A &  C);
Name          P05-02B;
Revision      01;
Date          05/21/96;
Designer      J. Carter;
Company       University of North Carolina, Charlotte;
Location      None;
Assembly      None;
Device        P16R4;
Format        j;

/************************************************************************/
/*   Practical Example 5.2b                                             */
/*   Three-Bit Counter using a SEQUENCE statement for the IFL           */
/************************************************************************/
/* COUNTER BITS */ PIN 17 = A;
                   PIN 16 = B;
                   PIN 15 = C;
       SEQUENCE[A,B,C] {
              PRESENT 'b'000      NEXT 'b'010;
              PRESENT 'b'001      NEXT 'b'000;
              PRESENT 'b'010      NEXT 'b'100;
              PRESENT 'b'011      NEXT 'b'000;
              PRESENT 'b'100      NEXT 'b'101;
              PRESENT 'b'101      NEXT 'b'111;
              PRESENT 'b'110      NEXT 'b'000;
              PRESENT 'b'111      NEXT 'b'011; }
```

Practical Exercise 5.3 CUPL Program of a State Machine

1. Write a CUPL program to implement of the state machine to the right on a 16R4 PAL using discrete logic expressions.

2. Write a CUPL program to implement of the state machine to the right a 16R4 PAL device SEQUENCE statements.

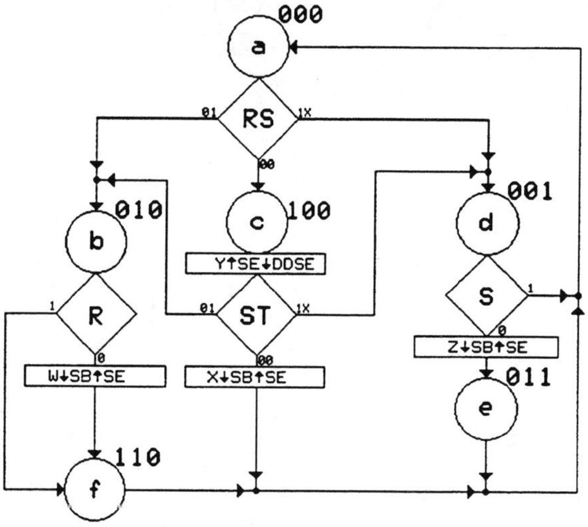

Illustrated below is the state table and output forming logic used to create the first solution

Following the table and schematics are the two programs. Note how the output forming logic was implemented for the second program.

	PRES ABC	INPUT RST	NEXT ABC	D_A	D_B	D_C	W	X	Y	Z
a	000	00*	100	$\bar R\bar S$	$\bar R S$	R	1	1	0	1
		01*	010				1	1	0	1
		10*	001				1	1	0	1
		11*	001				1	1	0	1
d	001	*0*	011	0	$\bar S$	$\bar S$	1	1	0	↓SB↑SE
		1	000				1	1	0	0
b	010	0**	110	1	1	0	↓SB↑SE	1	0	1
		1**	110				1	1	0	1
e	011	***	000	0	0	0	1	1	0	1
c	100	*00	000	0	$\bar S T$	S	1	↓SB↑SE	↑SE↓DDSE	1
		*01	010				1	1	↑SE↓DDSE	1
		*10	001				1	1	↑SE↓DDSE	1
		*11	001				1	1	↑SE↓DDSE	1
	101	***	***	*	*	*	*	*	*	*
f	110	***	000	0	0	0	1	1	0	
	111	***	***	*	*	*	*	*	*	*

$$D_A = \bar A\bar B\bar C\bar R\bar S + \bar A B\bar C$$

$$D_B = \bar A\bar B\bar C\bar R S + \bar B C\bar S + A\bar B\bar S T + \bar A B\bar C$$

$$D_C = \bar A\bar B\bar C R + \bar B C\bar S + A\bar B S$$

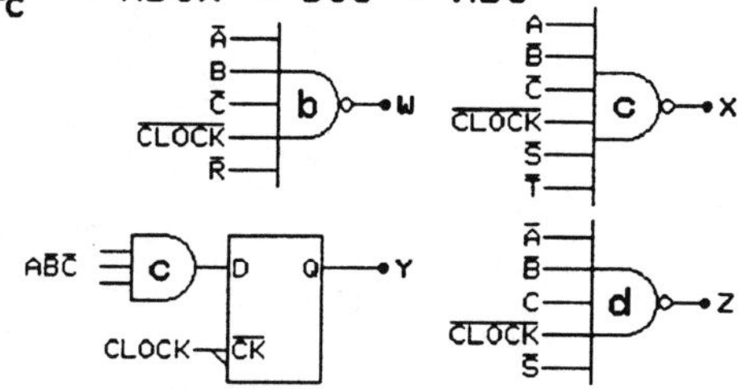


```
Name            P05-03A;
Revision        01;
Date            05/21/96;
Designer        J. Carter;
Location        None;
Assembly        None;
Device          P16R4;
Format          j;
/*************************************************************************/
/*   Practical Example 5.3a                                              */
/*   Three-Bit State Machine using Logic Expressions for IFL             */
/*************************************************************************/
/* CLOCK INPUT */       PIN  2 = CLK;
/* INPUT LOGIC */       PIN 17 = A;
                        PIN 16 = B;
                        PIN 15 = C;
/* OUTPUT LOGIC */      PIN 19 = W;
                        PIN 18 = X;
                        PIN 14 = Y;
                        PIN 13 = Z;
/* EXPRESSIONS */
        A.D =   (!A&!B&!C&!R&!S) # (!A& B&!C);
        B.D =   (!A&!B&!C&!R& S) # (!B& C&!S) # (A&!B&!S& T)#(!A& B&!C));
        C.D =   (!A&!B&!C& R)    # (!B& C&!S) # (A&!B& S);
        W    = !(!A& B&!C& !R     &!CLK);
        X    = !( A&!B&!C& !S&!T &!CLK);
        Y.D =   ( A&!B&!C);
        Z    = !(!A&!B& C& !S     &!CLK);
```

```
Name            P05-03B;
Revision        01;
Date            05/21/96;
Designer        J. Carter;
Location        None;
Assembly        None;
Device          P16R4;
Format          j;
/**************************************************************************/
/*   Practical Example 5.3b                                               */
/*   Three-Bit State Machine using a SEQUENCE statement for the IFL       */
/**************************************************************************/
/* CLOCK INPUT */      PIN  2 = CLK;
/* INPUT LOGIC */      PIN 17 = A;
                       PIN 16 = B;
                       PIN 15 = C;
/* OUTPUT LOGIC */     PIN 19 = W;
                       PIN 18 = X;
                       PIN 14 = Y;
                       PIN 13 = Z;
/* EXPRESSIONS */
   SEQUENCE[A,B,C] {
      PRESENT 'b'000                               /* STATE A */
              IF !R &  S       NEXT 'b'010;
              IF !R & !S       NEXT 'b'100;
              DEFAULT          NEXT 'b'001;
      PRESENT 'b'001                               /* STATE D */
              IF !S            NEXT 'b'011;
              DEFAULT          NEXT 'b'000;
      PRESENT 'b'010                               /* STATE B */
              IF !R            NEXT 'b'110;
              DEFAULT          NEXT 'b'110;
      PRESENT 'b'011           NEXT 'b'000;              /* STATE E */
      PRESENT 'b'100 OUT Y                         /* STATE C */
              IF !S&!T         NEXT 'b'000;
              IF !S& T         NEXT 'b'010;
              DEFAULT          NEXT 'b'001;
      PRESENT 'b'110           NEXT 'b'000;  }           /* STATE F */

      W   = !(!A& B&!C& !R    &!CLK);
      X   = !( A&!B&!C& !S&!T &!CLK);
      Y.D =  ( A&!B&!C);
      Z   = !(!A&!B& C& !S    &!CLK);
```

Practical Exercise 5.4: Simple Traffic Intersection Controller

Consider the simple vehicle traffic intersection illustrated in Figure 5.17. Design a PAL-based controller which will provide the logic signals needed to drive the lamps.

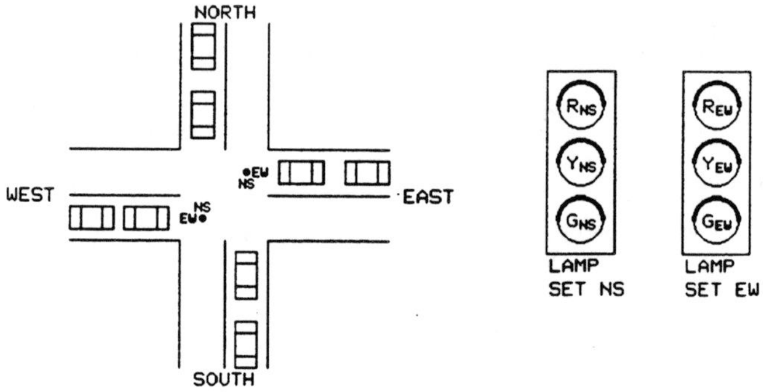

First, a state diagram is generated which identifies each of the states. Traffic flows N-S for three states, traffic is stopped on N-S for one state, traffic flows E-W for three states, traffic is stopped on E-W for one state. Consider the following state diagram. A glitch-free sequence was chosen since the counter is so simple.

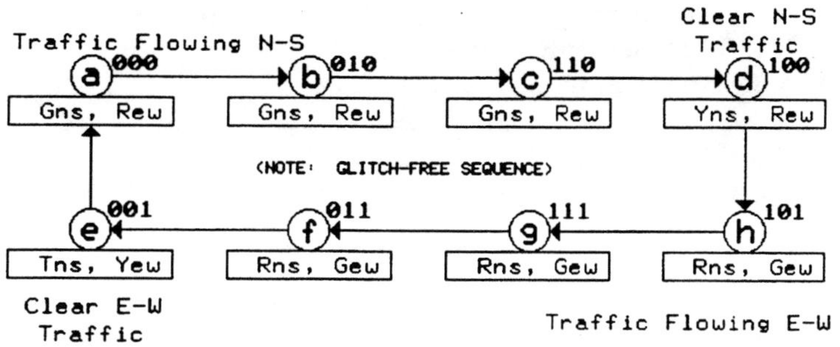


```
Name            P05-04;
Revision        01;
Date            05/21/96;
Designer        J. Carter;
Location        None;
Assembly        None;
Device          G22V10;
Format          j;
/*********************************************************************/
/*  Practical Example 5.4                                            */
/*  Simple Traffic Intersection Controller                           */
/*********************************************************************/
/* INPUT LOGIC */     PIN 23 = A;
                      PIN 22 = B;
                      PIN 21 = C;
/* OUTPUT LOGIC */    PIN 19 = RNS;        North-South Traffic Signal
                      PIN 18 = YNS;
                      PIN 17 = GNS;
                      PIN 16 = REW;        East-West Traffic Signal
                      PIN 15 = YEW;
                      PIN 14 = GEW;
/* EXPRESSIONS */
   SEQUENCE[A,B,C] {
        PRESENT 'b'000  NEXT 'b'010 OUT GNS OUT REW;      /* STATE A */
        PRESENT 'b'010  NEXT 'b'110 OUT GNS OUT REW;      /* STATE B */
        PRESENT 'b'110  NEXT 'b'100 OUT GNS OUT REW;      /* STATE C */
        PRESENT 'b'100  NEXT 'b'101 OUT YNS OUT REW;      /* STATE D */
        PRESENT 'b'101  NEXT 'b'111 OUT RNS OUT GEW;      /* STATE E */
        PRESENT 'b'111  NEXT 'b'011 OUT RNS OUT GEW;      /* STATE F */
        PRESENT 'b'011  NEXT 'b'001 OUT RNS OUT GEW;      /* STATE G */
        PRESENT 'b'001  NEXT 'b'000 OUT GNS OUT YEW; }    /* STATE H */
```

Practical Exercise 5.5: Moderate Traffic Intersection Controller

Consider the following vehicle traffic intersection. There is a transducer in each of the turn lanes which indicates that there is traffic waiting to turn. When this logic level is TRUE, and advanced green should be provided. Assume that the time delay between clock pulses is determined outside of the state machine. Design a PAL-based state machine which generates the logic signals needed to drive the lamps.

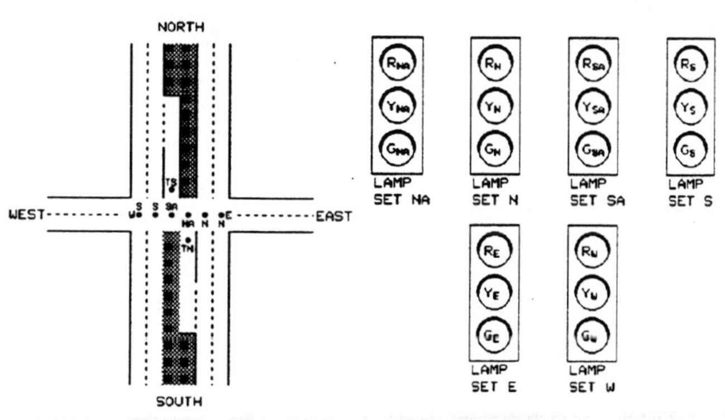

First, the common lamps are noted. Set NA = Set SA, Set N = Set S, Set E = Set W. Therefore we need only to control Set NA, Set N, and Set E, connecting Set SA, Set S and Set W to the same control signals. With this in mind, the following state diagram is determined:

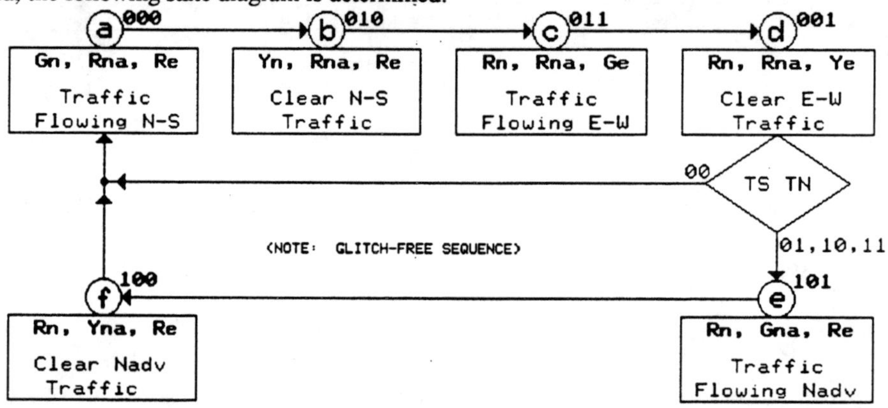

We now have two input signals, T_S and T_N. We have twelve output signals, G_N, Y_N, R_N, G_{NA}, Y_{NA}, R_{NA}, G_E, Y_E, R_E, and the three counter bits. Therefore, we will need to use two PAL devices. Let's use a 16R4 for the counter bits and the advanced turn signals, Set NA. Let's also use a 16L8 to generate the signals for the remaining traffic signals, Set N and Set S. The use of the 16L8 disallows the use of the sequence statement, so the logic expressions must be calculated. Therefore the following state table and expressions are determined:

	PRES ABC	TT SN	NEXT ABC	D_A	D_B	D_C	Gs Gn	Ys Yn	Rs Rn	Gsa Gna	Ysa Yna	Rsa Rna	Gw Ge	Yw Ye	Re
a	000	**	010	0	1	0	1	0	0	0	0	1	0	0	1
d	001	00	000	S+N	0	S+N	0	0	1	0	0	1	0	1	0
		01	101				0	0	1	0	0	1	0	1	0
		10	101				0	0	1	0	0	1	0	1	0
		11	101				0	0	1	0	0	1	0	1	0
b	010	**	011	0	1	1	0	1	0	0	0	1	0	0	1
c	011	**	001	0	0	1	0	0	1	0	0	1	1	0	0
f	100	**	000	0	0	0	0	0	1	0	1	0	0	0	1
e	101	**	100	1	0	0	0	0	1	1	0	0	0	0	1
	110	**	***	*	*	*	*	*	*	*	*	*	*	*	*
	111	**	***	*	*	*	*	*	*	*	*	*	*	*	*

$DA = \bar{A}\bar{B}CS + \bar{A}\bar{B}CN + AC$

$DB = \bar{A}\bar{C}$

$DC = \bar{A}\bar{B}CS + \bar{A}\bar{B}CN + B$

$GN = GS = \bar{A}\bar{B}\bar{C}$; $GE = GW = \bar{A}BC$

$YN = YS = \bar{A}B\bar{C}$; $YE = YW = \bar{A}\bar{B}C$

$RN = RS = A+C$; $RE = RW = A+\bar{C}$

$GNA = GSA = A\bar{B}C$

$YNA = YSA = A\bar{B}\bar{C}$

$RNA = RSA = \bar{A}$

With this information, the following CUPL program can be written.

```
Name            P05-05a;
Revision        01;
Date            05/21/96;
Designer        J. Carter;
Location        None;
Assembly        None;
Device          P16R4;
Format          j;
/************************************************************************/
/*   Practical Example 5.5                                              */
/*   Four-Way Traffic Controller with N-S Advanced Turn Lane  Chip 1 of 2 */
/************************************************************************/
/* INPUTS */              PIN  2 = S;      /* Sensor TS  (South turn lane) */
                          PIN  3 = N;      /* Sensor TN  (North turn lane) */
/* INPUT LOGIC */         PIN 17 = A;
                          PIN 16 = B;
                          PIN 15 = C;
/* OUTPUT LOGIC */        PIN 19 = GNA;    /* Lamp Sets NA and SA */
                          PIN 18 = YNA;
                          PIN 13 = RNA;
/* EXPRESSIONS */
       A.D = (!A&!B& C& S) # (!A&!B& C& N) # (A & C);
       B.D = (!A&!C);
       C.D = (!A&!B& C& S) # (!A&!B& C& N) # B;
       GNA = ( A&!B& C);
       YNA = ( A&!B&!C);
       RNA = !A;
```

```
Name            P05-05b;
Revision        01;
Date            05/21/96;
Designer        J. Carter;
Location        None;
Assembly        None;
Device          P16L8;
Format          j;
/*************************************************************************/
/*   Practical Example 5.5                                               */
/*   Four-Way Traffic Controller with N-S Advanced Turn Lane  Chip 2 of 2 */
/*************************************************************************/
/* INPUTS */               PIN  2 = A;    /* Counter bits from Chip 1    */
                           PIN  3 = B;
                           PIN  4 = C;
/* OUTPUT LOGIC */         PIN 19 = GN;   /* Lamp Sets N and S */
                           PIN 18 = YN;
                           PIN 17 = RN;
                           PIN 16 = GE;   /* Lamp Sets E and W */
                           PIN 15 = YE;
                           PIN 16 = RE;
/* EXPRESSIONS */
       GN = (!A&!B&!C);
       YN = (!A& B&!C);
       RN = A #  C;
       GE = (!A& B& C);
       YE = (!A&!B& C);
       RE = A # !C;
```

Practical Exercise 5.6: Complete Traffic Intersection Controller

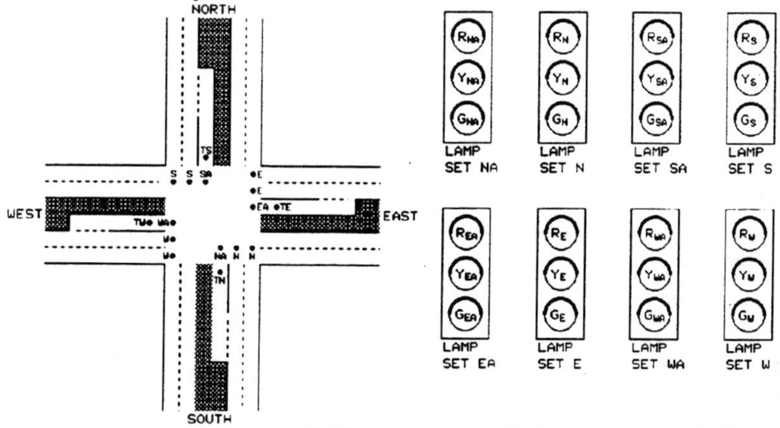

Consider the following traffic intersection. This intersection is very similar to that described in Practical Exercise 5.5, except turn lanes have been added for the East-West Traffic. <u>If we assume that opposite traffic patterns are similar</u>, as was done in Practical Exercise 5.5, the state diagram to the right is determined. Note the addition of the decision to State B, and the addition of the advanced turn signal for E-W traffic.

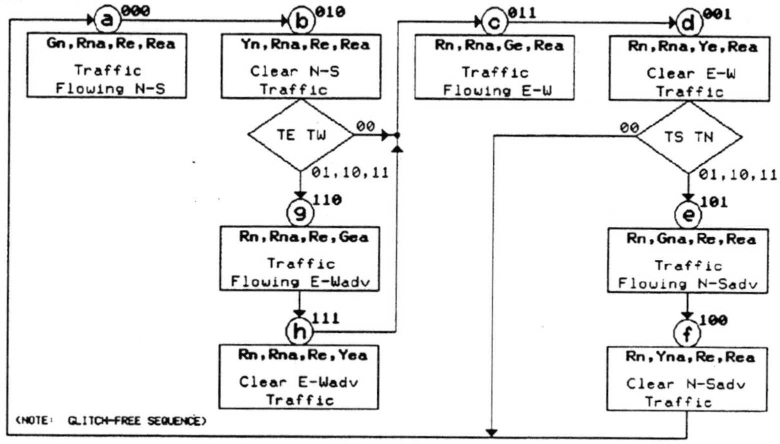

The following is the resulting state table. Again, note the addition of the two states and the addition of the E-W advance turn signal set. The state table and CUPL program are illustrated below.

	PRES ABC	TTTT SNEW	NEXT ABC	D_A	D_B	D_C	Gs Gn	Ys Yn	Rs Rn	Gsa Gna	Ysa Yna	Rsa Rna	Gw Ge	Yw Ye	Rw Re	Gwa Gea	Ywa Yea	Rwa Rea
a	000	****	010	0	1	0	1	0	0	0	0	1	0	0	1	0	0	1
d	001	00** 01** 10** 11**	000 101 101 101	S+N	0	S+N	0	0	1	0	0	1	0	1	0	0	0	1
b	010	**00 **01 **10 **11	011 110 110 110	E+W	1	$\overline{E}\overline{W}$	0	1	0	0	0	1	0	0	1	0	0	1
c	011	****	001	0	0	1	0	0	1	0	0	1	1	0	0	0	0	1
f	100	****	000	0	0	0	0	0	1	0	1	0	0	0	1	0	0	1
e	101	****	100	1	0	0	0	0	1	1	0	0	0	0	1	0	0	1
g	110	****	111	1	1	1	0	0	1	0	0	1	*	*	1	1	0	0
h	111	****	011	0	1	1	0	0	1	0	0	1	*	*	1	0	1	0

$$DA = \overline{A}\overline{B}CS+\overline{A}\overline{B}CN+\overline{A}B\overline{C}E+\overline{A}B\overline{C}W+A\overline{B}C+AB\overline{C}$$

$$DB = \overline{A}\overline{C} + AB$$

$$DC = \overline{A}\overline{B}CS + \overline{A}\overline{B}CN + \overline{A}B\overline{C}\overline{E}\overline{W} + BC + AB$$

$GN = GS = \overline{A}\overline{B}\overline{C}$ | $GE = GW = \overline{A}BC$

$YN = YS = \overline{A}B\overline{C}$ | $YE = YW = \overline{A}\overline{B}C$

$RN = RS = A+C$ | $RE = RW = A+\overline{C}$

$GNA = GSA = A\overline{B}C$ | $GEA = GWA = AB\overline{C}$

$YNA = YSA = A\overline{B}\overline{C}$ | $YEA = YWA = ABC$

$RNA = RSA = \overline{A}+B$ | $REA = RWA = \overline{A}+\overline{B}$

```
Name            P05-06a;
Revision        01;
Date            05/23/96;
Designer        J. Carter;
Location        None;
Assembly        None;
Device          P16R4;
Format          j;
/***********************************************************************/
/*  Practical Example 5.6                                              */
/*  Four-Way Traffic Controller with NS/EW Advanced Turn Lane  Chip 1 of 2  */
/*      Opposite traffic pattern controls are similar                  */
/***********************************************************************/
/* INPUTS */                PIN  2 = S;     /* Sensor TS  (South turn lane) */
                            PIN  3 = N;     /* Sensor TN  (North turn lane) */
                            PIN  4 = E;     /* Sensor TE  (East turn lane)  */
                            PIN  5 = W;     /* Sensor TW  (West turn lane)  */
/* INPUT LOGIC */           PIN 17 = A;
                            PIN 16 = B;
                            PIN 15 = C;
/* OUTPUT LOGIC */          PIN 19 = GNA;   /* Lamp Sets NA and SA */
                            PIN 18 = YNA;
                            PIN 13 = RNA;
                            PIN 12 = GEA;   /* Lamp Sets EA and WA */
/* EXPRESSIONS */
 A.D = (!A&!B& C& S) # (!A&!B& C& N) # (!A& B&!C& E) # !(A& B&!C& W) # (A&!B& C) # (A& B&!C);
 B.D = (!A&!C) & (A & B);
 C.D = (!A&!B& C& S) # (!A&!B& C& N) # (!A& B&!C&!E&!W) # (B &  C) # (A &  B);
 GNA = ( A&!B& C);
 YNA = ( A&!B&!C);
 RNA =  !A # B;
 GEA = ( A& B&!C);

Name            P05-06b;
Revision        01;
Date            05/23/96;
Designer        J. Carter;
Location        None;
Assembly        None;
Device          P16L8;
Format          j;
/***********************************************************************/
/*  Practical Example 5.6                                              */
/*  Four-Way Traffic Controller with NS/EW Advanced Turn Lane   Chip 2 of 2 */
/*      Opposite traffic pattern controls are similar                  */
/***********************************************************************/
/* INPUTS */                PIN  2 = A;     /* Counter bits from Chip 1     */
                            PIN  3 = B;
                            PIN  4 = C;
/* OUTPUT LOGIC */          PIN 19 = GN;    /* Lamp Sets N and S */
                            PIN 18 = YN;
                            PIN 17 = RN;
                            PIN 16 = GE;    /* Lamp Sets E and W */
                            PIN 15 = YE;
                            PIN 14 = RE;
                            PIN 13 = YEA;   /* Lamp Sets EA and WA */
                            PIN 12 = REA;
/* EXPRESSIONS */
        GN  = (!A &!B&!C);   YN  = (!A & B&!C);   RN  =   A # C;
        GE  = (!A & B& C);   YE  = (!A &!B& C);   RE  =   A # !C;
                             YEA = ( A & B& C);   REA =  !A # !B;
```

Again, the previous solution presupposes that opposite traffic signal controls are similar. If they are not, then the problem becomes considerably more complex. For example, if a vehicle is waiting in the North Advance turn lane when there is none in the South advance turn lane, it is more efficient to allow Southbound traffic to flow while the vehicle is taking advantage of the advanced turn. If this is applied to all four directions, the following state diagram may result. This solution has 16 states, enabling the use of a four-bit counter. Also, casual observation of the traffic patterns reveals that there are no common opposite control signals, so each of the 24 lamps must be independently controlled, requiring additional PAL outputs in the design.

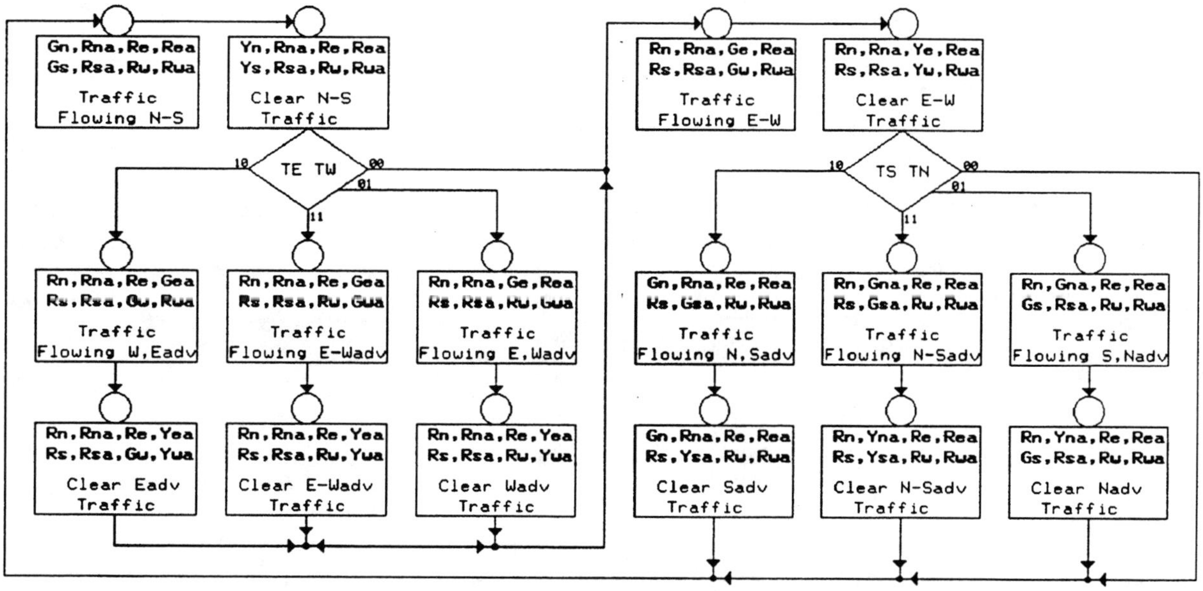

Practical Exercise 7.1: Simple ROM-Based Counter.

Consider the state machine defined by the following figure. Draw the state table, ROM program, and a schematic diagram of a ROM-based state machine which exhibits these characteristics.

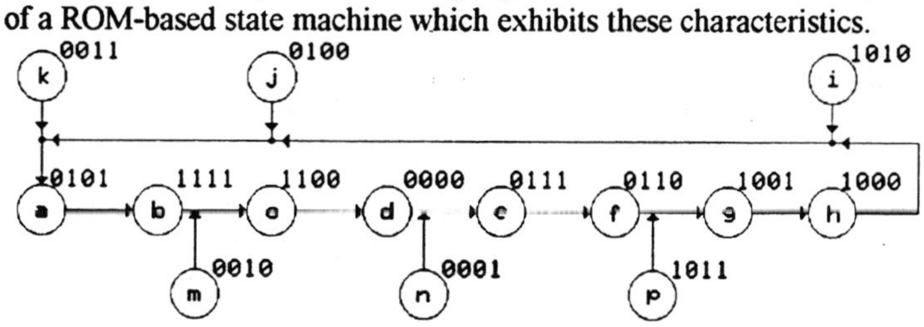

Solution:

STATE TABLE PRESENT				NEXT				ROM PROGRAM ADDRESS				DATA			
A	B	C	D	A	B	C	D	A_3	A_2	A_1	A_0	D_3	D_2	D_1	D_0
0	0	0	0	0	1	1	1	0	0	0	0	0	1	1	1
0	0	0	1	0	1	1	1	0	0	0	1	0	1	1	1
0	0	1	0	1	1	0	0	0	0	1	0	1	1	0	0
0	0	1	1	0	1	0	1	0	0	1	1	0	1	0	1
0	0	1	0	1	1	0	0	0	0	1	0	1	1	0	0
0	0	1	1	0	1	0	1	0	0	1	1	0	1	0	1
0	1	0	0	0	1	0	1	0	1	0	0	0	1	0	1
0	1	0	1	1	1	1	1	0	1	0	1	1	1	1	1
0	1	1	0	1	0	0	1	0	1	1	0	1	0	0	1
0	1	1	1	0	1	1	0	0	1	1	1	0	1	1	0
1	0	0	0	0	1	0	1	1	0	0	0	0	1	0	1
1	0	0	1	1	0	0	0	1	0	0	1	1	0	0	0
1	0	1	0	0	1	0	1	1	0	1	0	0	1	0	1
1	0	1	1	1	0	0	1	1	0	1	1	1	0	0	1
1	1	0	0	0	0	0	0	1	1	0	0	0	0	0	0
1	1	0	1	X	X	X	X	1	1	0	1	0	0	0	0
1	1	1	0	X	X	X	X	1	1	1	0	0	0	0	0
1	1	1	1	1	1	0	0	1	1	1	1	1	1	0	0

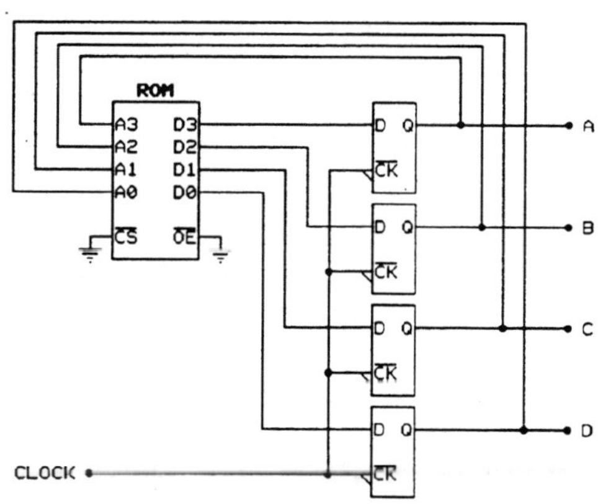

Practical Exercise 7.2. Handling Multiple Outputs in ROM-Based Designs.

Consider the state machine defined by Figure 7.9. Develop the state diagram, schematic diagram, state table and ROM program for a ROM-based solution for this state machine.

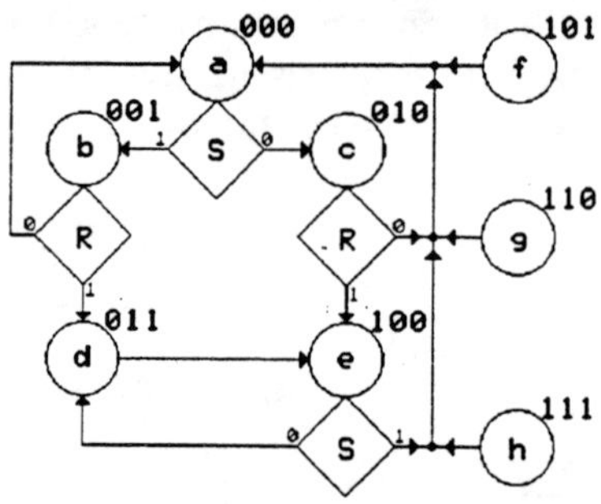

Solution: Note that each state in the state table requires four ROM addresses, since there are two input stimuli, requiring four permutations of input possibilities. Stored at each o the four subsequent addresses is the output data for each specific permutation. Note how the use of the "don't care" states in the CUPL program were used to simplify the coding.

STATE TABLE

	PRES ABC	INPUT RS	NEXT ABC
a	000	*0	010
		*1	001
b	001	0*	000
		1*	011
c	010	*0	000
		*1	100
d	011	**	100
e	100	*0	011
		*1	000
f	101	**	000
g	110	**	000
h	111	**	000

ROM PROGRAM

	A A_4	B A_3	C A_2	R A_1	S A_0	A D_2	B D_1	C D_0
a	0	0	0	0	0	0	1	0
	0	0	0	0	1	0	0	1
	0	0	0	1	0	0	1	0
	0	0	0	1	1	0	0	1
b	0	0	1	0	0	0	0	0
	0	0	1	0	1	0	0	0
	0	0	1	1	0	0	1	1
	0	0	1	1	1	0	1	1
c	0	1	0	0	0	0	0	0
	0	1	0	0	1	1	0	0
	0	1	0	1	0	0	0	0
	0	1	0	1	1	1	0	0
d	0	1	1	0	0	1	0	0
	0	1	1	0	1	1	0	0
	0	1	1	1	0	1	0	0
	0	1	1	1	1	1	0	0

ROM PROGRAM

	A A_4	B A_3	C A_2	R A_1	S A_0	A D_2	B D_1	C D_0
e	1	0	0	0	0	0	1	1
	1	0	0	0	1	0	0	0
	1	0	0	1	0	0	1	1
	1	0	0	1	1	0	0	0
f	1	0	1	0	0	0	0	0
	1	0	1	0	1	0	0	0
	1	0	1	1	0	0	0	0
	1	0	1	1	1	0	0	0
g	1	1	0	0	0	0	0	0
	1	1	0	0	1	0	0	0
	1	1	0	1	0	0	0	0
	1	1	0	1	1	0	0	0
h	1	1	1	0	0	0	0	0
	1	1	1	0	1	0	0	0
	1	1	1	1	0	0	0	0
	1	1	1	1	1	0	0	0

```
Name       P07-02;
Date       05/24/96;
Designer   J.W. Carter;
Device     RA8P4;
/*********************************************************************/
/* Practical Exercise 7.2                                           */
/*      State machine with three inputs, uses DM74S387 256x4 PROM */
/*      TABLE construct is built from the state table directly.    */
/*********************************************************************/
/**  Inputs  **/
PIN [2..1]    = [A1..A0];          /* Inputs R, S                  */
PIN [5..3]    = [A4..A2];          /* Present State, A, B, C       */
/**  Outputs  **/
PIN [11..9]   = [D2..D0];          /* Next State                   */
FIELD input  = [A4..A0];
FIELD output = [D2..D0];
TABLE input =>output  {
/* State a */   'b'000X0 => 'b'010;
                'b'000X1 => 'b'001;
/* State b */   'b'0010X => 'b'000;
                'b'0011X => 'b'011;
/* State c */   'b'010X0 => 'b'000;
                'b'010X1 => 'b'100;
/* State d */   'b'001XX => 'b'100;
/* State e */   'b'100X0 => 'b'011;
                'b'100X1 => 'b'000;
/* State f */   'b'101XX => 'b'000;
/* State g */   'b'110XX => 'b'000;
/* State h */   'b'111XX => 'b'000; }
```

Practical Exercise 7.3. A Simple Soda Machine Controller.

Consider the state machine defined by Practical Example 7.4. Develop the CUPL program from the ROM program. This is probably an exercise in tedium, but the effort will help to hone the skill of program writing.

```
Name        P07-03;
Date        05/24/96;
Designer    J.W. Carter;
Device      RA8P8;
/*******************************************************************/
/* Practical Exercise 7.3                                          */
/*     Soda Machine Controller using a DM74S471 256 x 8 PROM       */
/*******************************************************************/
/**  Inputs  **/
        PIN [5..1]   = [A4..A0];  /* Present State & Inputs */
        PIN [18..17] = [A6..A5];
/**  Outputs  **/
        PIN [9..6]   = [D3..D0];  /* Next State             */
        PIN [12..11] = [D5..D4];  /* Outputs                */
FIELD input =  [A6..A0];
FIELD output = [D5..D0];
TABLE input =>output  {
/* State a */      'b'0000000 => 'b'000000;   'b'0000001 => 'b'000000;
                   'b'0000010 => 'b'000000;   'b'0000011 => 'b'000000;
                   'b'0000100 => 'b'000000;   'b'0000101 => 'b'000000;
                   'b'0000100 => 'b'000000;   'b'0000111 => 'b'000000;
                   'b'0001000 => 'b'001000;   'b'0001001 => 'b'001000;
                   'b'0001010 => 'b'001000;   'b'0001011 => 'b'001000;
                   'b'0001100 => 'b'001000;   'b'0001101 => 'b'001000;
                   'b'0001110 => 'b'001000;   'b'0001111 => 'b'001000;
/* State b */      'b'0010000 => 'b'001000;   'b'0010001 => 'b'001000;
                   'b'0010010 => 'b'001000;   'b'0010011 => 'b'001000;
                   'b'0010100 => 'b'100000;   'b'0010101 => 'b'100000;
                   'b'0010110 => 'b'100000;   'b'0010111 => 'b'100000;
                   'b'0011000 => 'b'010000;   'b'0011001 => 'b'010000;
                   'b'0011010 => 'b'010000;   'b'0011011 => 'b'010000;
/* State c */      'b'0100000 => 'b'010000;   'b'0100001 => 'b'000001;
                   'b'0100010 => 'b'000010;   'b'0100100 => 'b'100100;
                   'b'0101000 => 'b'010000;   'b'0101001 => 'b'000001;
                   'b'0101010 => 'b'000010;   'b'0101100 => 'b'100100;
/* State d */      'b'1000000 => 'b'000100;   'b'1000001 => 'b'000100;
                   'b'1000010 => 'b'000100;   'b'1000011 => 'b'000100;
                   'b'1000X00 => 'b'000100;   'b'1000101 => 'b'000100;
                   'b'1000110 => 'b'000100;   'b'1000111 => 'b'000100;
                   'b'1001000 => 'b'000100;   'b'1001001 => 'b'000100;
                   'b'1001010 => 'b'000100;   'b'1001011 => 'b'000100;
                   'b'1001100 => 'b'000100;   'b'1001101 => 'b'000100;
                   'b'1001110 => 'b'000100;   'b'1001111 => 'b'000100;  }
```

Practical Exercise 7.4: A Simple Candy Machine Controller.

This is the same problem as Practical Exercise 4.6. Implement with a ROM. OE connected to !CLOCK

	$A_{8..6}$ PRES ABC	A_5 NA N	A_4 Qi Q	A_3 CRi R	A_2 CDi C	A_1 PCi P	A_0 CGi G	$D_{6..4}$ NEXT ABC	D_3 Qo	D_2 CBo	D_1 PCo	D_0 CGo
a	000	1	0	0	0	0	0	000	0	0	0	0
		0	1	0	0	0	0	001	0	0	0	0
b	001	1	0	0	0	0	0	001	0	0	0	0
		0	1	0	0	0	0	010	0	0	0	0
		0	0	1	0	0	0	100	0	0	0	0
		0	0	0	0	0	1	000	0	0	0	1
c	010	1	0	0	0	0	0	010	0	0	0	0
		0	1	0	0	0	0	010	1	0	0	0
		0	0	1	0	0	0	100	1	0	0	0
		0	0	0	1	0	0	000	0	1	0	0
		0	0	0	0	1	0	000	0	0	1	0
		0	0	0	0	0	1	100	0	0	0	1
d	100	X	X	X	X	X	X	000	1	0	0	0

```
Name        P07-04;
Date        05/24/96;
Designer    J.W. Carter;
Device      RA9P8;
/*******************************************************************/
/* Practical Exercise 7.4                                          */
/*     Candy Machine Controller using a DM74S473 512 x 8 PROM      */
/*     Pin 15, Output Enable connected to !CLOCK                   */
/*******************************************************************/
/**  Inputs  **/
PIN [3..1]   = [A8..A6];           /* Present State A, B, C     */
```

```
PIN 4 = A5;                         /* NA, No Activity            */
PIN 5 = A4;                         /* Qi, Quarter Input          */
PIN 16 = A3;                        /* CRi, Coin Return Input     */
PIN 17 = A2;                        /* CBi, Chocolate Bar Input   */
PIN 18 = A1;                        /* PCi, Potato Chip Input     */
PIN 19 = A0;                        /* CGi, Chewing Gum Input     */
/**  Outputs  **/
PIN [8..6]   = [D6..D4];        /* Next State                 */
PIN 9 = D3;                         /* Qo, Quarter Output         */
PIN 11 = D2;                        /* CBo, Chocolate Bar Output  */
PIN 12 = D1;                        /* PCo, Potato Chip Output    */
PIN 13 = D0;                        /* CGo, Chewing Gum Output    */
FIELD input =  [A8..A0];
FIELD output = [D6..D0];
TABLE input => output  {
/*                   A876543210       D6543210  */
/* State a */       'b'000100000 => 'b'0000000;
                    'b'000010000 => 'b'0010000;
/* State b */       'b'001100000 => 'b'0010000;
                    'b'001010000 => 'b'0100000;
                    'b'001001000 => 'b'1000000;
                    'b'001000001 => 'b'0000001;
/* State c */       'b'010100000 => 'b'0100000;
                    'b'010010000 => 'b'0101000;
                    'b'010001000 => 'b'1001000;
                    'b'010000100 => 'b'0000100;
                    'b'010000010 => 'b'0000010;
                    'b'010000001 => 'b'1000001;
/* State d */       'b'100XXXXXX => 'b'0001000; }
```

Practical Exercise 7.5: A Moderate Candy Machine Controller

(Note this problem is similar to Practical Exercise 4.7)

The state table for the solution is illustrated on the next page. This design requires the use of a ROM which has 12 address lines and 12 data lines. Or, two ROMs can be combined to provide the12-bit wide data word. For example a 4K x 4 ROM can be used to implement the four-bit counter, and a 4K x 8 ROM can be used to implement the output forming logic. The resulting CUPL table structure is illustrated below:

```
FIELD input =  [A11..A0];
FIELD output = [D11..D0];
TABLE input => output  {
/*                   A109876543210         D109876543210                A109876543210          D109876543210 */
/* State a */       'b'000010000000 => 'b'000000000000;           'b'000001000000 => 'b'101000000000;
                    'b'000000100000 => 'b'011000000000;           'b'000000010000 => 'b'000100000000;
/* State q */       'b'0001XXXXXXXX => 'b'000100000000;
/* State f */       'b'001010000000 => 'b'001000000000;           'b'001001000000 => 'b'001010000000;
                    'b'001000100000 => 'b'001001000000;           'b'001000010000 => 'b'000100000000;
                    'b'001000001000 => 'b'000010001000;           'b'001000000100 => 'b'100000000100;
                    'b'001000000010 => 'b'000000000010;           'b'001000000001 => 'b'000000000001;
/* State h */       'b'001110000000 => 'b'001100000000;           'b'001101000000 => 'b'110100000000;
                    'b'001100100000 => 'b'010000000000;           'b'001100010000 => 'b'000100000000;
                    'b'001100001000 => 'b'100000001000;
/* State j */       'b'010010000000 => 'b'010000000000;           'b'010001000000 => 'b'110101000000;
                    'b'010000100000 => 'b'011100000000;           'b'010000010000 => 'b'000100000000;
                    'b'010000001000 => 'b'100001001000;           'b'010000000100 => 'b'101100000100;
/* State d */       'b'010110000000 => 'b'010100000000;           'b'010101000000 => 'b'011100000000;
                    'b'010100100000 => 'b'100100000000;           'b'010100010000 => 'b'000100000000;
                    'b'010100001000 => 'b'101100001000;
/* State b */       'b'011010000000 => 'b'011000000000;           'b'011001000000 => 'b'001100000000;
                    'b'011000100000 => 'b'110000000000;           'b'011000010000 => 'b'000100000000;
/* State k */       'b'0111XXXXXXXX => 'b'001000000000;
/* State p */       'b'1000XXXXXXXX => 'b'000000000000;
/* State e */       'b'100110000000 => 'b'100100000000;           'b'100101000000 => 'b'011101000000;
                    'b'100100100000 => 'b'001000000000;           'b'100100010000 => 'b'000100000000;
                    'b'100100001000 => 'b'100000101000;           'b'100100000100 => 'b'000000000100;
/* State g */       'b'101010000000 => 'b'101000000000;           'b'101001000000 => 'b'001000000000;
                    'b'101000100000 => 'b'001100000000;           'b'101000010000 => 'b'000100000000;
                    'b'101000001000 => 'b'000000001000;
/* State m */       'b'1011XXXXXXXX => 'b'000000100000;
/* State c */       'b'110010000000 => 'b'110000000000;           'b'110001000000 => 'b'010000000000;
                    'b'110000100000 => 'b'010100000000;           'b'110000010000 => 'b'000100000000;
/* State n */       'b'1101XXXXXXXX => 'b'001001000000; }
```

	$A_{11..8}$ PRES ABCD	A_7 NA	A_6 Qi	A_5 Di	A_4 CRi	A_3 CGi	A_2 CCi	A_1 CBi	A_0 PCi	$D_{11..8}$ NEXT ABCD	D_7 Qo	D_6 Do	D_5 No	D_4 CRo	D_3 CGo	D_2 CCo	D_1 CBo	D_0 PCo
a	0000	1	0	0	0	0	0	0	0	0000	0	0	0	0	0	0	0	0
		0	1	0	0	0	0	0	0	1010	0	0	0	0	0	0	0	0
		0	0	1	0	0	0	0	0	0110	0	0	0	0	0	0	0	0
		0	0	0	1	0	0	0	0	0001	0	0	0	0	0	0	0	0
q	0001	*	*	*	*	*	*	*	*	0001	0	0	0	1	0	0	0	0
f	0010	1	0	0	0	0	0	0	0	0010	0	0	0	0	0	0	0	0
		0	1	0	0	0	0	0	0	0010	1	0	0	0	0	0	0	0
		0	0	1	0	0	0	0	0	0010	0	1	0	0	0	0	0	0
		0	0	0	1	0	0	0	0	0001	0	0	0	0	0	0	0	0
		0	0	0	0	1	0	0	0	0000	1	0	0	0	1	0	0	0
		0	0	0	0	0	1	0	0	1000	0	0	0	0	0	1	0	0
		0	0	0	0	0	0	1	0	0000	0	0	0	0	0	0	1	0
		0	0	0	0	0	0	0	1	0000	0	0	0	0	0	0	0	1
h	0011	1	0	0	0	0	0	0	0	0011	0	0	0	0	0	0	0	0
		0	1	0	0	0	0	0	0	1101	0	0	0	0	0	0	0	0
		0	0	1	0	0	0	0	0	0100	0	0	0	0	0	0	0	0
		0	0	0	1	0	0	0	0	0001	0	0	0	0	0	0	0	0
		0	0	0	0	1	0	0	0	1000	0	0	0	0	1	0	0	0
j	0100	1	0	0	0	0	0	0	0	0100	0	0	0	0	0	0	0	0
		0	1	0	0	0	0	0	0	1101	0	1	0	0	0	0	0	0
		0	0	1	0	0	0	0	0	0111	0	0	0	0	0	0	0	0
		0	0	0	1	0	0	0	0	0001	0	0	0	0	0	0	0	0
		0	0	0	0	1	0	0	0	1000	0	1	0	0	1	0	0	0
		0	0	0	0	0	1	0	0	1011	0	0	0	0	0	1	0	0
d	0101	1	0	0	0	0	0	0	0	0101	0	0	0	0	0	0	0	0
		0	1	0	0	0	0	0	0	0111	0	0	0	0	0	0	0	0
		0	0	1	0	0	0	0	0	1001	0	0	0	0	0	0	0	0
		0	0	0	1	0	0	0	0	0001	0	0	0	0	0	0	0	0
		0	0	0	0	1	0	0	0	1011	0	0	0	0	1	0	0	0
b	0110	1	0	0	0	0	0	0	0	0110	0	0	0	0	0	0	0	0
		0	1	0	0	0	0	0	0	0011	0	0	0	0	0	0	0	0
		0	0	1	0	0	0	0	0	1100	0	0	0	0	0	0	0	0
		0	0	0	1	0	0	0	0	0001	0	0	0	0	0	0	0	0
k	0111	*	*	*	*	*	*	*	*	0010	0	0	1	0	0	0	0	0
p	1000	*	*	*	*	*	*	*	*	0000	0	1	0	0	0	0	0	0
e	1001	1	0	0	0	0	0	0	0	1001	0	0	0	0	0	0	0	0
		0	1	0	0	0	0	0	0	0111	0	1	0	0	0	0	0	0
		0	0	1	0	0	0	0	0	0010	0	0	0	0	0	0	0	0
		0	0	0	1	0	0	0	0	0001	0	0	0	0	0	0	0	0
		0	0	0	0	1	0	0	0	1000	0	0	1	0	1	0	0	0
		0	0	0	0	0	1	0	0	0000	0	0	0	0	0	1	0	0
g	1010	1	0	0	0	0	0	0	0	1010	0	0	0	0	0	0	0	0
		0	1	0	0	0	0	0	0	0010	0	0	0	0	0	0	0	0
		0	0	1	0	0	0	0	0	0011	0	0	0	0	0	0	0	0
		0	0	0	1	0	0	0	0	0001	0	0	0	0	0	0	0	0
		0	0	0	0	1	0	0	0	0000	0	0	0	0	1	0	0	0
m	1011	*	*	*	*	*	*	*	*	0000	0	0	1	0	0	0	0	0
c	1100	1	0	0	0	0	0	0	0	1100	0	0	0	0	0	0	0	0
		0	1	0	0	0	0	0	0	0100	0	0	0	0	0	0	0	0
		0	0	1	0	0	0	0	0	0101	0	0	0	0	0	0	0	0
		0	0	0	1	0	0	0	0	0001	0	0	0	0	0	0	0	0
n	1101	*	*	*	*	*	*	*	*	0010	0	1	0	0	0	0	0	0
	1110	*	*	*	*	*	*	*	*	****	*	*	*	*	*	*	*	*
	1111	*	*	*	*	*	*	*	*	****	*	*	*	*	*	*	*	*

Practical Exercise 7.6: Cutter Controller, Pattern 1 Algorithm.

State 0:	CUTTER_UP; next 1;		
State 1:	CUTTER_UP;		
	GO_UP; GO_RIGHT;	IF (MAX_UP = 1) or (MAX_RIGHT = 1) then next 11 else next 2;	
State 2:	CUTTER_DOWN;		
	GO_UP;	IF MAX_UP=1 then next 11	else next 3;
State 3:	GO_RIGHT;	IF MAX_RIGHT=1 then next 11	else next 4;
State 4:	GO_UP;	IF MAX_UP=1 then next 11	else next 5;
State 5:	GO_RIGHT;	IF MAX_RIGHT=1 then next 11	else next 6;
State 6:	GO_DOWN;	IF MAX_DOWN=1 then next 11	else next 7;
State 7:	GO_DOWN;	IF MAX_DOWN=1 then next 11	else next 8;
State 8:	GO_LEFT;	IF MAX_LEFT=1 then next 11	else next 9;
State 9:	GO_LEFT;	IF MAX_LEFT=1 then next 11	else next 10;
State 10:	GO LEFT; GO DOWN;	next 11;	
State 11:	CUTTER_UP;	next 11;	

Practical Exercise 7.7: Cutter Controller, Pattern 1 State Diagram.

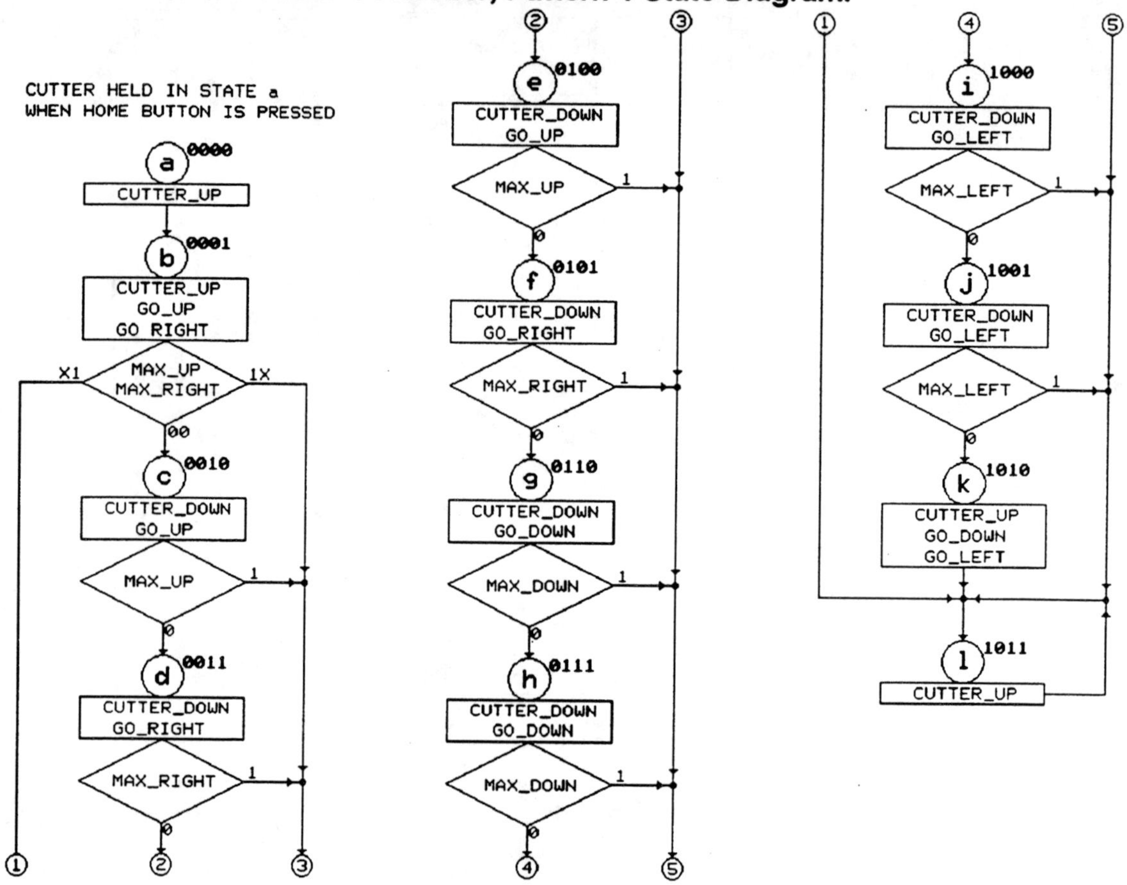

Practical Exercise 7.8: Cutter Controller Pattern 2.

Develop the state diagram, state table, and CUPL program for the pattern illustrated in the following figure.

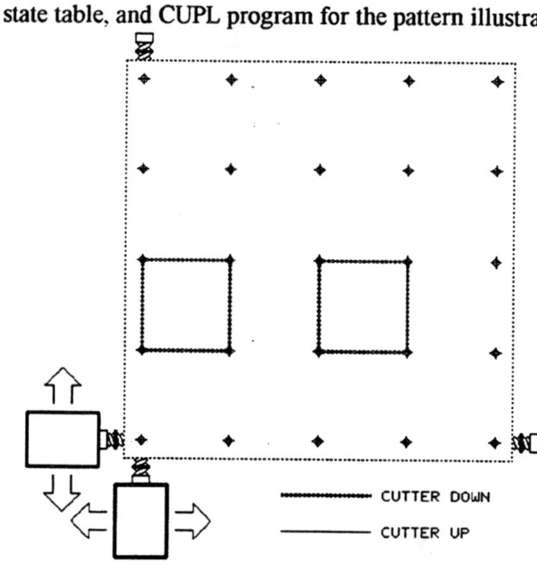

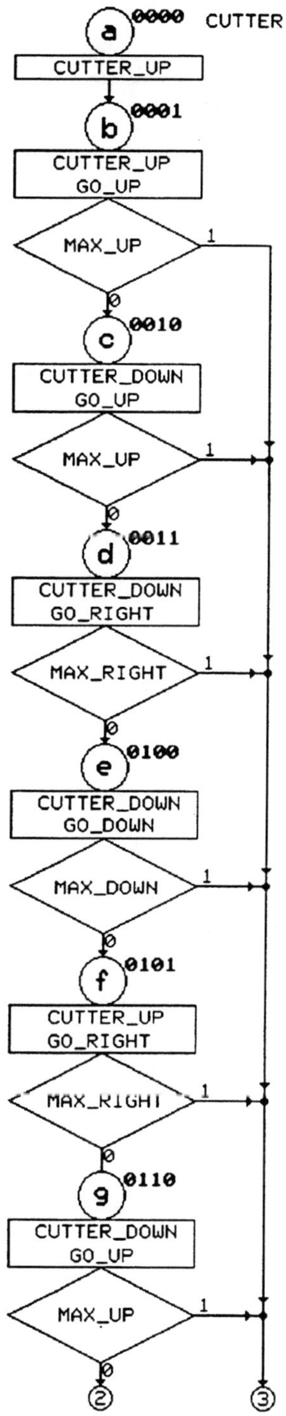

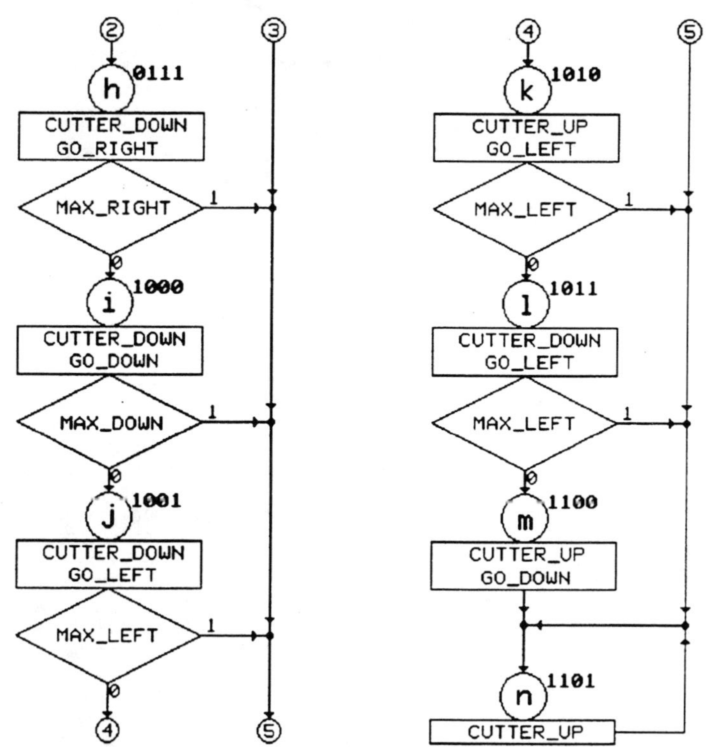

	PARTITION P3210	PRESENT ABCD	INPUTS S3210	NEXT ABCD	CUTTER C_DN	MOTORS G_UP	G_DN	G_RT	G_LF
a	0010	0000	XXXX	0001	0	0	0	0	0
b	0010	0001	0XXX 1XXX	0010 1101	0	1	0	0	0
c	0010	0010	0XXX 1XXX	0011 1101	1	1	0	0	0
d	0010	0011	XX0X XX1X	0100 1101	1	0	0	1	0
e	0010	0100	X0XX X1XX	0101 1101	1	0	1	0	0
f	0010	0101	XX0X XX1X	0110 1101	1	0	0	1	0
g	0010	0110	0XXX 1XXX	0111 1101	1	1	0	0	0
h	0010	0111	XX0X XX1X	1000 1101	1	0	0	1	0
i	0010	1000	X0XX X1XX	1001 1101	1	0	1	0	0
j	0010	1001	XXX0 XXX1	1010 1101	1	0	0	0	1
k	0010	1010	XXX0 XXX1	1011 1101	0	0	0	0	1
l	0010	1011	XXX0 XXX1	1100 1101	1	0	0	0	0
m	0010	1100	XXXX	1101	0	0	1	0	0
n	0010	1101	XXXX	1011	0	0	0	0	0

```
Name      P07-08;
Date      05/25/96;
Designer  J.W. Carter;
Device    RA12P9;
/*****************************************************************/
/* Practical Exercise 7.8: Pattern Cutter using a 4096 x 9 PROM */
/*****************************************************************/
/**  Inputs  **/
PIN [01..04] = [A11..A08];      /* Partition Inputs P3..P0       */
PIN [05..08] = [A07..A04];      /* Counter Bits A, B, C, D       */
PIN [09..10] = [A03..A02];      /* Input Stimulus S3..S0         */
PIN [13..14] = [A01..A00];
/**  Outputs  **/
PIN [15..19] = [D00..D04];      /* Output Forming Logic          */
PIN [20..23] = [D05..D08];      /* Next Address Bits             */
FIELD input  = [A11..A00];
FIELD output = [D08..D00];
TABLE input  => output  {
*/ PATTERN 2    Note: X-Y motors disabled in an error condition. */
       /*                 PPPP    SSSS           DDDDD  */
       /*                 3210ABCD3210           ABCD43210  */
       /* State a */  'b'00100000XXXX => 'b'000100000;
       /* State b */  'b'001000010XXX => 'b'001001000;
                      'b'001000011XXX => 'b'110100000;
       /* State c */  'b'001000100XXX => 'b'001111000;
                      'b'001000101XXX => 'b'110100000;
       /* State d */  'b'00100011XX0X => 'b'010010010;
                      'b'00100011XX1X => 'b'110100000;
       /* State e */  'b'00100100X0XX => 'b'010110100;
                      'b'00100100X1XX => 'b'110100000;
       /* State f */  'b'00100101XX0X => 'b'011010010;
                      'b'00100101XX1X => 'b'110100000;
       /* State g */  'b'001001100XXX => 'b'011111000;
                      'b'001001101XXX => 'b'110100000;
       /* State h */  'b'00100111XX0X => 'b'100010010;
                      'b'00100111XX1X => 'b'110100000;
       /* State i */  'b'00101000X0XX => 'b'100110100;
                      'b'00101000X1XX => 'b'110100000;
       /* State j */  'b'00101001XXX0 => 'b'101010001;
                      'b'00101001XXX1 => 'b'110100000;
       /* State k */  'b'00101010XXX0 => 'b'101100001;
                      'b'00101010XXX1 => 'b'110100000;
       /* State l */  'b'00101011X0XX => 'b'110010000;
                      'b'00101011X1XX => 'b'110100000;
       /* State m */  'b'00101100XXXX => 'b'110100100;
       /* State n */  'b'00101101XXXX => 'b'110100000;
```

Practical Exercise 7.9: Cutter Controller Pattern 3.

Develop the state diagram, state table, and CUPL program for the pattern illustrated in the following figure.

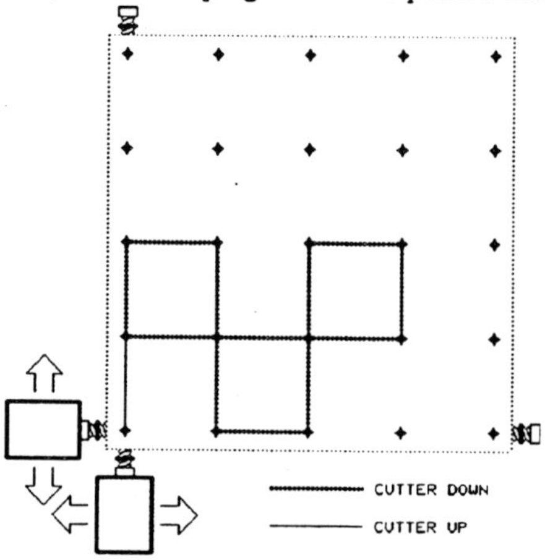

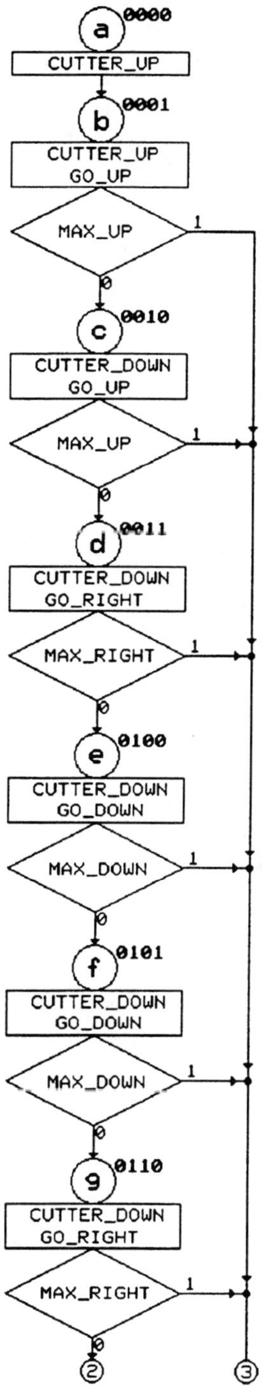

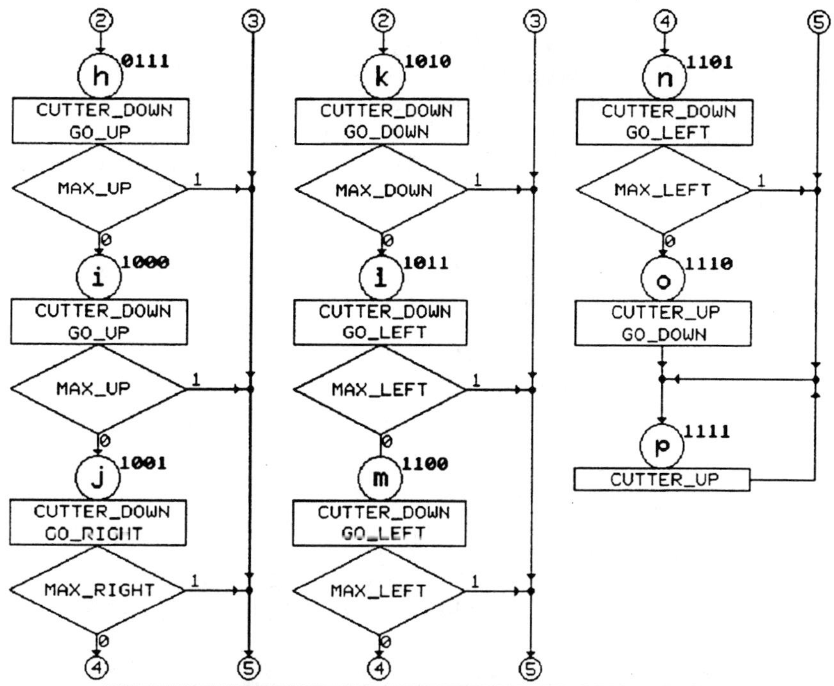

	PARTITION P3210	PRESENT ABCD	INPUTS S3210	NEXT ABCD	CUTTER C_DN	MOTORS G_UP	G_DN	G_RT	G_LF
a	0011	0000	XXXX	0001	0	0	0	0	0
b	0011	0001	0XXX 1XXX	0010 1111	0	1	0	0	0
c	0011	0010	0XXX 1XXX	0011 1111	1	1	0	0	0
d	0011	0011	XX0X XX1X	0100 1111	1	0	0	1	0
e	0011	0100	X0XX X1XX	0101 1111	1	0	1	0	0
f	0011	0101	X0XX X1XX	0110 1111	1	0	1	0	0
g	0011	0110	XX0X XX1X	0111 1111	1	0	0	1	0
h	0011	0111	0XXX 1XXX	1000 1111	1	1	0	0	0
i	0011	1000	0XXX 1XXX	1001 1111	1	1	0	0	0
j	0011	1001	XX0X XX1X	1010 1111	1	0	0	1	0
k	0011	1010	X0XX X1XX	1011 1111	1	0	1	0	0
l	0011	1011	XXX0 XXX1	1100 1111	1	0	0	0	1
m	0011	1100	XXX0 XXX1	1101 1111	1	0	0	0	1
n	0011	1101	XXX0 XXX1	1110 1111	1	0	0	0	1
o	0011	1110	XXXX	1110	0	0	1	0	0
p	0011	1111	XXXX	1111	0	0	0	0	0

```
Name      P07-09;
Date      05/25/96;
Designer  J.W. Carter;
Device    RA12P9;
/*****************************************************************/
/* Practical Exercise 7.9: Pattern Cutter using a 4096 x 9 PROM */
/*****************************************************************/
/**  Inputs  **/
PIN [01..04] = [A11..A08];      /* Partition Inputs P3..P0      */
PIN [05..08] = [A07..A04];      /* Counter Bits A, B, C, D      */
PIN [09..10] = [A03..A02];      /* Input Stimulus S3..S0        */
PIN [13..14] = [A01..A00];
/**  Outputs  **/
PIN [15..19] = [D00..D04];      /* Output Forming Logic         */
PIN [20..23] = [D05..D08];      /* Next Address Bits            */
FIELD input  = [A11..A00];
FIELD output = [D08..D00];
TABLE input  => output  {
*/ PATTERN 3   Note: X-Y motors disabled in an error condition. */
      /*              PPPP    SSSS          DDDDD  */
      /*              3210ABCD3210        ABCD43210  */
      /* State a */ 'b'00110000XXXX => 'b'000100000;
      /* State b */ 'b'001100010XXX => 'b'001001000;
                    'b'001100011XXX => 'b'111100000;
      /* State c */ 'b'001100100XXX => 'b'001111000;
                    'b'001100101XXX => 'b'111100000;
      /* State d */ 'b'00110011XX0X => 'b'010010010;
                    'b'00110011XX1X => 'b'111100000;
      /* State e */ 'b'00110100X0XX => 'b'010110100;
                    'b'00110100X1XX => 'b'111100000;
      /* State f */ 'b'00110101X0XX => 'b'011010100;
                    'b'00110101X1XX => 'b'111100000;
      /* State g */ 'b'00110110XX0X => 'b'011110010;
                    'b'00110110XX1X => 'b'111100000;
      /* State h */ 'b'001101110XXX => 'b'100011000;
                    'b'001101111XXX => 'b'111100000;
      /* State i */ 'b'001110000XXX => 'b'100111000;
                    'b'001110001XXX => 'b'111100000;
      /* State j */ 'b'00111001XX0X => 'b'101010010;
                    'b'00111001XX1X => 'b'111100000;
      /* State k */ 'b'00111010X0XX => 'b'101110100;
                    'b'00111010X1XX => 'b'111100000;
      /* State l */ 'b'00111011XXX0 => 'b'110010001;
                    'b'00111011XXX1 => 'b'111100000;
      /* State m */ 'b'00111100XXX0 => 'b'110110001;
                    'b'00111100XXX1 => 'b'111100000;
      /* State n */ 'b'00111101XXX0 => 'b'111010001;
                    'b'00111101XXX1 => 'b'111100000;
      /* State o */ 'b'00111110XXXX => 'b'111100100;
      /* State p */ 'b'00111111XXXX => 'b'111100000;
```

Practical Exercise 8.1: TAB, Transfer Accumulators, Inherent

NAME: PRACT. EXERCISE 8.1 DATE:
MNEMONIC: TAB OPCODE: $16

		μP ROM		COMPUTER CONTROL UNIT										ARITHMETIC LOGIC UNIT									STATUS AND SHIFT							PROGRAM CONTROL UNIT										
		ADDRESS		LOAD CTRL		BRANCH ADDRESS								LDB	LDA	SOURCE			CI	FCTN			R/L	SE	LD	FLAG SELECT			OE	IY	IX	VMA	R/W	SP	PC	IDB			DBB	IR
				I1	I0	A7	A6	A5	A4	A3	A2	A1	A0			I2	I1	I0		I2	I1	I0				A2	A1	A0								S2	S1	S0		
				36	35	34	33	32	31	30	29	28	27	26	25	24	23	22	21	20	19	18	17	16	15	14	13	12	11	10	09	08	07	06	05	04	03	02	01	00
T0	ALU - PAR A, (B)←(ALU), [Z]	0	4	0	0									0		0	0	1	0	0	0	0																		
				Z										LDB		A				PAR																				

Practical Exercise 8.2: SBA, Subtract Accumulators, Inherent

NAME: PRACT. EXERCISE 8.2 DATE:
MNEMONIC: SBA OPCODE: $10

State	Description	µP ROM Address	Load Ctrl (T1 T0)	Branch Address (A7–A0)	LDB	LDA	Source (T2 T1 T0)	CI	FCTN (T2 T1 T0)	R/L	SE	LD	Flag Select (A2 A1 A0)	OE	IY	IX	VMA	R/W	SP	PC	IDB (S2 S1 S0)	DBB	IR
T0	ALU = SUB BA, (A)←(ALU). SET FLAGS. [Z]	06	0 0			0	1 1 0	1	0 1 1			0											
			Z			LDA	BA	C	SUB			L											

Practical Exercise 8.3: ADDB #$NN, Add to Accumulator, Immediate

NAME: PRACT. EXERCISE 8.3 DATE:
MNEMONIC: ADDB #$NN OPCODE: $CB

State	Description	µP ROM Address	Load Ctrl (T1 T0)	Branch Address (A7–A0)	LDB	LDA	Source (T2 T1 T0)	CI	FCTN (T2 T1 T0)	R/L	SE	LD	Flag Select (A2 A1 A0)	OE	IY	IX	VMA	R/W	SP	PC	IDB (S2 S1 S0)	DBB	IR
T0	(DBB)←(EDB), PC→IDB Path, ALU = PAR D+1, [µPC]	0A	0 1				0 0 0	1	0 0 0												0 0 1	0	
			µPC				D	C	PAR												PC	D	
T1	(PC)←(PC)+1, DBB→IDB Path, SET FLAGS, ALU = ADD BD, (B)←(ALU), [Z]	0B	0 0		0		1 1 1	0	0 1 0			0								0	0 0 0		
			Z		LDB		BD		ADD			L								P	DBB		

Practical Exercise 8.4: LDS #$NN, Load Stack Pointer, Immediate

NAME: PRACT. EXERCISE 8.4 DATE:
MNEMONIC: LDS #$NN OPCODE: 8E

State	Description	µP ROM Address	Load Ctrl (T1 T0)	Branch Address (A7–A0)	LDB	LDA	Source (T2 T1 T0)	CI	FCTN (T2 T1 T0)	R/L	SE	LD	Flag Select (A2 A1 A0)	OE	IY	IX	VMA	R/W	SP	PC	IDB (S2 S1 S0)	DBB	IR
T0	(DBB)←(EDB), PC→IDB Path, ALU = PAR D+1, [µPC]	0C	0 1				0 0 0	1	0 0 0												0 0 1	0	
			µPC				D	C	PAR												PC	D	
T1	(PC)←(PC)+1, DBB→IDB Path, ALU = PAR D, [µPC]	0D	0 1				0 0 0	0	0 0 0											0	0 0 0		
			µPC				D		PAR											P	DBB		
T2	(SP)←(ALU), [Z]	0E	0 0																0				
			Z																S				

NOTE THAT A CLOCK CYCLE COULD BE SAVED BY FETCHING THE NEXT OPCODE IN T2, AND THEN JUMPING TO $01 INSTEAD OF [Z].

Practical Exercise 8.5: ADDB $NN, Add to Accumulator, Direct

NAME: PRACT. EXERCISE 8.5 DATE:
MNEMONIC: ADDB $NN OPCODE: $DB

State	Description	µP ROM Address	Load Ctrl (T1 T0)	Branch Address (A7–A0)	LDB	LDA	Source (T2 T1 T0)	CI	FCTN (T2 T1 T0)	R/L	SE	LD	Flag Select (A2 A1 A0)	OE	IY	IX	VMA	R/W	SP	PC	IDB (S2 S1 S0)	DBB	IR
T0	(DBB)←(EDB), PC→IDB Path, ALU = PAR D+1, [µPC]	14	0 1				0 0 0	1	0 0 0												0 0 1	0	
			µPC				D	C	PAR												PC	D	
T1	(PC)←(PC)+1, DBB→IDB Path, Memory Read, [µPC]	15	0 1														1	1		0	0 0 0	1	
			µPC														RD			P	DBB		
T2	(DBB)←(EDB), DBB→IDB Path, SET FLAGS, ALU = PAR D, (A)←(ALU), [Z]	16	0 0		0		1 1 1	0	0 1 0			0									0 0 0	0	
			Z		LDB		BD		ADD			L									DBB	D	

Practical Exercise 8.6: LDS $NN, Load Stack Pointer, Direct

NAME: PRACT. EXERCISE 8.6 DATE:
MNEMONIC: LDS $NN OPCODE: $9E

State	Description	µP ROM Address	Load Ctrl (T1 T0)	Branch Address (A7–A0)	LDB	LDA	Source (T2 T1 T0)	CI	FCTN (T2 T1 T0)	R/L	SE	LD	Flag Select (A2 A1 A0)	OE	IY	IX	VMA	R/W	SP	PC	IDB (S2 S1 S0)	DBB	IR
T0	(DBB)←(EDB), PC→IDB Path, ALU = PAR D+1, [µPC]	18	0 1				0 0 0	1	0 0 0												0 0 1	0	
			µPC				D	C	PAR												PC	D	
T1	(PC)←(PC)+1, DBB→IDB Path, Memory Read, [µPC]	19	0 1														1	1		0	0 0 0	1	
			µPC														RD			P	DBB		
T2	(DBB)←(EDB), DBB→IDB Path, ALU = PAR D, [µPC]	1A	0 1				0 0 0	0	0 0 0												0 0 0	0	
			µPC				D		PAR												DBB	D	
T3	(SP)←(ALU), [Z]	1B	0 0																0				
			Z																S				

NOTE THAT A CLOCK CYCLE COULD BE SAVED BY FETCHING THE NEXT OPCODE IN T3, AND THEN JUMPING TO $01 INSTEAD OF [Z].

Practical Exercise 8.7: ADDB $NN,Y, Add to Accumulator, Indexed

NAME: PRACT. EXERCISE 8.7 DATE:
MNEMONIC: ADDB $NN,Y OPCODE: A6

		µP ROM	Computer Control Unit										Arithmetic Logic Unit									Status and Shift							Program Control Unit										
Step	Description	Address	Load Ctrl I1 (36)	Load Ctrl I0 (35)	A7 (34)	A6 (33)	A5 (32)	A4 (31)	A3 (30)	A2 (29)	A1 (28)	A0 (27)	LDB (26)	LDA (25)	Source I2 (24)	Source I1 (23)	Source I0 (22)	CI (21)	FCTN I2 (20)	FCTN I1 (19)	FCTN I0 (18)	R/L (17)	SE (16)	LD (15)	Flag Select A2 (14)	Flag Select A1 (13)	Flag Select A0 (12)	OE (11)	IY (10)	IX (09)	VMA (08)	R/W (07)	SP (06)	PC (05)	IDB S2 (04)	IDB S1 (03)	IDB S0 (02)	DBB (01)	IR (00)
T0	(DBB)←(EDB), PC→IDB Path.	20	0	1											0	0	0	1	0	0	0														0	0	1	0	
	ALU - PAR D+1, [µPC]		µPC												D			C	PAR																PC			D	
T1	(PC)←(PC)+1, Y+OPERAND→IDB Path.	21	0	1																											1	1		0	1	0	0	1	
	Memory Read, [µPC]		µPC																												RD			P	IY+OP				
T2	(DBB)←(EDB), DBB→IDB Path, SET	22	0	0									0		1	1	1	0	0	1	0			0											0	0	0	0	
	FLAGS, ALU - ADD BD, (B)←(ALU), [Z]		Z										LDB		BD				ADD					L											DBB			D	

Practical Exercise 8.8: LDS $NN,Y, Load Stack Pointer, Indexed

NAME: PRACT. EXERCISE 8.8 DATE:
MNEMONIC: LDS $NN,Y OPCODE: $AE

		µP ROM	Computer Control Unit										Arithmetic Logic Unit									Status and Shift							Program Control Unit										
Step	Description	Address	Load Ctrl I1 (36)	Load Ctrl I0 (35)	A7 (34)	A6 (33)	A5 (32)	A4 (31)	A3 (30)	A2 (29)	A1 (28)	A0 (27)	LDB (26)	LDA (25)	Source I2 (24)	Source I1 (23)	Source I0 (22)	CI (21)	FCTN I2 (20)	FCTN I1 (19)	FCTN I0 (18)	R/L (17)	SE (16)	LD (15)	Flag Select A2 (14)	Flag Select A1 (13)	Flag Select A0 (12)	OE (11)	IY (10)	IX (09)	VMA (08)	R/W (07)	SP (06)	PC (05)	IDB S2 (04)	IDB S1 (03)	IDB S0 (02)	DBB (01)	IR (00)
T0	(DBB)←(EDB), PC→IDB Path.	24	0	1											0	0	0	1	0	0	0														0	0	1	0	
	ALU - PAR D+1, [µPC]		µPC												D			C	PAR																PC			D	
T1	(PC)←(PC)+1, Y+OPERAND→IDB Path.	25	0	1																											1	1		0	1	0	0	1	
	Memory Read, [µPC]		µPC																												RD			P	IY+OP				
T2	(DBB)←(EDB), DBB→IDB Path,	26	0	1											0	0	0	0	0	0	0														0	0	0	0	
	ALU - PAR D, [µPC]		µPC												D				PAR																DBB			D	
T3	(SP)←(ALU), [Z]	27	0	0																													0						
			Z																														S						

NOTE THAT A CLOCK CYCLE COULD BE SAVED BY FETCHING THE NEXT OPCODE IN T3, AND THEN JUMPING TO $01 INSTEAD OF [Z].

Practical Exercise 8.9: BCC *+$NN, Branch if Carry Clear, Relative

NAME: PRACT. EXERCISE 8.9 DATE:
MNEMONIC: BCC *+$NN OPCODE: 24

		µP ROM	Computer Control Unit										Arithmetic Logic Unit									Status and Shift							Program Control Unit										
Step	Description	Address	Load Ctrl I1 (36)	Load Ctrl I0 (35)	A7 (34)	A6 (33)	A5 (32)	A4 (31)	A3 (30)	A2 (29)	A1 (28)	A0 (27)	LDB (26)	LDA (25)	Source I2 (24)	Source I1 (23)	Source I0 (22)	CI (21)	FCTN I2 (20)	FCTN I1 (19)	FCTN I0 (18)	R/L (17)	SE (16)	LD (15)	Flag Select A2 (14)	Flag Select A1 (13)	Flag Select A0 (12)	OE (11)	IY (10)	IX (09)	VMA (08)	R/W (07)	SP (06)	PC (05)	IDB S2 (04)	IDB S1 (03)	IDB S0 (02)	DBB (01)	IR (00)
T0	(DBB)←(EDB), PC→IDB Path.	37	0	1	0	0	1	1	1	0	0	1			0	0	0	1	0	0	0				0	0	1	1							0	0	1	0	
	ALU - PAR D+1, cc=CC, PL=T2, [µPC]		µPC		JPL $39 if C=0										D			C	PAR						cc=CC										PC			D	
T1	(PC)←(ALU), [Z]	38	0	0																														0				1	
			Z																															P					
T2	DBB→IDB Path, ALU - ADD DP, [µPC]	39	0	1											1	0	0	1	0	1	0														0	0	0	0	
			µPC												DP			C	ADD																DBB			D	
T3	(PC)←(ALU), [Z]	3A	0	0																														0					
			Z																															P					

Practical Exercise 10.1: Address Bus Decoding No. 1.

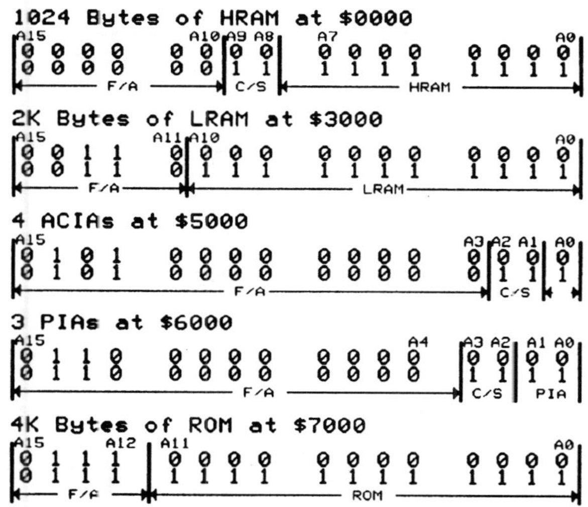

1024 Bytes of HRAM at $0000

HRAM	A9 A8	$\overline{CS1}$	$\overline{CS0}$
0	0 0	A9	Q0
1	0 1	A9	Q1
2	1 0	$\overline{A9}$	Q0
3	1 1	$\overline{A9}$	Q1

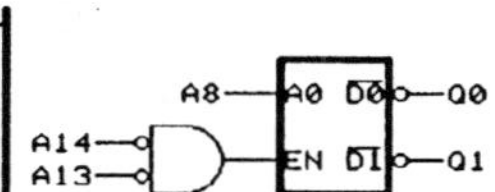

2K Bytes of LRAM at $3000

LRAM	$\overline{CS2}$	CS1	CS0
0	$\overline{A14}$	A13	1

4 ACIAs at $5000

ACIA	A2 A1	CS0	$\overline{CS1}$	$\overline{CS2}$
0	0 0	A2	A1	$\overline{A14 \cdot \overline{A13}}$
1	0 1	A2	$\overline{A1}$	
2	1 0	$\overline{A2}$	A1	
3	1 1	$\overline{A2}$	$\overline{A1}$	

3 PIAs at $6000

PIA	A3 A2	CS0	$\overline{CS1}$	$\overline{CS2}$
0	0 0	$\overline{A3}$	A2	$\overline{A14 \cdot \overline{A12}}$
1	0 1	$\overline{A3}$	$\overline{A2}$	
2	1 0	$\overline{A3}$	A2	

4K Bytes of ROM at $7000

ROM	$\overline{OE2}$	OE1
0	0	A14•A13•A12

FREE-ALLOCATION TABLE

```
       A14 A13 A12
HRAM | 0 (0)(0) 0    0 0                       |¬A14 • ¬A13
LRAM | 0 (0)(1) 1    0                         |¬A14 • A13
ACIA | 0 (1)(0) 1    0 0 0 0   0 0 0 0   0     |A14 • ¬A13
PIA  | 0 (1) 1 (0)   0 0 0 0   0 0 0 0         |A14 • ¬A12
ROM  | 0 (1)(1)(1)                             |A14 • A13 • A12
```

```
/* PRACTICAL EXERCISE 10.1  */
NAME       PE10-01;
PARTNO     ADDRESS BUS DECODE;
REVISION   01;
DATE       04/19/96;
DESIGNER   J.W. CARTER;
COMPANY    UNC CHARLOTTE;
DEVICE     P16L8;
/* INPUTS */
PIN 1 = A1;
PIN 2 = A2;
PIN 3 = A3;
PIN 4 = A8;
PIN 5 = A12;
PIN 7 = A13;
PIN 8 = A14;
/* OUTPUTS */
PIN 12 = !(!A14 & !A13 & !A8); /* HRAM #0, HRAM #2: !CS0 */
PIN 13 = !(!A14 & !A13 &  A8); /* HRAM #1, HRAM #3: !CS0 */
PIN 14 = !( A14 & !A13);       /* ALL ACIAS: !CS2  */
PIN 15 = !( A14 & !A12);       /* ALL PIAS:   CS2  */
PIN 16 =  A14 & A13 & A12;     /* ROM: OE1 */
PIN 17 = !A1;                  /* ACIA #1, ACIA #3: !CS1 */
PIN 18 = !A2;                  /* ACIA #2, ACIA #3:  CS0 */
                               /*     AND  PIA #1:  CS1 */
PIN 19 = !A3;                  /* PIA #0,  PIA #1:   CS0 */

/* USE ADDITIONAL INVERTER FOR !A9 & !A14 */
```

Practical Exercise 10.2: Address Bus Decoding No. 2.

```
1 ACIA at $1000
A15                                     A1  A0
0 0 0 1   0 0 0 0   0 0 0 0   0 0 0 | 0
0 0 0 1   0 0 0 0   0 0 0 0   0 0 0 | 1
|<----------------- F/A ---------------->| AC

2 PIAs at $1008
A15                               A3 A2  A1 A0
0 0 0 1   0 0 0 0   0 0 0 0   1 | 0 | 0 0
0 0 0 1   0 0 0 0   0 0 0 0   1 | 1 | 1 1
|<----------------- F/A ------------>| C |<PIA>|

2 BUFFERs at $1010
A15                                     A1  A0
0 0 0 1   0 0 0 0   0 0 0 1   0 0 0 | 0
0 0 0 1   0 0 0 0   0 0 0 1   0 0 0 | 1
|<----------------- F/A ---------------->|<B>|

8K of LRAM at $2000
A15   A12                                 A0
0 0 1 | 0 | 0 0 0 0   0 0 0 0   0 0 0 0
0 0 1 | 1 | 1 1 1 1   1 1 1 1   1 1 1 1
|<F/A>|CS|<------------- LRAM ------------>|

16K of ROM at $4000
A15A14A13                                 A0
0 1 | 0 0   0 0 0 0   0 0 0 0   0 0 0 0
0 1 | 1 1   1 1 1 1   1 1 1 1   1 1 1 1
|<F/A>|<-------------- ROM --------------->|
```

FREE-ALLOCATION TABLE

	A15	A14	A13	A12								A4	A3			
ACIA	0	⓪	⓪	1	0 0 0 0				0	0	0	⓪	⓪	0	0	$\overline{A14} \cdot \overline{A13} \cdot \overline{A4} \cdot \overline{A3}$
PIA	0	⓪	⓪	1	0 0 0 0				0	0	0	0	①			$\overline{A14} \cdot \overline{A13} \cdot A3$
BUFF	0	⓪	⓪	1	0 0 0 0				0	0	0	①	0	0	0	$\overline{A14} \cdot \overline{A13} \cdot A4$
LRAM	0	⓪	①													$\overline{A14} \cdot A13$
ROM	0	①														$A14$

1 ACIA at $1000

ACIA	CS0	$\overline{CS1}$	$\overline{CS2}$
0	$\overline{A14} \cdot \overline{A13}$	A4	A3

2 PIAs at $1008

PIA	A2	CS0	$\overline{CS1}$	$\overline{CS2}$
0	0	$\overline{A2}$	A14	$\overline{\overline{A13} \cdot A3}$
1	1	A2	A14	$\overline{\overline{A13} \cdot A3}$

2 BUFFERs at $1010

BUFF	A0	$\overline{G2}$	$\overline{G1}$
0	0	$\overline{A0}$	$\overline{\overline{A14} \cdot \overline{A13} \cdot A4}$
1	1	A0	$\overline{\overline{A14} \cdot \overline{A13} \cdot A4}$

8K Bytes of LRAM at $3000

LRAM	A12	$\overline{OE1}$
0	0	Q0
1	0	Q1

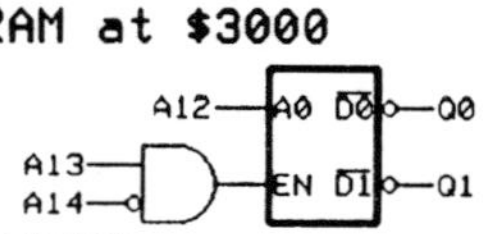

16K of ROM at $4000

ROM	$\overline{OE}$
0	$\overline{A14}$

```
/* PRACTICAL EXERCISE 10.2  */
NAME       PE10-02;
PARTNO     ADDRESS BUS DECODE;
REVISION   01;
DATE       04/19/96;
DESIGNER   J.W. CARTER;
COMPANY    UNC CHARLOTTE;
DEVICE     P16L8;

/* INPUTS */

PIN 1 = A0;
PIN 2 = A2;
PIN 3 = A3;
PIN 4 = A4;
PIN 5 = A12;
PIN 7 = A13;
PIN 8 = A14;

/* OUTPUTS */

PIN 12 = !(!A14 & !A13 & !A12); /* LRAM #0, !OE1 */
PIN 13 = !(!A14 & !A13 &  A12); /* LRAM #1, !OE1 */
PIN 14 = !A14 & !A13;           /* ACIA: CS0  */
PIN 15 = !(!A14 & !A13 &  A3);  /* ALL PIAS:  !CS2  */
PIN 16 = !(!A14 & !A13 &  A4);  /* BUFF #0, BUFF #1: !G2 */
PIN 17 = !A0;                   /* BUFF #1: !G1 */
PIN 18 = !A2;                   /* PIA #0:  CS0 */
```

Practical Exercise 10.3: Address Bus Decoding No. 3.

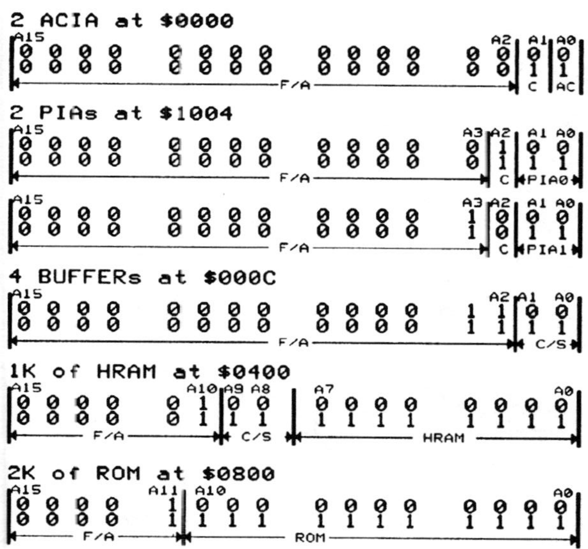

2 ACIAs at $0000

ACIA	A1	$\overline{CS0}$	CS1	$\overline{CS2}$
0	0	$\overline{A1}$	$\overline{A11}\cdot\overline{A10}$	$\overline{A3}\cdot\overline{A2}$
1	1	A1	$\overline{A11}\cdot\overline{A10}$	$\overline{A3}\cdot\overline{A2}$

2 PIAs at $1008

PIA0	A2	CS0	$\overline{CS1}$	$\overline{CS2}$
0	1	A2	$\overline{A11}\cdot\overline{A10}$	A3

PIA1	A2	CS0	$\overline{CS1}$	$\overline{CS2}$
0	0	A3	$\overline{A11}\cdot\overline{A10}$	A2

4 BUFFERs at $000C

BUFF	A1	A0	$\overline{G2}$	$\overline{G1}$
0	0	0	$\overline{A1}$	Q0
1	0	1	$\overline{A1}$	Q1
2	1	0	A1	Q0
3	1	1	A1	Q1

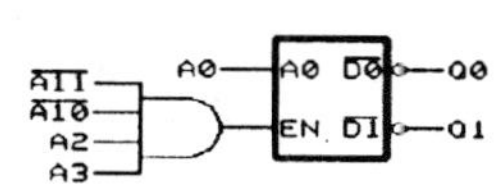

1K Bytes of HRAM at $0400

HRAM	A9	A8	$\overline{CS1}$	$\overline{CS0}$
0	0	0	$\overline{A9}$	Q0
1	0	1	$\overline{A9}$	Q1
2	1	0	A9	Q0
3	1	1	A9	Q1

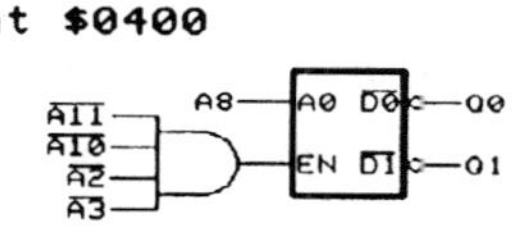

2K of ROM at $0800

ROM	$\overline{OE2}$	OE1
0	0	A11

FREE-ALLOCATION TABLE

					A11	A10							A3	A2	
ACIA	0	0	0	0	⓪	⓪	0	0	0	0	0	0	⓪	⓪	$\overline{A11}\cdot\overline{A10}\cdot\overline{A3}\cdot\overline{A2}$
PIA0	0	0	0	0	⓪	⓪	0	0	0	0	0	0	⓪	①	$\overline{A11}\cdot\overline{A10}\cdot\overline{A3}\cdot A2$
PIA1	0	0	0	0	⓪	⓪	0	0	0	0	0	0	①	⓪	$\overline{A11}\cdot\overline{A10}\cdot A3\cdot\overline{A2}$
BUFF	0	0	0	0	⓪	⓪	0	0	0	0	0	0	①	①	$\overline{A11}\cdot\overline{A10}\cdot A3\cdot A2$
HRAM	0	0	0	0	⓪	①	0								$\overline{A11}\cdot A10$
ROM	0	0	0	0	①										A11

```
/* PRACTICAL EXERCISE 10.3  */
NAME        PE10-03;
PARTNO      ADDRESS BUS DECODE;
REVISION    01;
DATE        04/19/96;
DESIGNER    J.W. CARTER;
COMPANY     UNC CHARLOTTE;
DEVICE      P16L8;
/* INPUTS */
PIN 1 = A0;
PIN 2 = A1;
PIN 3 = A2;
PIN 4 = A8;
PIN 5 = A9;
PIN 7 = A10;
PIN 8 = A11;
/* OUTPUTS */
PIN 12 = !(!A11 & !A10);                          /* ACIA, PIA */
PIN 13 = !( !A3 &   A2);                           /* ACIA */
PIN 14 = !(!A11 & !A10 &  A3 &  A2 & !A0);        /* BUFF0,2 */
PIN 15 = !(!A11 & !A10 &  A3 &  A2 &  A0);        /* BUFF1,3 */
PIN 16 = !(!A11 & !A10 & !A3 & !A2 & !A8);        /* HRAM0,2 */
PIN 17 = !(!A11 & !A10 & !A3 & !A2 &  A8);        /* HRAM1,3 */
PIN 18 = !A1                                       /* BUFF0,1 */
PIN 19 = !A9                                       /* HRAM0,1 */
```

Practical Exercise 10.4: Address Bus Decoding No. 4.

768 Bytes of HRAM at $0400

4 ACIAs at $1000

1 PIAs at $1010

3 BUFFERs at $1020

2K of ROM at $2800

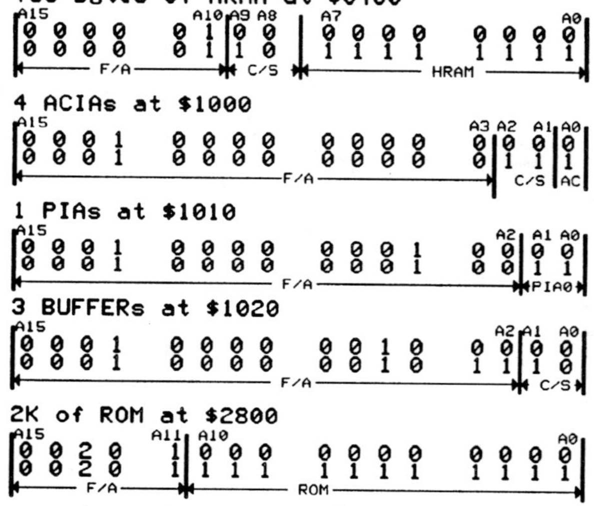

768 Bytes of HRAM at $0400

HRAM	A9	A8	$\overline{CS1}$	$\overline{CS0}$
0	0	0	$\overline{A9}$	Q0
1	0	1	$\overline{A9}$	Q1
2	1	0	A9	Q0

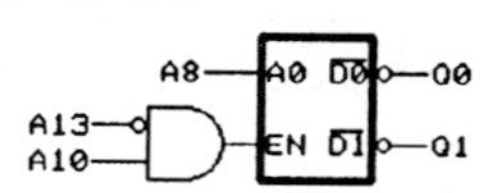

4 ACIAs at $1000

ACIA	A2	A1	CS0	$\overline{CS1}$	$\overline{CS2}$
0	0	0	$\overline{A2}$	A1	$A12 \cdot \overline{A5} \cdot \overline{A4}$
1	0	1	$\overline{A2}$	$\overline{A1}$	
2	1	0	A2	A1	
3	1	1	A2	$\overline{A1}$	

1 PIA at $1010

PIA0	A2	CS0	$\overline{CS1}$	$\overline{CS2}$
0	1	A12	$\overline{A4}$	0

3 BUFFERs at $1020

BUFF	A1	A0	$\overline{G2}$	$\overline{G1}$
0	0	0	$\overline{A1}$	Q0
1	0	1	$\overline{A1}$	Q1
2	1	0	A1	Q0

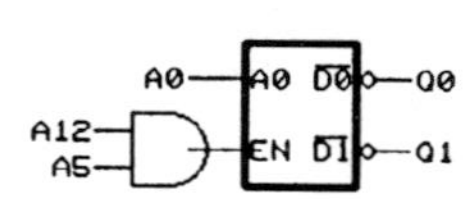

2K of ROM at $0800

ROM	$\overline{OE2}$	OE1
0	0	A13

FREE-ALLOCATION TABLE

HRAM	0 0 (0)(0)	0 1			$\overline{A13} \cdot \overline{A12}$
ACIA	0 0 0 (1)	0 0 0 0	0 0 (0)(0)	0	$A12 \cdot \overline{A5} \cdot \overline{A4}$
PIA	0 0 0 (1)	0 0 0 0	0 0 0 (1)	0 0	$A12 \cdot A4$
BUFF	0 0 0 (1)	0 0 0 0	0 0 (1) 0	0 0	$A12 \cdot A5$
ROM	0 0 (1) 0	1			A13

(Circled bits: A13 A12; A5 A4)

```
/* PRACTICAL EXERCISE 10.4   */
NAME      PE10-04;
PARTNO    ADDRESS BUS DECODE;
REVISION  01;
DATE      04/19/96;
DESIGNER  J.W. CARTER;
COMPANY   UNC CHARLOTTE;
DEVICE    P16L8;
/* INPUTS */
PIN 1 = A0;
PIN 2 = A1;
PIN 3 = A2;
PIN 4 = A4;
PIN 5 = A5;
PIN 7 = A12;
PIN 8 = A13;
/* OUTPUTS */
PIN 12 = !(!A13 & !A12 & !A8);   /* HRAM0,2 */
PIN 13 = !(!A13 & !A12 &  A8);   /* HRAM1   */
PIN 14 = !( A12 & !A5 & !A4);    /* ACIAs   */
PIN 15 = !(A12 & A5 & !A0);      /* BUFF0,2 */
PIN 16 = !(A12 & A5 &  A0);      /* BUFF1   */
PIN 17 = !A1                     /* ACIA,BUFF */
PIN 18 = !A2                     /* ACIA    */
PIN 19 = !A4                     /* PIA     */
/*  USE INVERTER FOR !A9, NEEDED FOR HRAM */
```

Index to Transparency Masters

Chapter 1

Chapter 2

Chapter 3

Chapter 4

Chapter 5

Chapter 6

Chapter 7

Chapter 8

Chapter 9

Chapter 10

$$Y = A\overline{B}C + \overline{A}\,\overline{B}\,\overline{C} + \overline{A}B\overline{C} + \overline{A}\,\overline{B}C$$

EXPRESSION SET 1.1 Conventional Logic Expression

$$Y = f(A,B,C)$$

EXPRESSION SET 1.2 Domain of an Expression

$$Y = f(A,B,C) = A\overline{B}C + \overline{A}\,\overline{B}\,\overline{C} + \overline{A}B\overline{C} + \overline{A}\,\overline{B}C$$

EXPRESSION SET 1.3 Complete Form of a Conventional SOP Expression

Operator	Character
NOT	!
AND	&
OR	#
XOR	$

TABLE 1.1 Typographical Logical Operators.

$$Y = f(A,B,C) = (A\&!B\&C)\#(!A\&!B\&!C)\#(!A\&B\&!C)\#(!A\&!B\&C)$$

EXPRESSION SET 1.4 Typographical Form of a Logical Expression.

1. $Y = f(A,B,C) = A\overline{B}C + \overline{A}\,\overline{B}\,\overline{C} + \overline{A}B\overline{C} + \overline{A}\,\overline{B}C$
2. $Y = f(A,B,C) = 101 + 000 + 010 + 001$
3. $Y = f(A,B,C) = \sum(5,0,2,1)$
4. $Y = f(A,B,C) = \sum(0,1,2,5)$

EXPRESSION SET 1.5 Summation Form of a Logical Expression

$$W = f(A,B,C) = \overline{A}\,\overline{B}C + A\overline{B}\,\overline{C} + ABC$$

$$X = f(A,B,C,D) = \overline{A}CD + AB\overline{C} + \overline{C}\,\overline{D}$$

$$Y = f(A,B,C) = \overline{A} + B\overline{C} + \overline{A}C$$

EXPRESSION SET 1.6 Conventional SOP Expressions

```
W = (!A&!B&C)#(A&!B&!C)#(A&B&C)

X = (!A&C&D)#(A&B&!C)#(!C&!D)

Y = !A #(B&!C)#(!A&C)
```

EXPRESSION SET 1.7 Typographic SOP Expressions

1. $Y = AB\overline{C}D + AB\overline{C}\,\overline{D}$
2. $Y = AB\overline{C}(D + \overline{D})$
3. $Y = AB\overline{C}(1)$
4. $Y = AB\overline{C}$

EXPRESSION SET 1.8 Elimination of Redundant Term in an SOP Expression

$$Y = f(A,B,C,D) = AB\overline{C} = AB\overline{C}D + AB\overline{C}\,\overline{D}$$

EXPRESSION SET 1.9 Expansion of SOP Expression by Adding Redundant Terms

$$Y = f(A,B,C,D) = AB$$

$$Y = f(A,B,C,D) = AB\overline{C}\,\overline{D} + AB\overline{C}D + ABC\overline{D} + ABCD$$

EXPRESSION SET 1.10 Expansion of SOP Expression to Include Redundant Terms

$$W = f(A,B,C) = \overline{A}\,\overline{B}C + A\overline{B}\,\overline{C} + ABC$$

$$X = f(A,B,C,D) = \overline{A}CD + AB\overline{C} + \overline{C}\,\overline{D}$$

$$Y = f(A,B,C) = \overline{A} + B\overline{C} + \overline{A}C$$

EXPRESSION SET 1.11 Conventional SOP Expressions

$$W = f(A,B,C) = \overline{A}\,\overline{B}C + A\overline{B}\,\overline{C} + ABC$$

$$W = f(A,B,C) = \sum(001,100,111)$$

$$W = f(A,B,C) = \sum(1,4,7)$$

EXPRESSION SET 1.12 Conversion from Conventional to Summation Form

1. $X = f(A,B,C,D) = \overline{A}CD + AB\overline{C} + \overline{C}\,\overline{D}$
2. $X = f(A,B,C,D) = \overline{A}CD(B + \overline{B}) + AB\overline{C}(D + \overline{D}) + \overline{C}\,\overline{D}(\overline{A}\,\overline{B} + \overline{A}B + A\overline{B} + AB)$
3. $X = f(A,B,C,D) = \overline{A}BCD + \overline{A}\,\overline{B}CD + AB\overline{C}D + AB\overline{C}\,\overline{D} + \overline{A}\,\overline{B}\,\overline{C}\,\overline{D} + \overline{A}B\overline{C}\,\overline{D} + A\overline{B}\,\overline{C}\,\overline{D} + AB\overline{C}\,\overline{D}$
4. $X = f(A,B,C,D) = \sum(0111,0011,1101,1100,0000,0100,1000,1100)$
5. $X = f(A,B,C,D) = \sum(7,3,13,12,0,4,8,12)$
6. $X = f(A,B,C,D) = \sum(0,3,4,7,8,12,13)$

EXPRESSION SET 1.13 Conversion from Conventional to Summation Form

1. $Y = f(A,B,C) = \overline{A} + B\overline{C} + \overline{A}C$

2. $Y = f(A,B,C) = \overline{A}(\overline{B}\overline{C} + \overline{B}C + B\overline{C} + BC) + B\overline{C}(A + \overline{A}) + \overline{A}C(B + \overline{B})$

3. $Y = f(A,B,C) = \overline{A}\overline{B}\overline{C} + \overline{A}\overline{B}C + \overline{A}B\overline{C} + \overline{A}BC + AB\overline{C} + \overline{A}B\overline{C} + \overline{A}BC + \overline{A}\overline{B}C$

4. $Y = f(A,B,C) = \sum(000,001,010,011,110,010,011,001)$

5. $Y = f(A,B,C) = \sum(0,1,2,3,6,2,3,1)$

6. $Y = f(A,B,C) = \sum(0,1,2,3,6)$

EXPRESSION SET 1.14 Conversion from Conventional to Summation Form

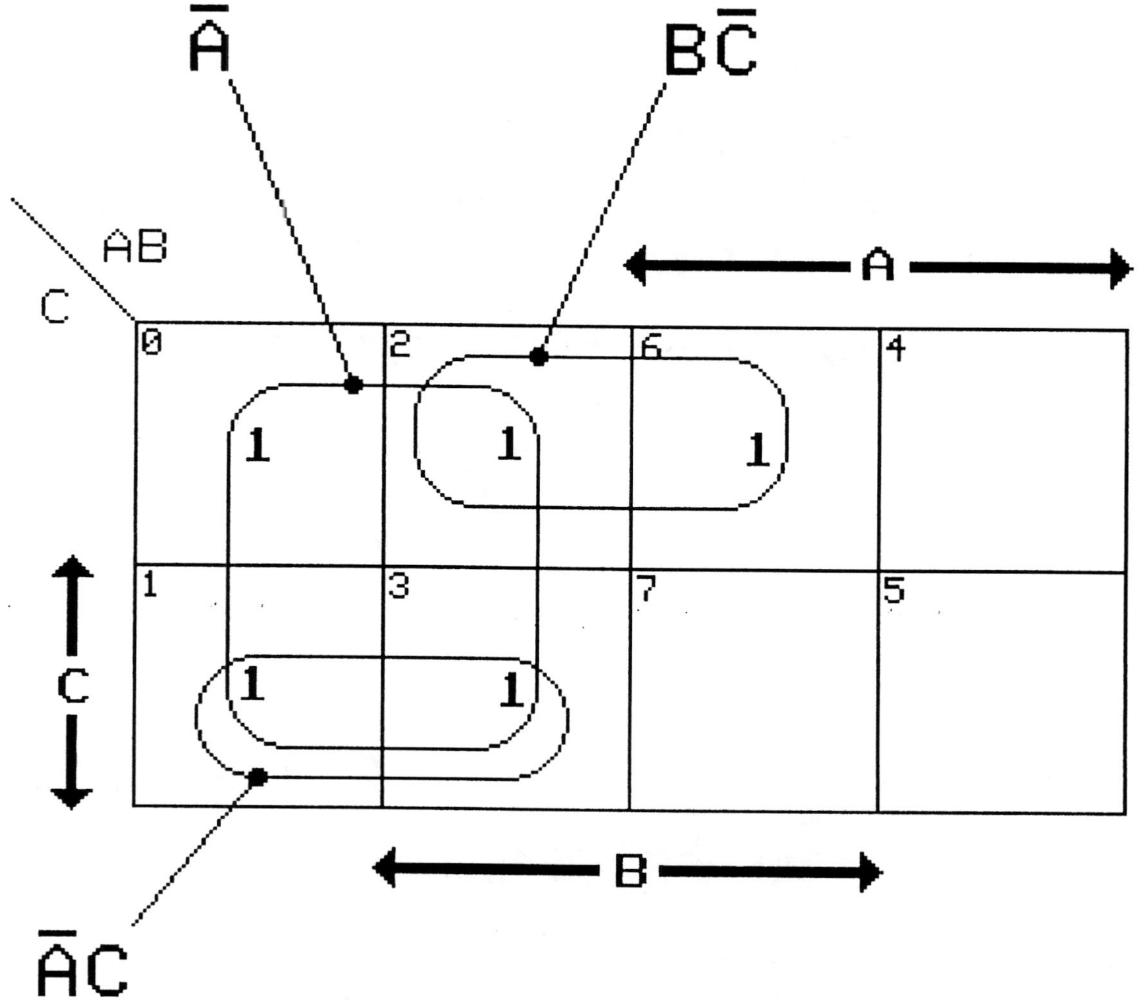

FIGURE 1.1 Karnaugh map for expression Y.

$$W = f(A,B,C) = \overline{A}\,\overline{B}\,\overline{C} + \overline{A}\,\overline{B}C + A\overline{B}\,\overline{C} + A\overline{B}C$$

$$X = f(A,B,C,D) = \overline{A}B + A\overline{B}C\overline{D} + \overline{B}CD$$

$$Y = f(A,B,C) = A\overline{B} + \overline{B}C$$

EXPRESSION SET 1.15 Practical Exercise 1.1 Conventional Expressions

$$T = f(A,B,C) = \sum(0,3,4,5)$$

$$U = f(A,B,C,D) = \sum(0,1,2,8,9,11,14,15)$$

$$V = f(A,B,C) = \sum(1,3,4,6,7)$$

EXPRESSION SET 1.16 Practical Exercise 1.1, Summation Expressions

$$Y = f(A,B,C) = A\overline{B}C + \overline{A}BC + \overline{A}\,\overline{B} + \overline{B}\,\overline{C}$$

EXPRESSION SET 1.17 Conventional SOP Expression

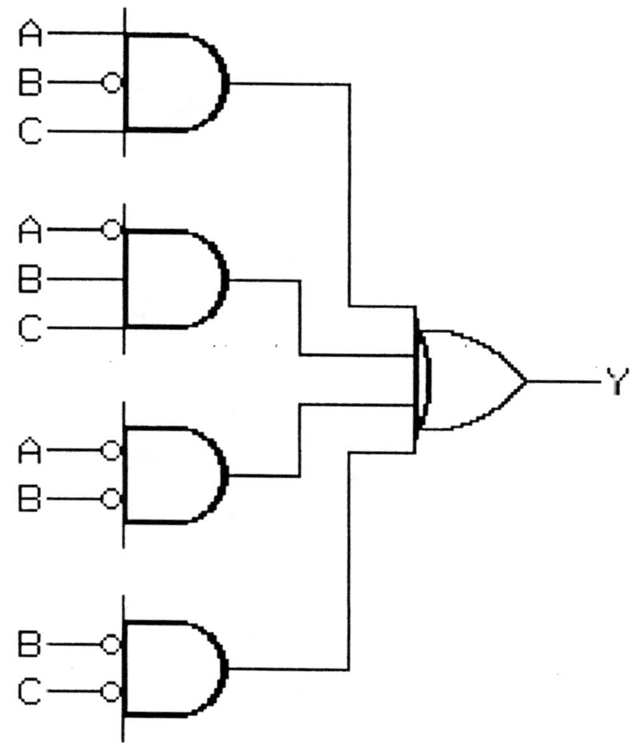

FIGURE 1.2 Discrete logic implementation of SOP expression.

$$A + B = \overline{\overline{A}\,\overline{B}}$$

EXPRESSION SET 1.18 DeMorgan's Theorem Applied to OR Logic

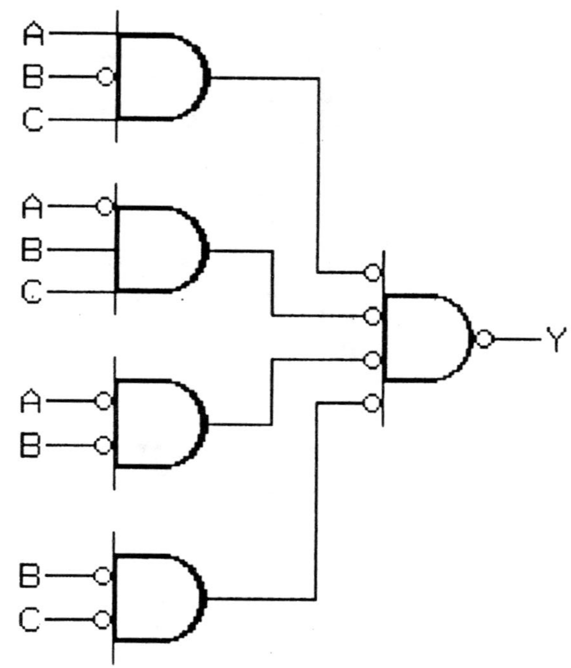

FIGURE 1.3 DeMorgan's equivalent NAND output buffer.

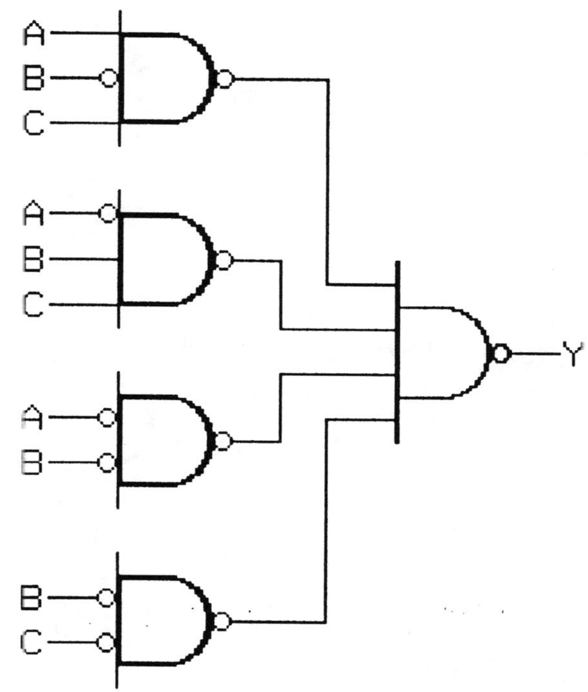

FIGURE 1.4 ***NAND implementation of conventional summation expression.***

$$Y = f(A,B,C) = A\overline{B}C + \overline{A}BC + \overline{A}\,\overline{B} + \overline{B}\,\overline{C}$$

EXPRESSION SET 1.19 ***Conventional SOP Expression***

$$Y = f(A,B,C) = A\overline{B}C + \overline{A}BC + \overline{A}\,\overline{B} + \overline{B}\,\overline{C}$$

$$Y = f(A,B,C) = A\overline{B}C + \overline{A}BC + \overline{A}\,\overline{B}(C + \overline{C}) + \overline{B}\,\overline{C}(A + \overline{A})$$

$$Y = f(A,B,C) = A\overline{B}C + \overline{A}BC + \overline{A}\,\overline{B}C + \overline{A}\,\overline{B}\,\overline{C} + A\overline{B}\,\overline{C} + \overline{A}\,\overline{B}\,\overline{C}$$

$$Y = f(A,B,C) = \sum(101,011,001,000,100,000)$$

$$Y = f(A,B,C) = \sum(5,3,1,0,4,0)$$

$$Y = f(A,B,C) = \sum(0,1,3,4,5)$$

EXPRESSION SET 1.20 Summation Solution for Multiplexer Implementation

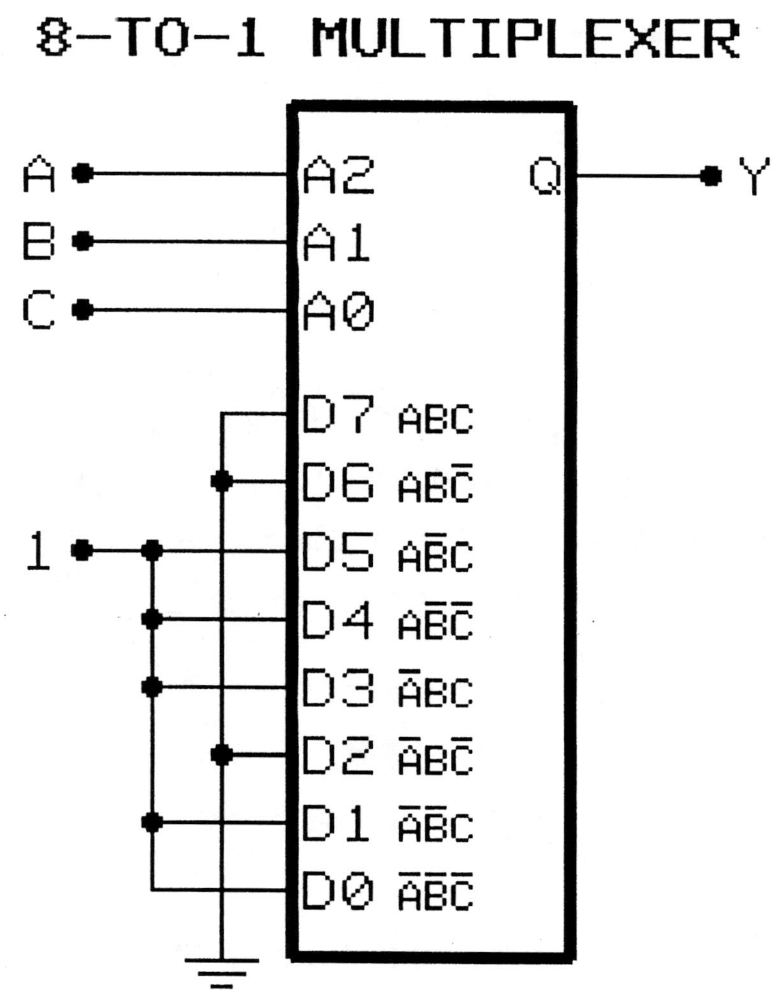

FIGURE 1.5 Multiplexer implementation of SOP expression.

$$X = f(A,B,C,D) = \overline{A}CD + AB\overline{C} + \overline{C}\,\overline{D}$$

EXPRESSION SET 1.21 Practical Example 1.2, Conventional Logical Expression

1. $X = f(A,B,C,D) = \overline{A}CD + AB\overline{C} + \overline{C}\,\overline{D}$
2. $X = f(A,B,C,D) = \overline{A}CD(B + \overline{B}) + AB\overline{C}(D + \overline{D}) + \overline{C}\,\overline{D}(\overline{A}\,\overline{B} + \overline{A}B + A\overline{B} + AB)$
3. $X = f(A,B,C,D) = \overline{A}BCD + \overline{A}\,\overline{B}CD + AB\overline{C}D + AB\overline{C}\,\overline{D} + \overline{A}\,\overline{B}\,\overline{C}\,\overline{D} + \overline{A}B\overline{C}\,\overline{D} + A\overline{B}\,\overline{C}\,\overline{D} + AB\overline{C}\,\overline{D}$
4. $X = f(A,B,C,D) = \sum(0111,0011,1101,1100,0000,0100,1000,1100)$
5. $X = f(A,B,C,D) = \sum(7,3,13,12,0,4,8,12)$
6. $X = f(A,B,C,D) = \sum(0,3,4,7,8,12,13)$

EXPRESSION SET 1.22 Practical Example 1.2, Summation Expression Conversion

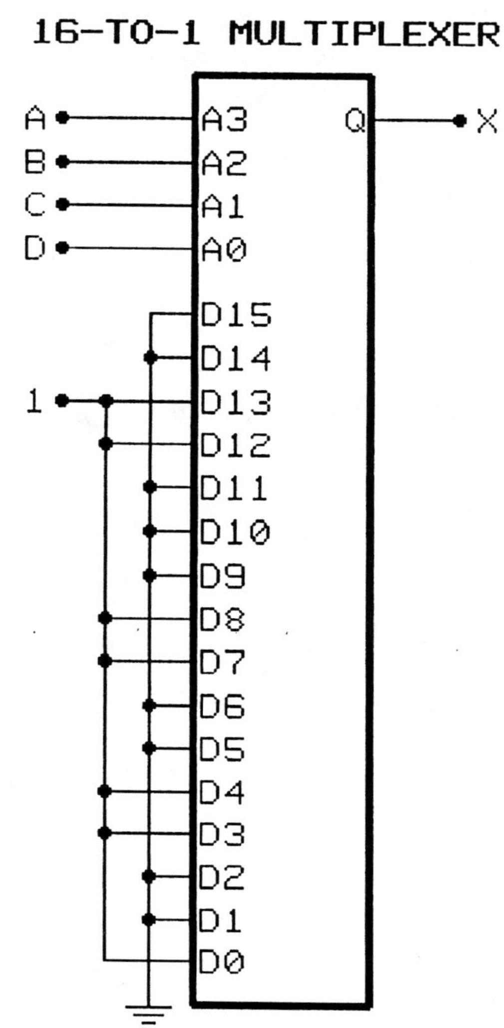

FIGURE 1.6 Multiplexer implementation of SOP expression.

$$W = f(A,B,C) = \overline{A}\,\overline{B}\,\overline{C} + \overline{A}\,\overline{B}C + A\overline{B}\,\overline{C} + A\overline{B}C$$

$$X = f(A,B,C,D) = \overline{A}B + A\overline{B}C\overline{D} + \overline{B}CD$$

$$Y = f(A,B,C) = A\overline{B} + \overline{B}C$$

EXPRESSION SET 1.23 Practical Exercise 1.2, Logic Functions

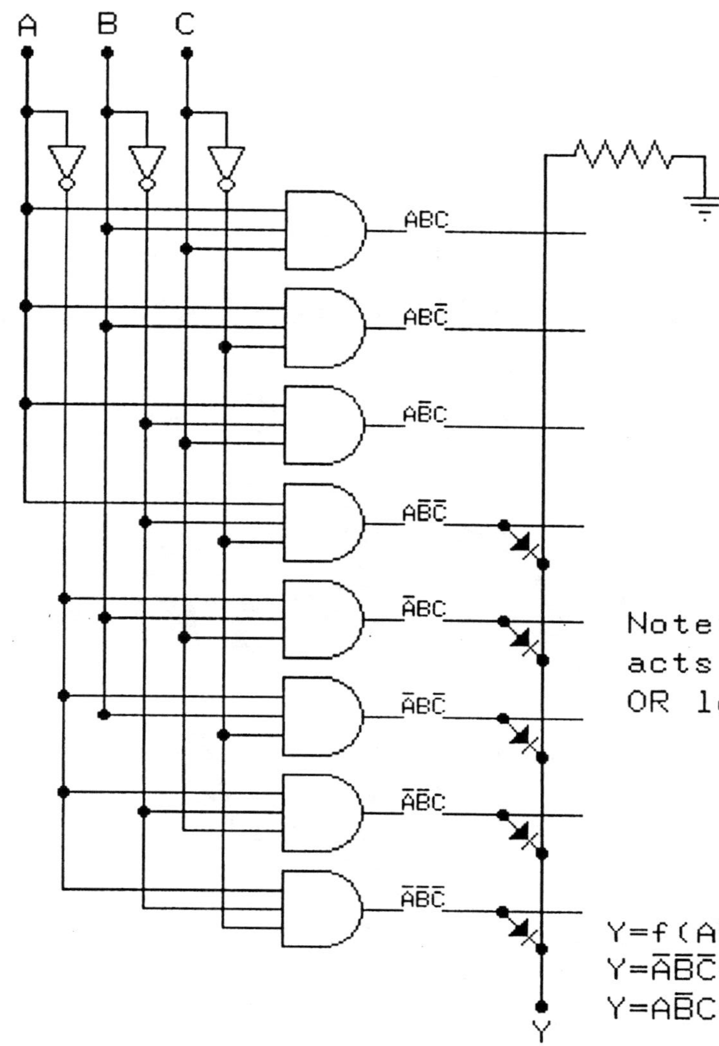

ADDRESS	DATA
7	0
6	0
5	0
4	1
3	1
2	1
1	1
0	1

Note: Diode Matrix acts as a 5-input OR logic gate.

$Y=f(ABC)=E(0,1,2,3,4)$

$Y=\bar{A}\bar{B}\bar{C}+\bar{A}\bar{B}C+\bar{A}B\bar{C}+\bar{A}BC+A\bar{B}\bar{C}$

$Y=A\bar{B}C+\bar{A}BC+\bar{A}\bar{B}+\bar{B}\bar{C}$

FIGURE 1.7 8-word by 1-bit read-only memory.

$$W = AB\overline{C} + A\overline{B}C + \overline{A}\,\overline{B}\,\overline{C}$$

$$X = A\overline{B} + \overline{A}C$$

$$Y = AB + \overline{A}\,\overline{B} + AC$$

$$Z = A\overline{B}\,\overline{C} + BC$$

EXPRESSION SET 1.24 ***Practical Example 1.3, Logic Functions***

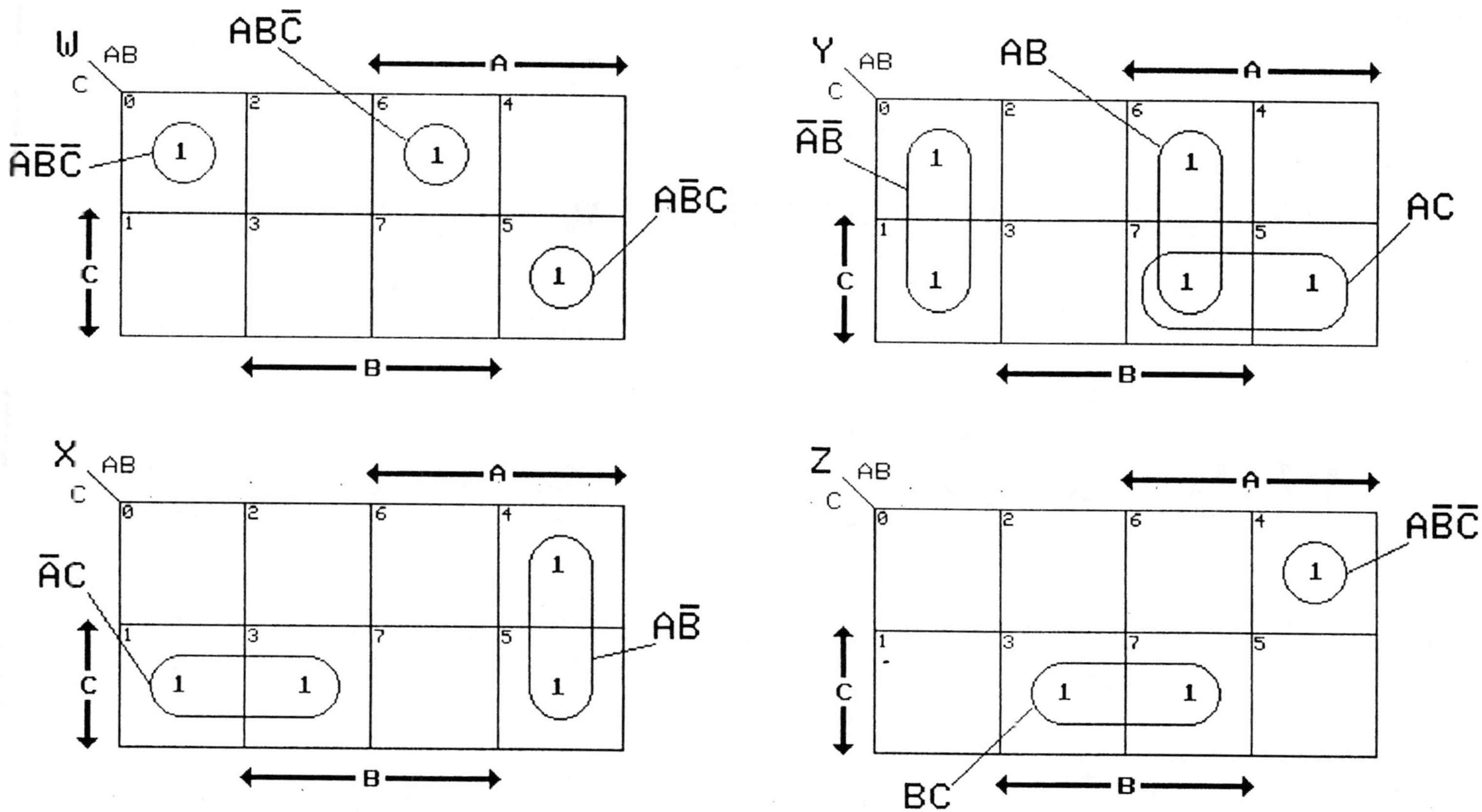

FIGURE 1.8 Karnaugh maps of logical functions W, X, Y, Z.

$$W = f(A,B,C) = \sum(0,5,6)$$

$$X = f(A,B,C) = \sum(1,3,4,5)$$

$$Y = f(A,B,C) = \sum(0,1,5,6,7)$$

$$Z = f(A,B,C) = \sum(3,4,7)$$

EXPRESSION SET 1.25 Practical Example 1.3, Summation Equations.

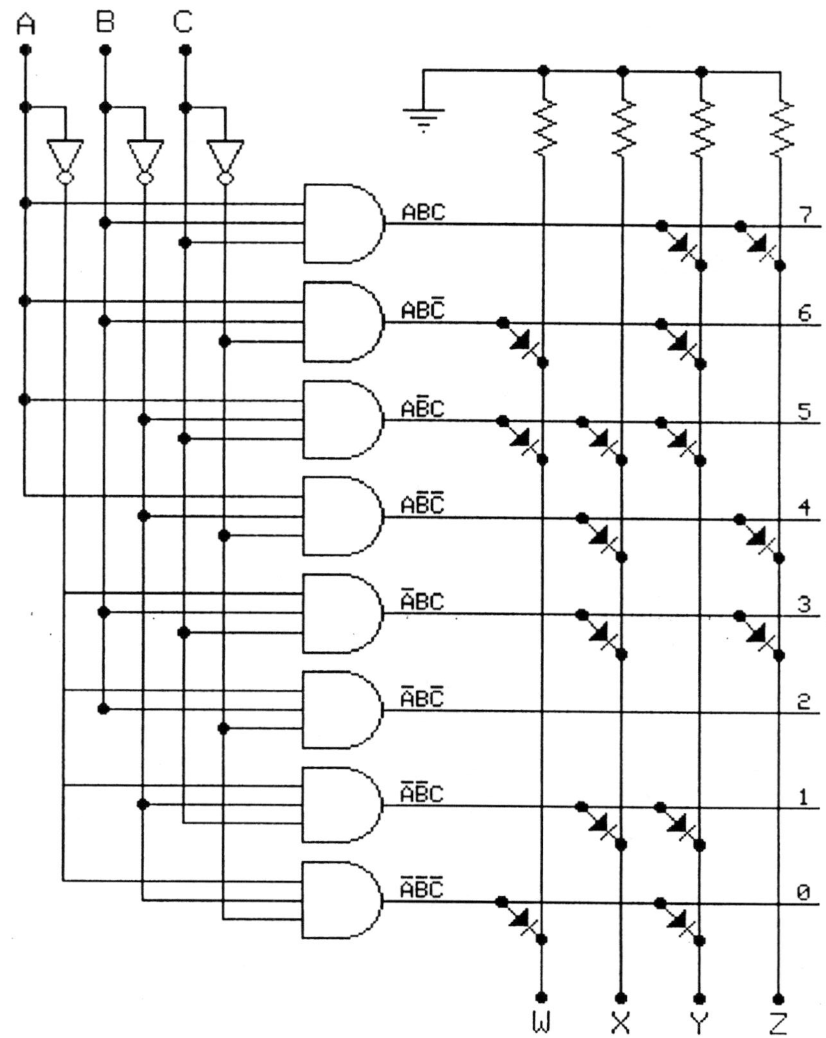

ADDRESS	DATA
ABC	WXYZ
7	0011
6	1010
5	1110
4	0101
3	0101
2	0000
1	0110
0	1010

FIGURE 1.9 Practical example 1.3, ROM circuit.

$$W = \overline{A}\,\overline{C} + A\overline{C}\,\overline{D} + A\overline{B}D$$

$$X = A\overline{C} + BD + AB + \overline{A}\,\overline{B}C\overline{D}$$

$$Y = \overline{C}\,\overline{D} + \overline{A}B + B\overline{C} + A\overline{B}D$$

$$Z = \overline{C}D + AB\overline{C} + \overline{A}B\overline{D} + A\overline{B}C\overline{D}$$

EXPRESSION SET 1.26 Practical Exercise 1.3, Logic Functions

$$W = B\overline{C} + ACD + \overline{A}\,\overline{B}C$$

$$X = \overline{A}\,\overline{B}C + BD + AB\overline{C}$$

$$Y = AC + \overline{A}\,\overline{C}$$

$$Z = B\overline{C} + BD$$

EXPRESSION SET 1.27 Logical Functions

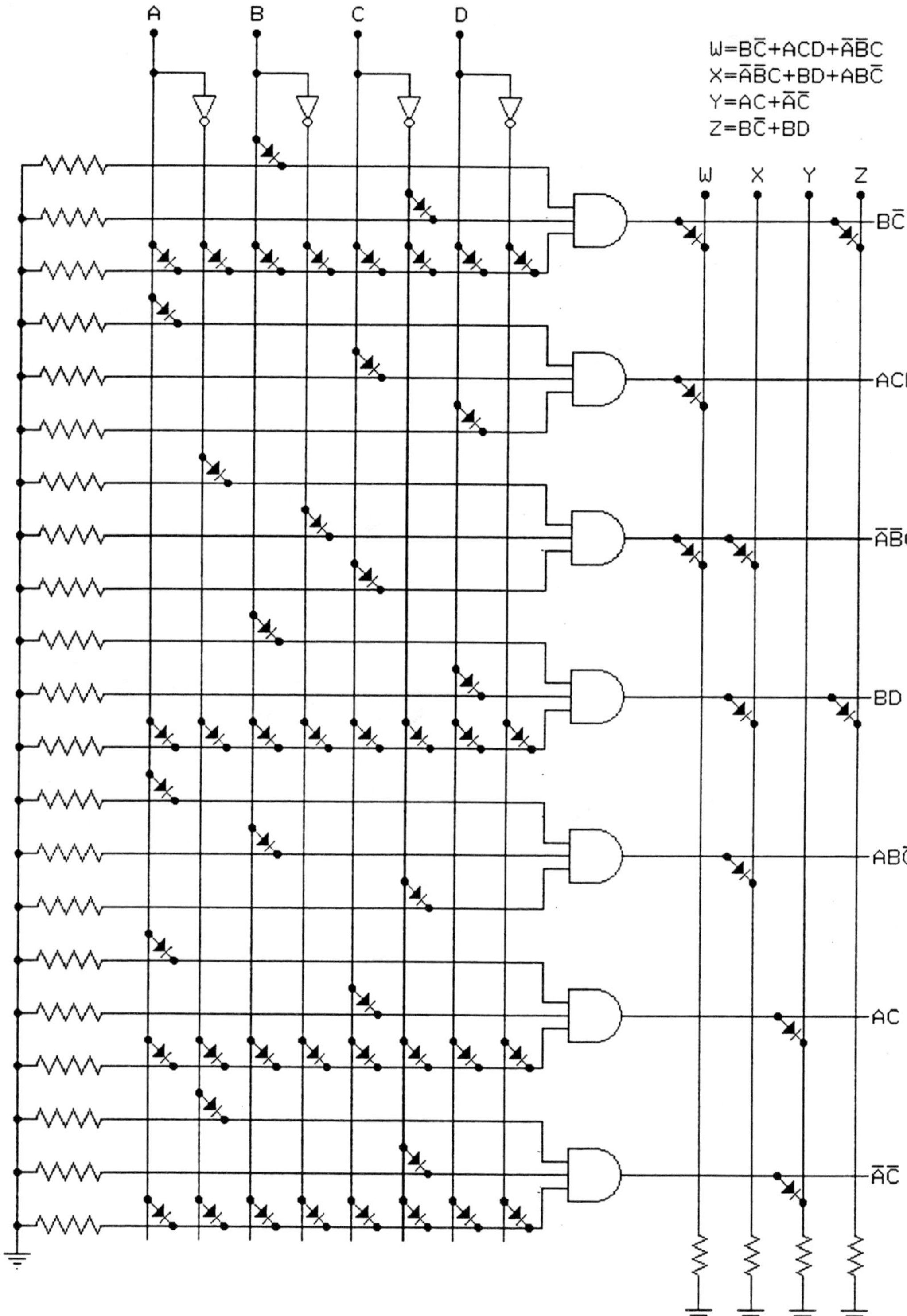

FIGURE 1.10 FPLA implementation of logical functions.

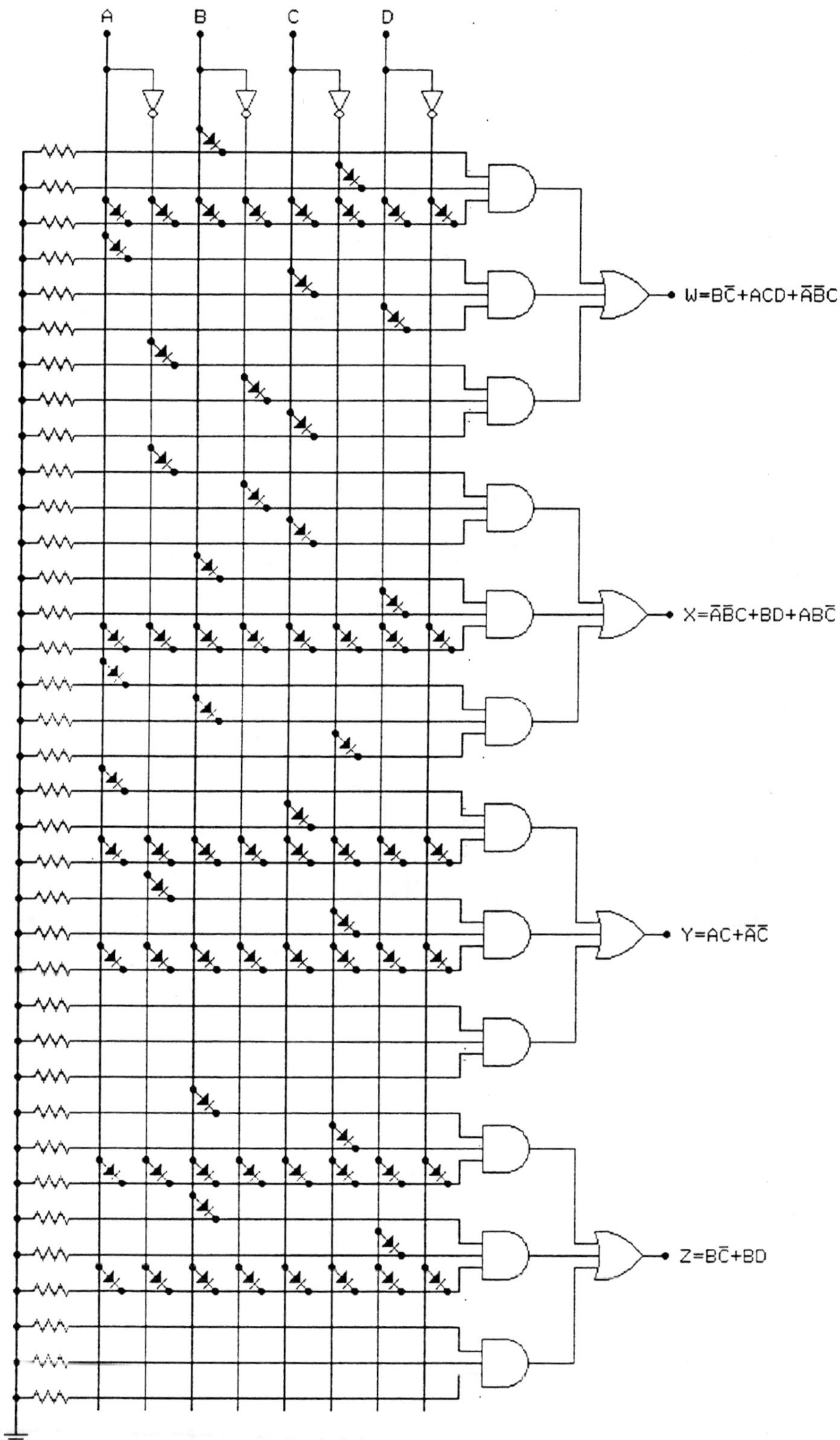

FIGURE 1.11 PAL implementation of logical functions.

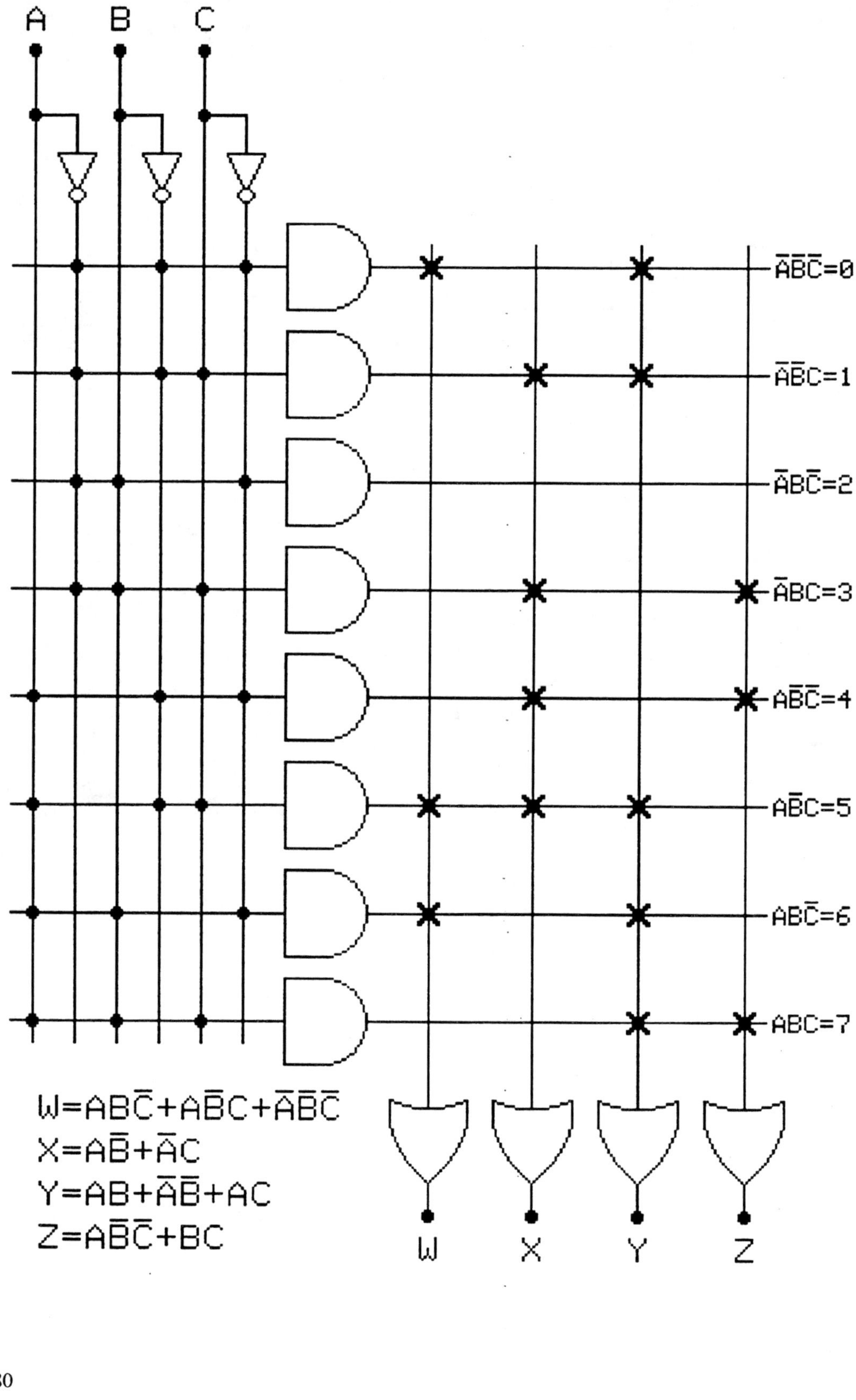

FIGURE 1.12 ROM schematic, simplified form.

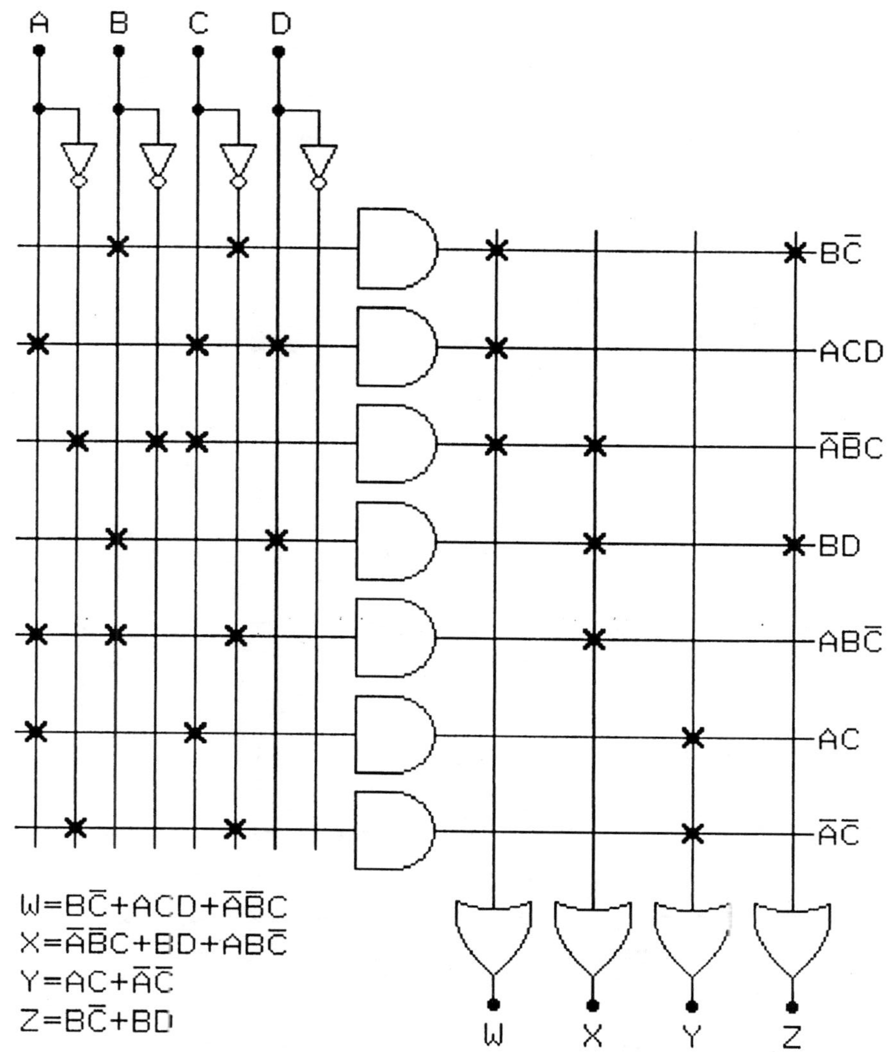

FIGURE 1.13 FPLA schematic, simplified form.

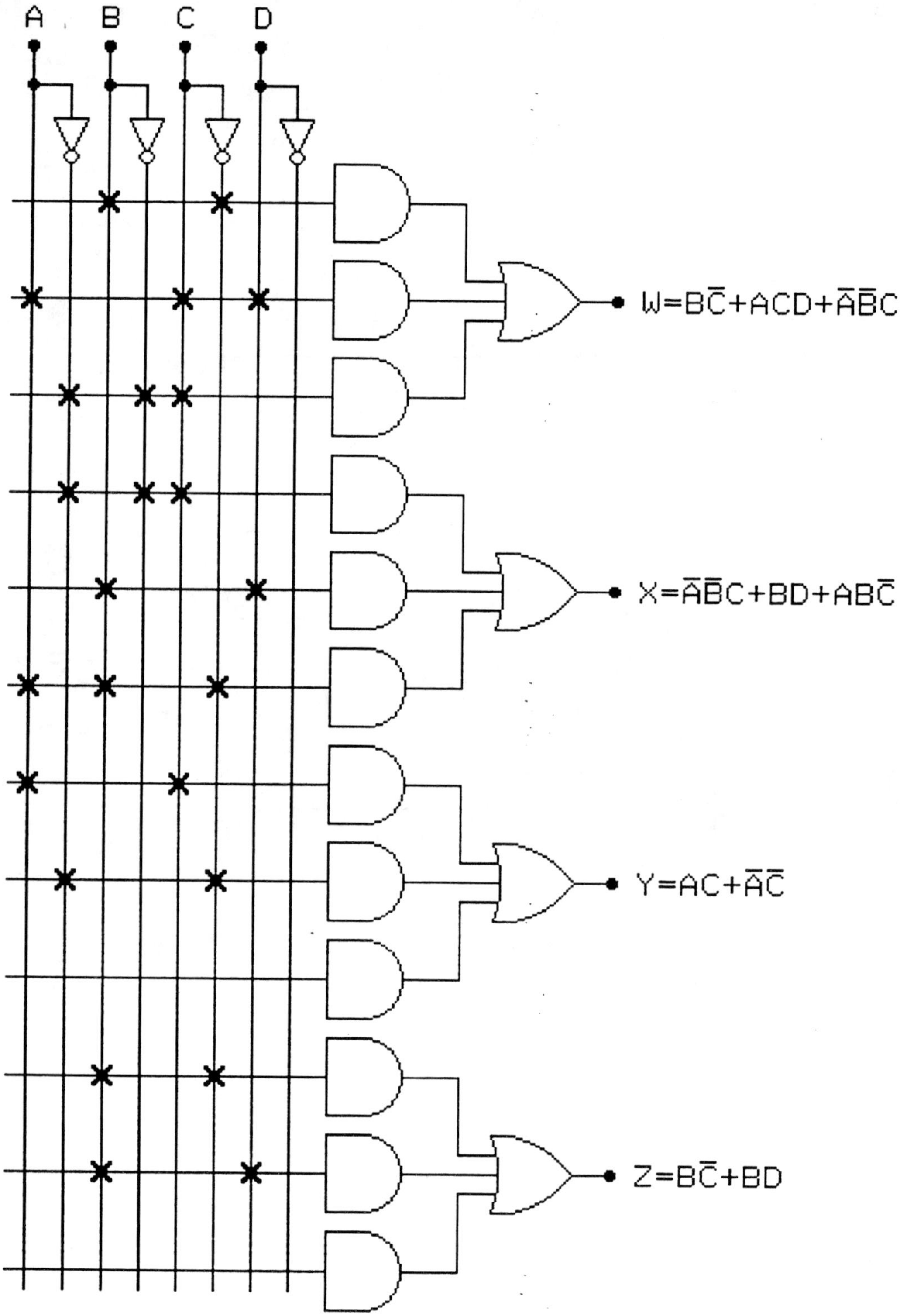

FIGURE 1.14 PAL schematic, first form.

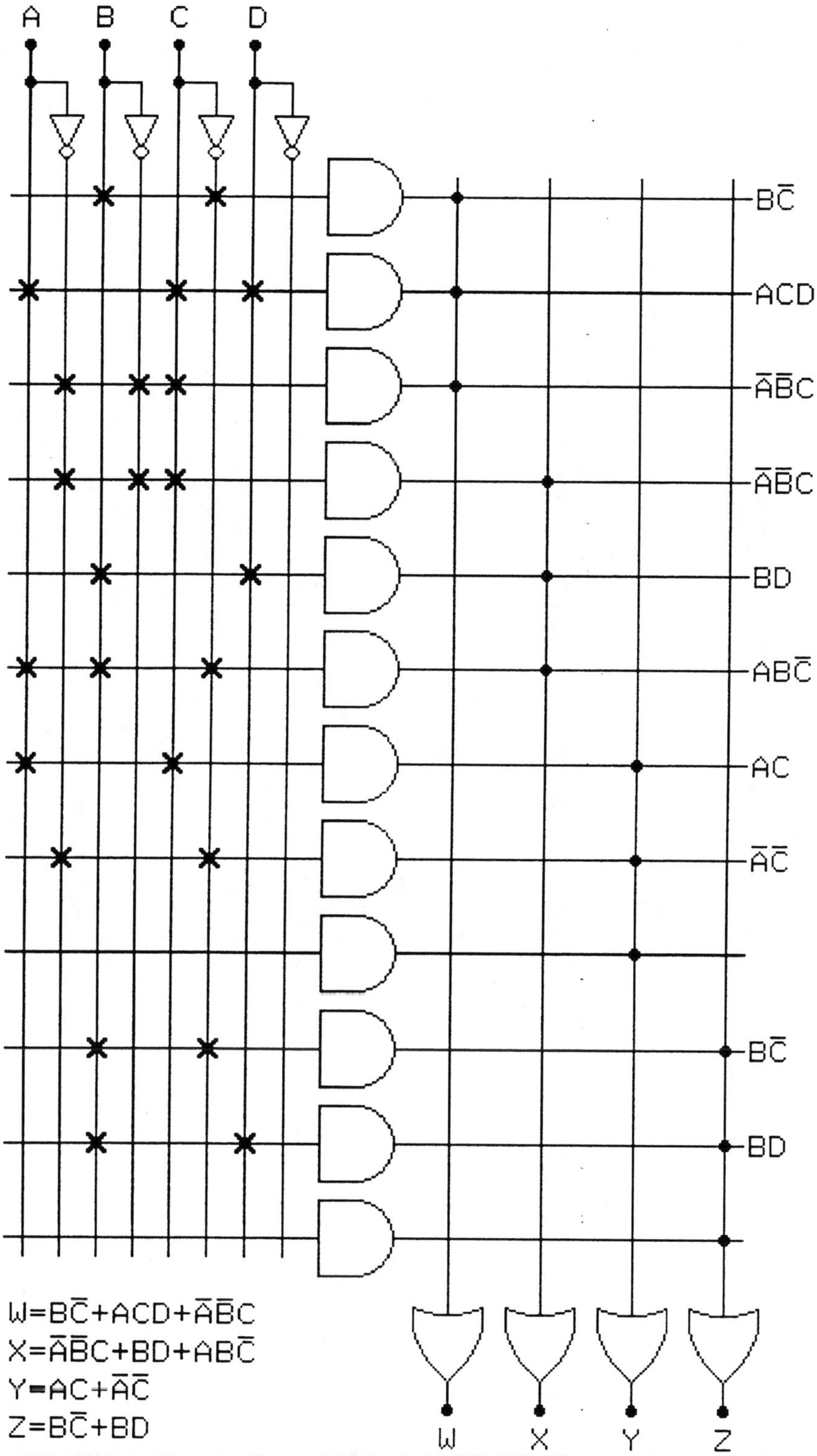

FIGURE 1.15 PAL schematic, second form

Device	Input Matrix	Output Matrix	Output Logic
ROM	*Hardwired*	*Programmable*	*Combinational, SOP only.*
FPLA	*Programmable*	*Programmable*	*Several output buffer types available.*
PAL	*Programmable*	*Hardwired*	*Several output buffer types available.*

TABLE 1.2 ROM/FPLA/PAL device summary.

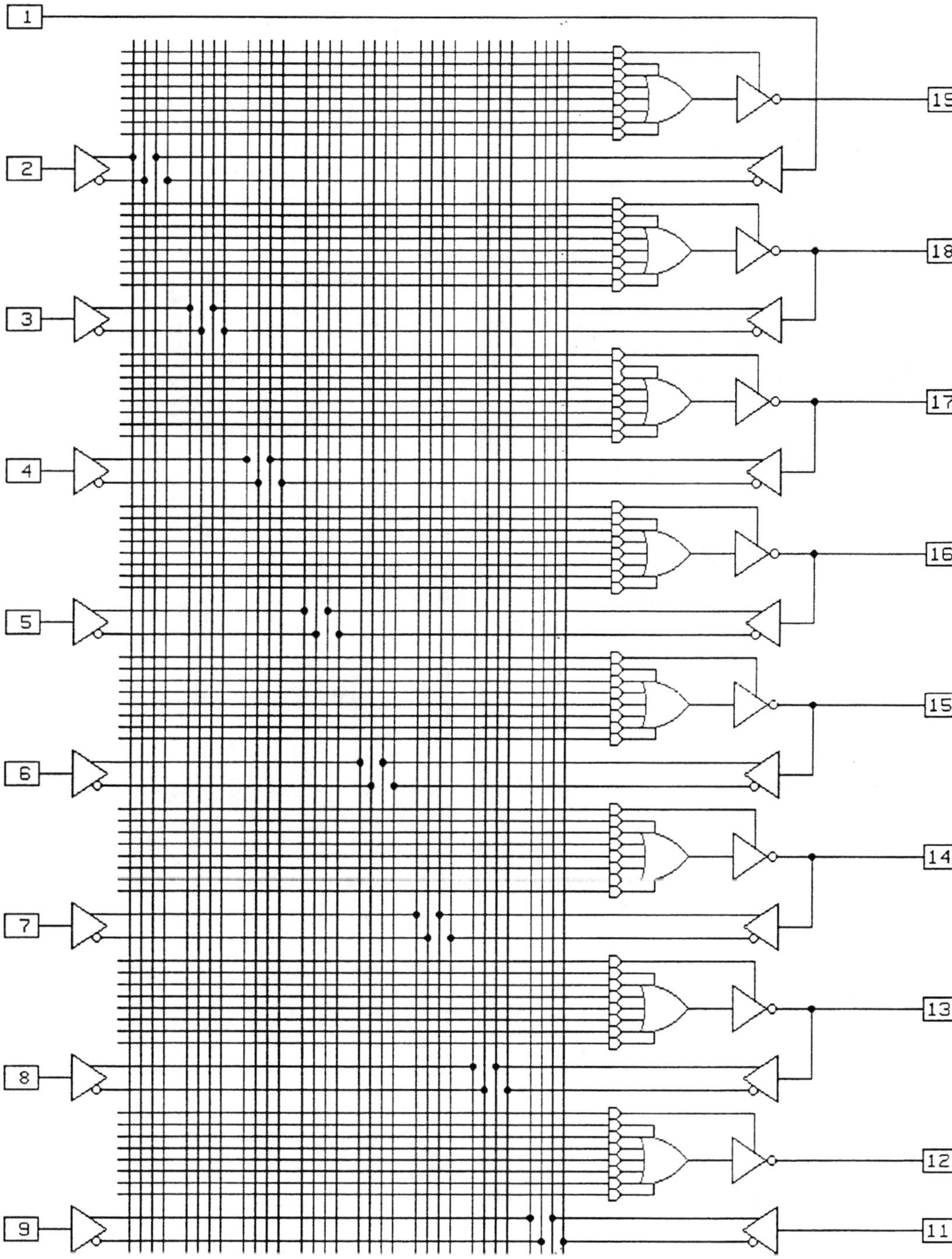

FIGURE 1.16 Unprogrammed 16L8 PAL circuit diagram.

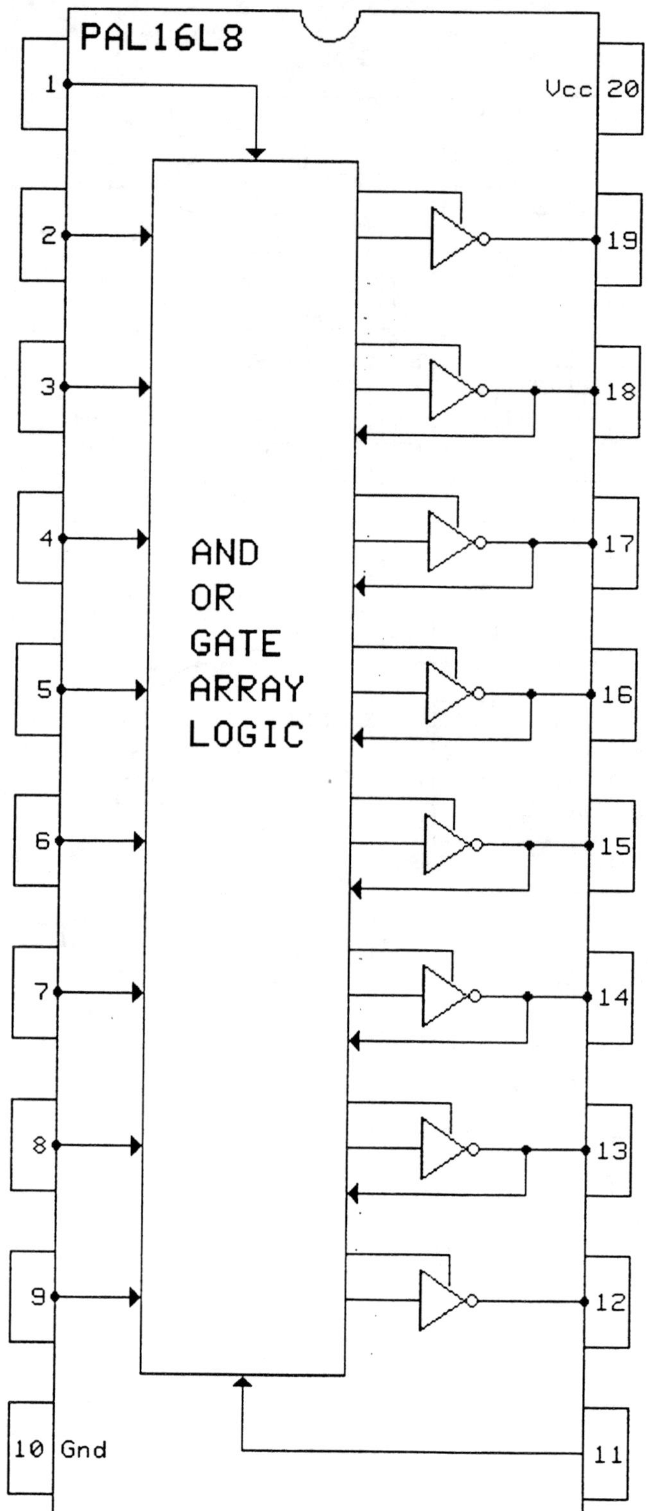

FIGURE 1.17 PAL16L8 device diagram.

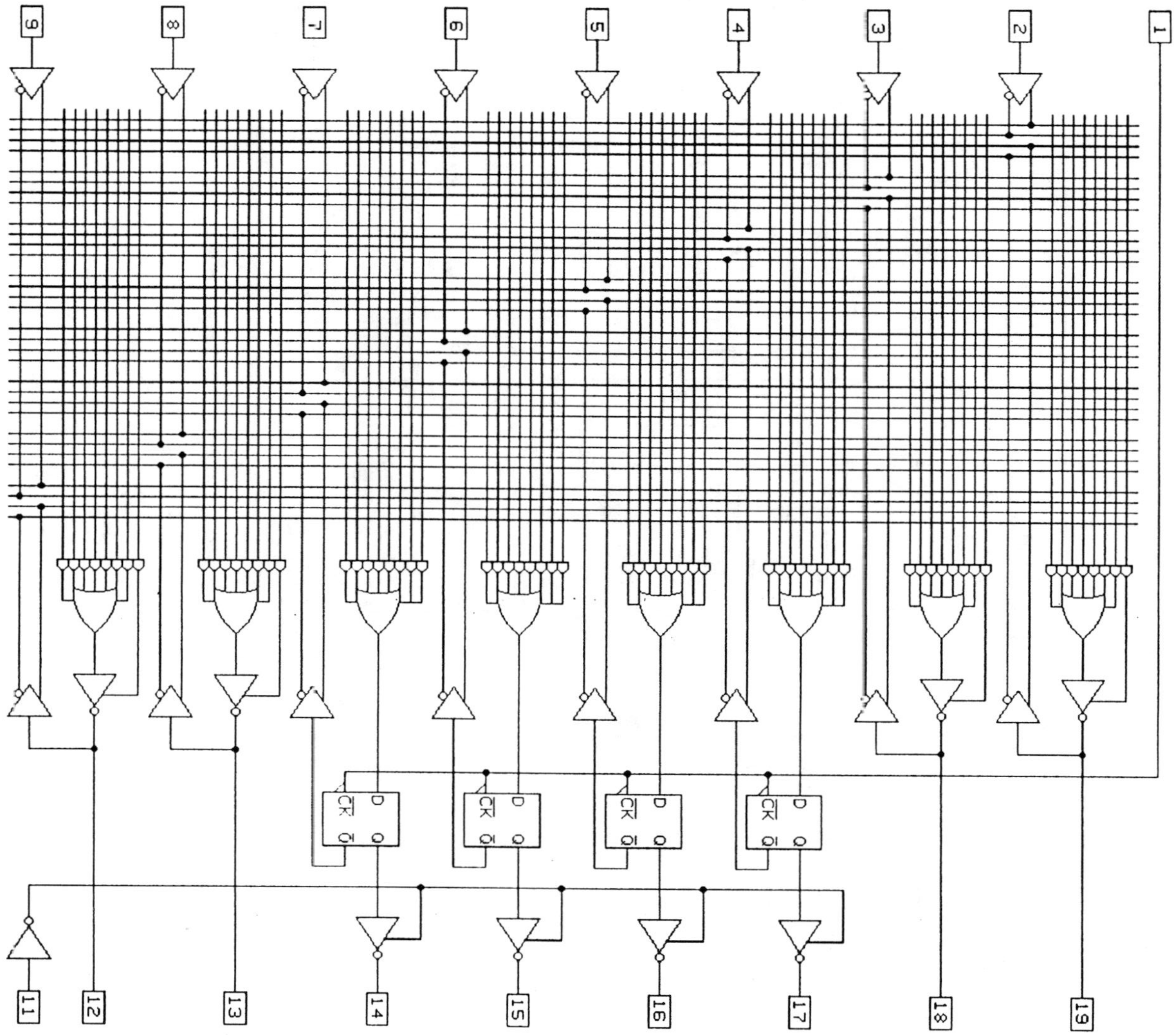

FIGURE 1.18 Unprogrammed PAL 16R4 circuit diagram.

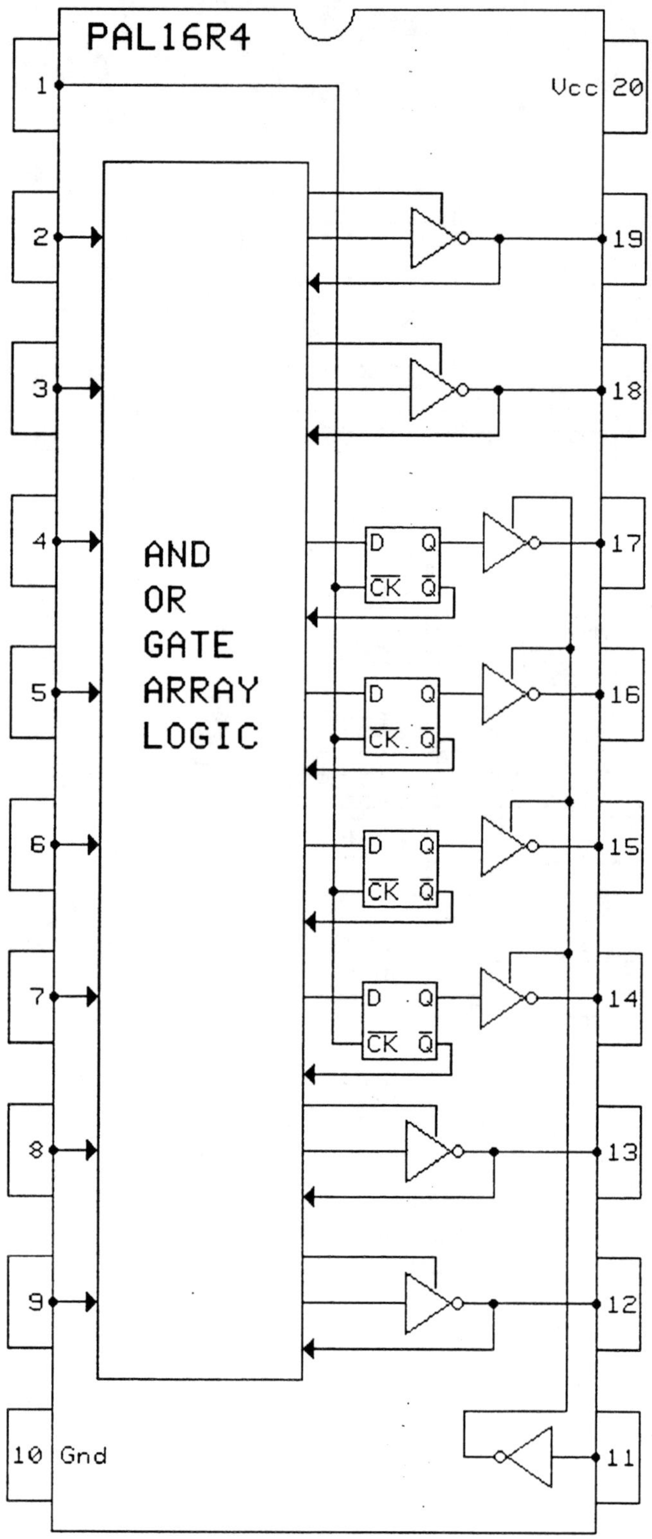

FIGURE 1.19 PAL16R4 device diagram.

V-TYPE MACROCELL

A1	A0	OUTPUT CONFIGURATION
0	0	REGISTERED/ACTIVE-LOW
0	1	REGISTERED/ACTIVE-HIGH
1	0	COMBINATORIAL/ACTIVE-LOW
1	1	COMBINATORIAL/ACTIVE-HIGH

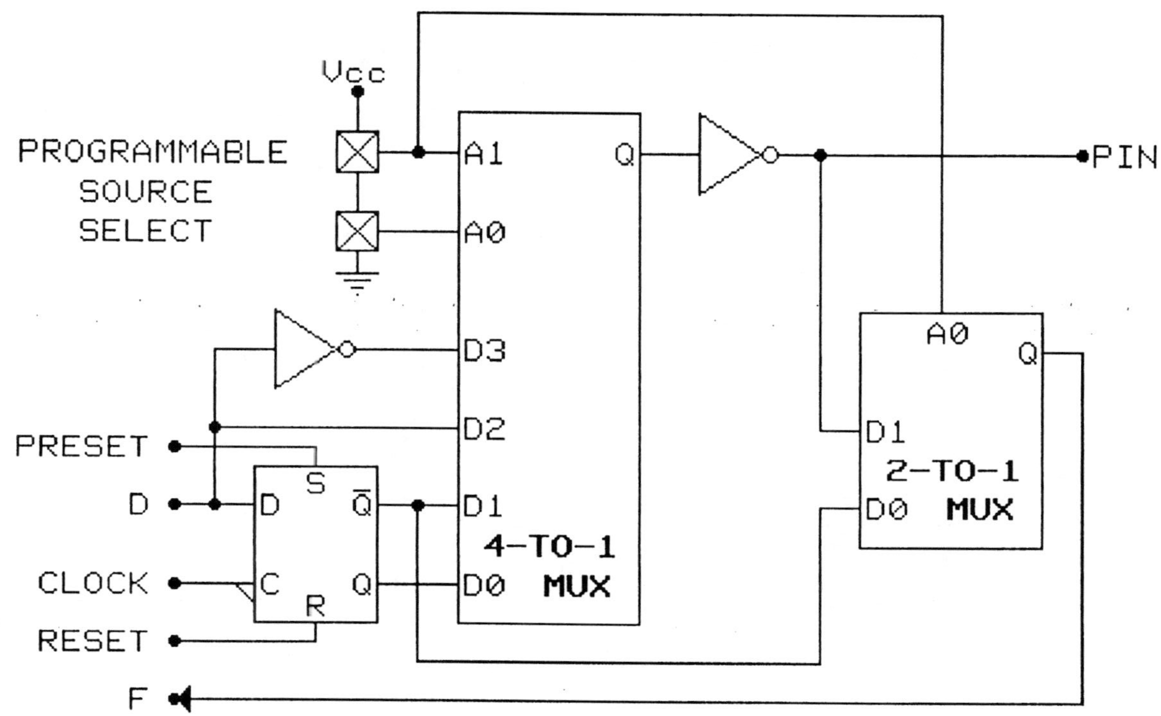

FIGURE 1.20 V-type macrocell.

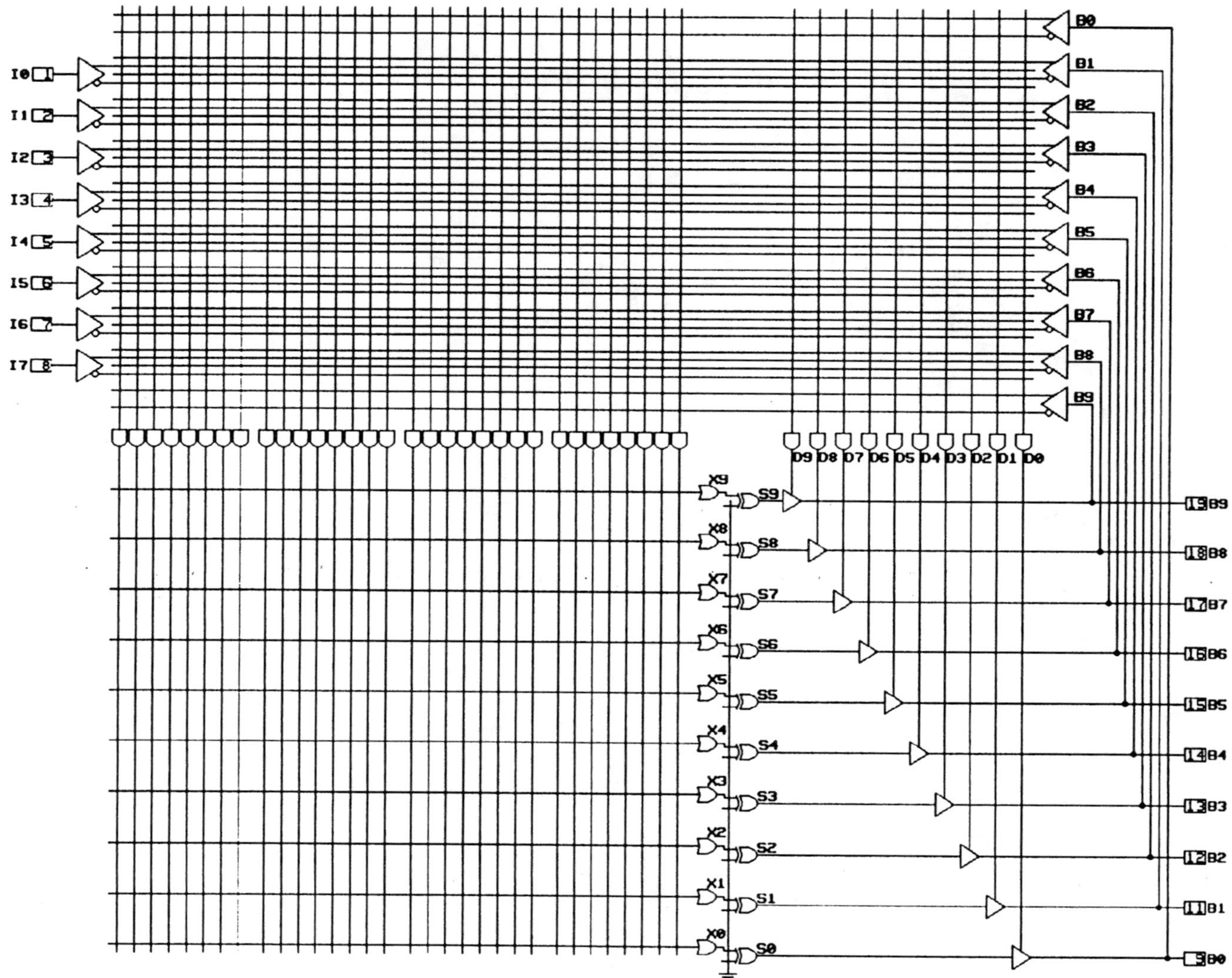

FIGURE 1.22 Philips PLS153 FPLA (18 x 32 x 10)

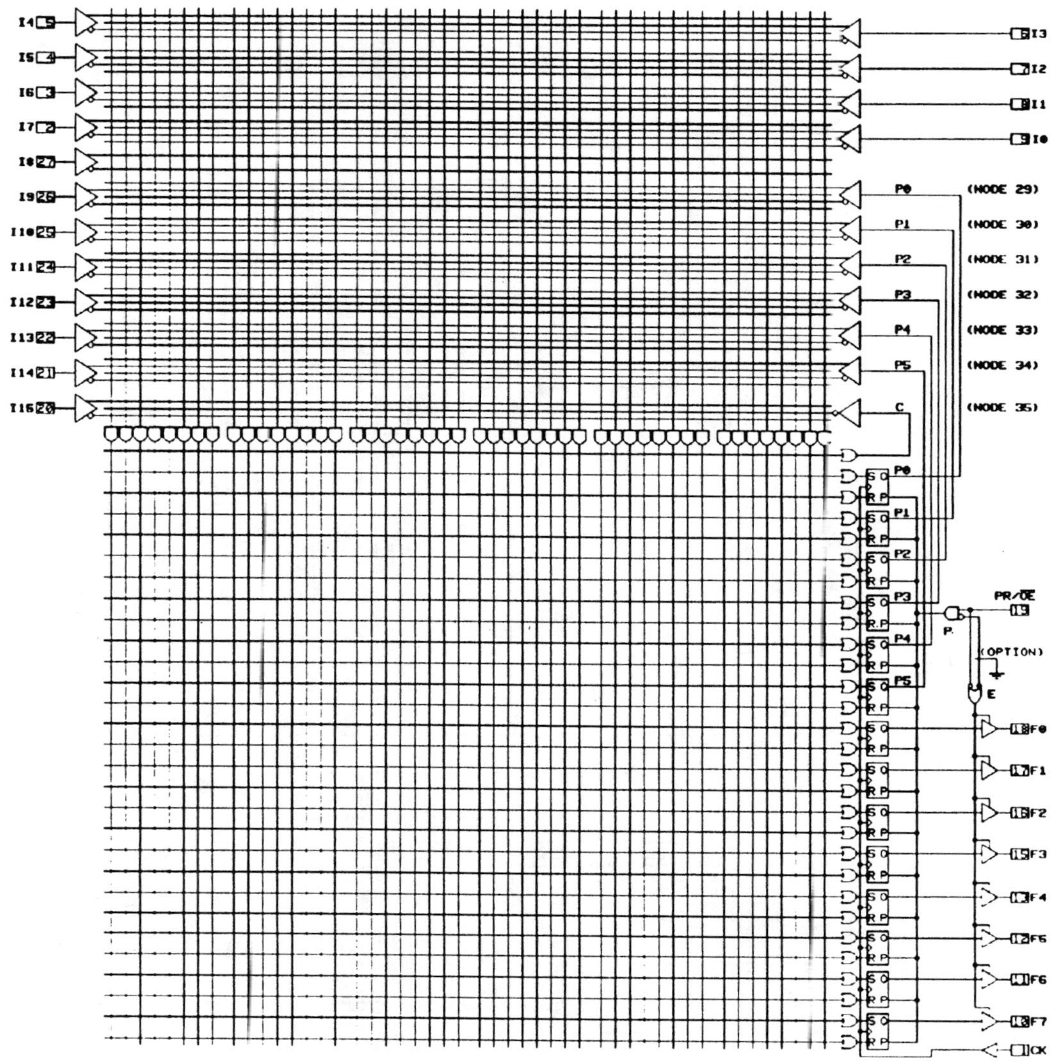

FIGURE 1.23 Philips PLS105 FPLS (16 x 48 x 8).

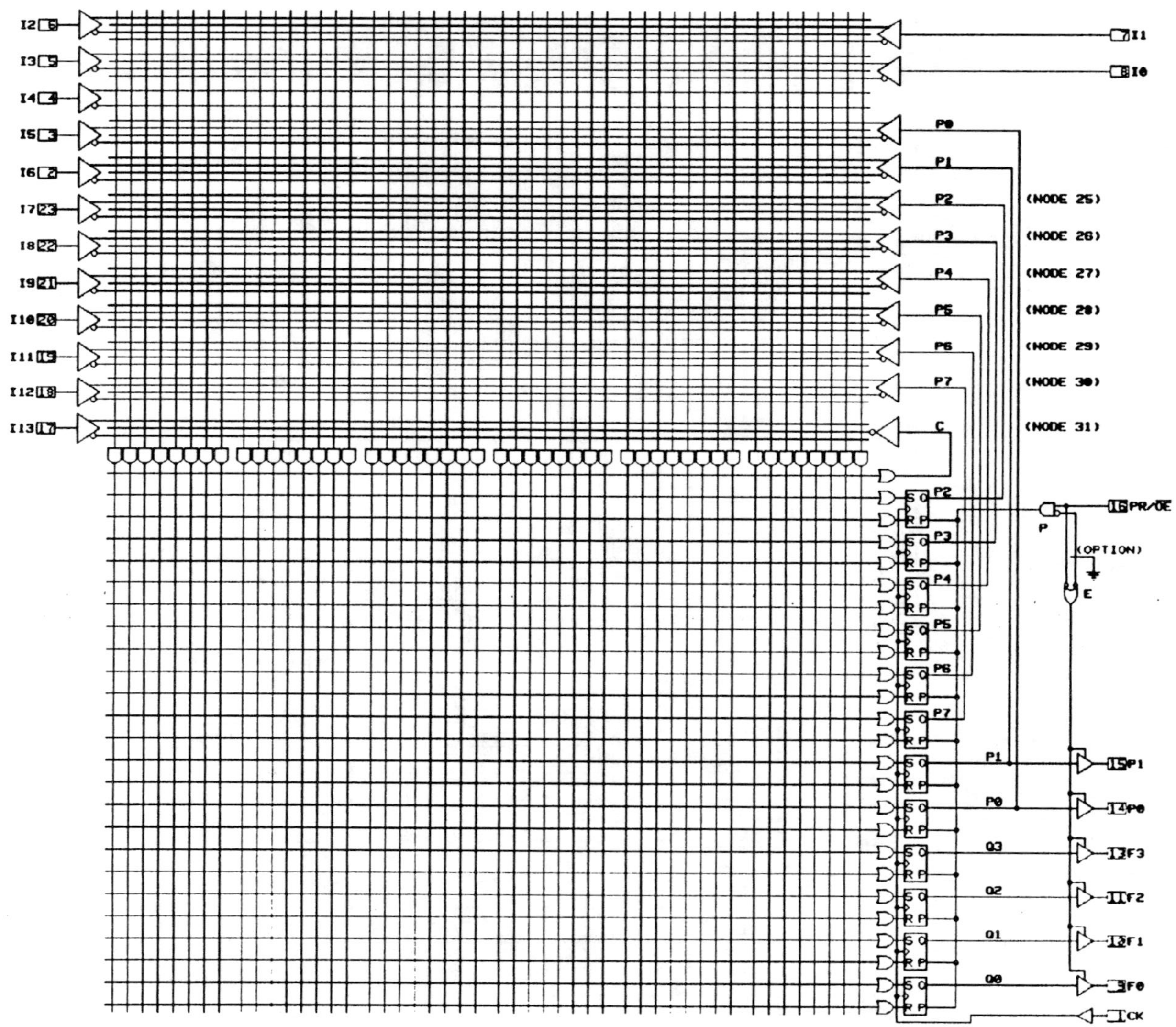

FIGURE 1.24 Philips PLS167 FPLS (14 x 48 x 6).

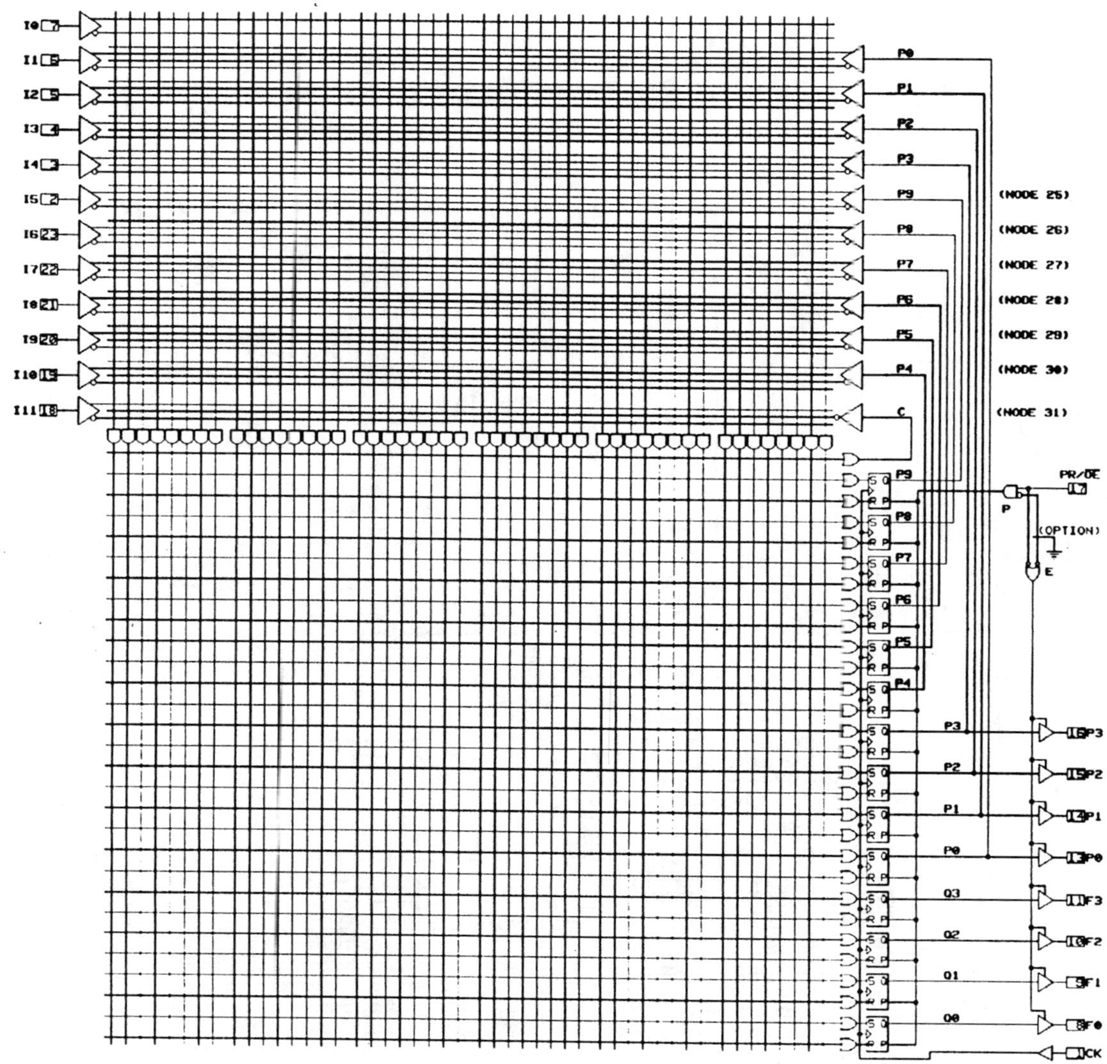

FIGURE 1.25 Philips/Signetics PLS168 FPLS (12 x 48 x 8).

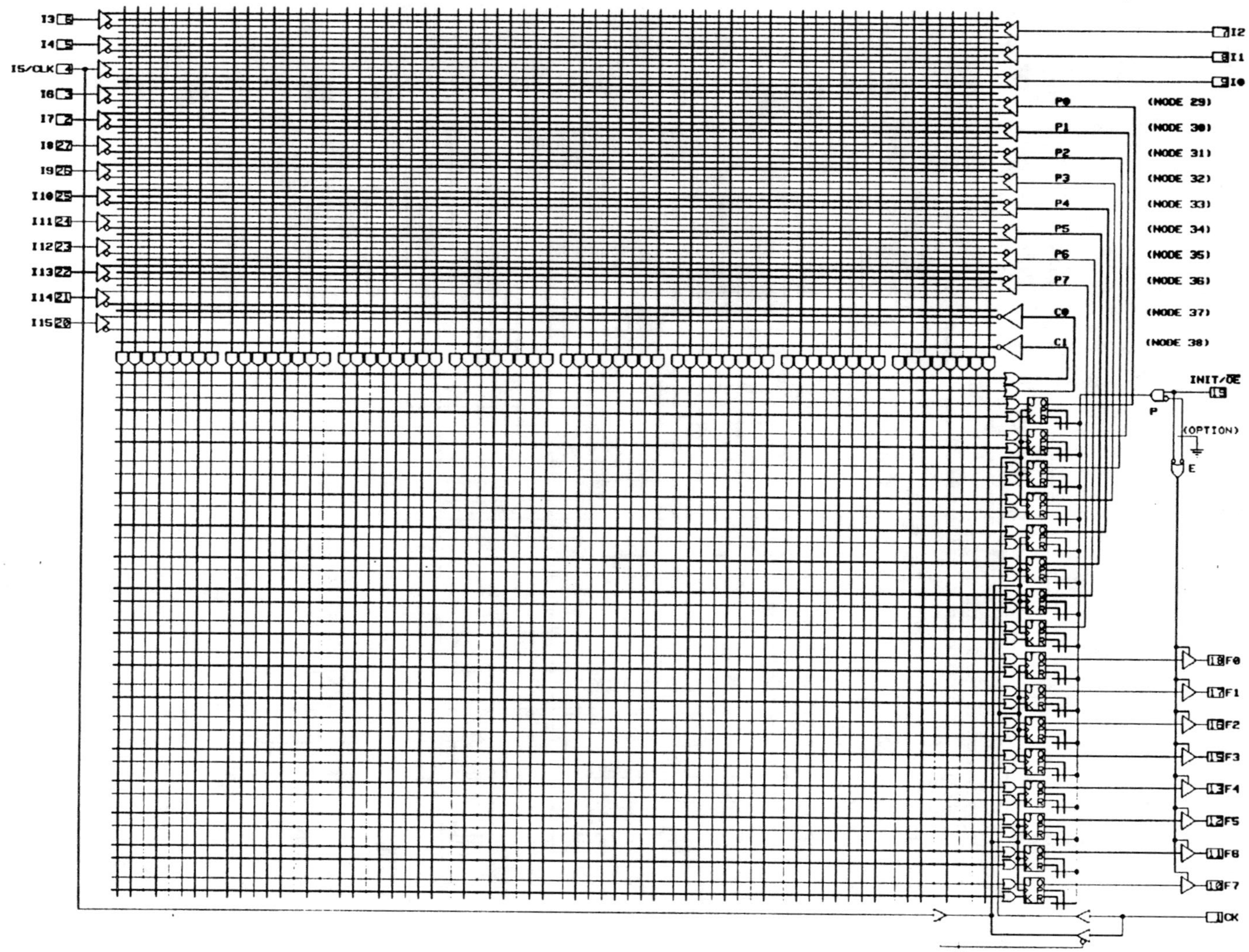

FIGURE 1.26   Philips/Signetics PLS405 FPLS (16 x 64 x 8).

Lattice GAL Devices							
16V8							
20V8							
20RA10							
20XV10							
22V10							
replace the following *Bipolar Programmable Logic Devices*							
10H8	14H4	16H2	16P2	16RP4	18P8	20L10*	20RP6
10L8	14H8	16H6	16P6	16RP6	18U8	20P8	20RP8
10P8	14L4	16H8	16P8	16RP8	20H2	20R4	20RA10
12H6	14L8	16L2	16R4	18H4	20H8	20R6	20X4*
12L6	14P4	16L6	16R6	18L4	20L2	20R8	20X8*
12P6	14P8	16L8	16R8	18P4	20L8	20RP4	20X10*
							22V10

* GAL20XV10 replaces 20L10, 20X10, 20X8 and 20X4.

TABLE 1.3 GAL/PAL replacement table.

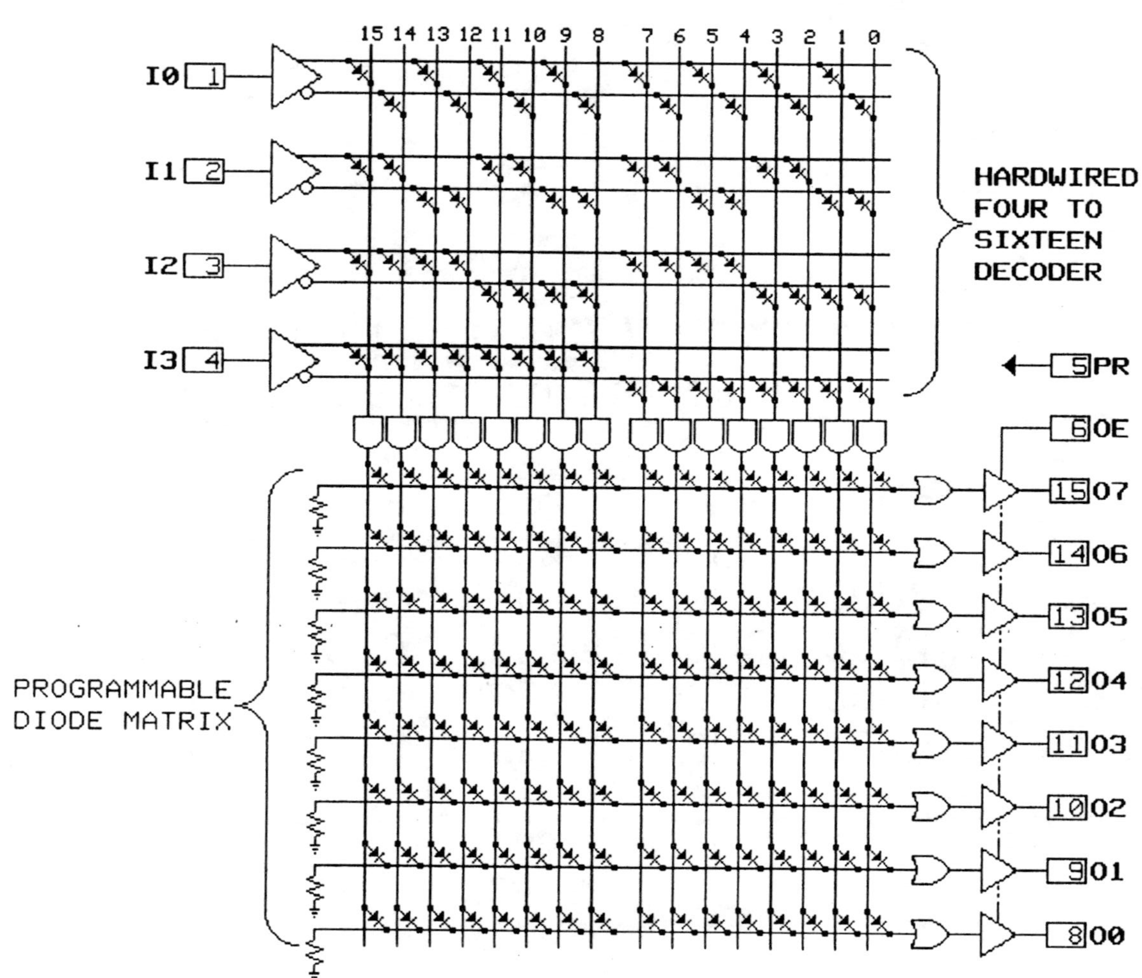

FIGURE 1.27 Unprogrammed 16 x 8 programmable ROM (PROM).

Binary	ASCII
0000	0011 0000
0001	0011 0001
0010	0011 0010
0011	0011 0011
0100	0011 0100
0101	0011 0101
0110	0011 0110
0111	0011 0111
1000	0011 1000
1001	0011 1001
1010	0100 0001
1011	0100 0010
1100	0100 0011
1101	0100 0100
1110	0100 0101
1111	0100 0110

TABLE 1.4 Binary to ASCII code conversion.

Binary	ASCII
Address I 3210	Data O 7654 3210
0000	0011 0000
0001	0011 0001
0010	0011 0010
0011	0011 0011
0100	0011 0100
0101	0011 0101
0110	0011 0110
0111	0011 0111
1000	0011 1000
1001	0011 1001
1010	0100 0001
1011	0100 0010
1100	0100 0011
1101	0100 0100
1110	0100 0101
1111	0100 0110

TABLE 1.5 PROM-based binary to ASCII code conversion.

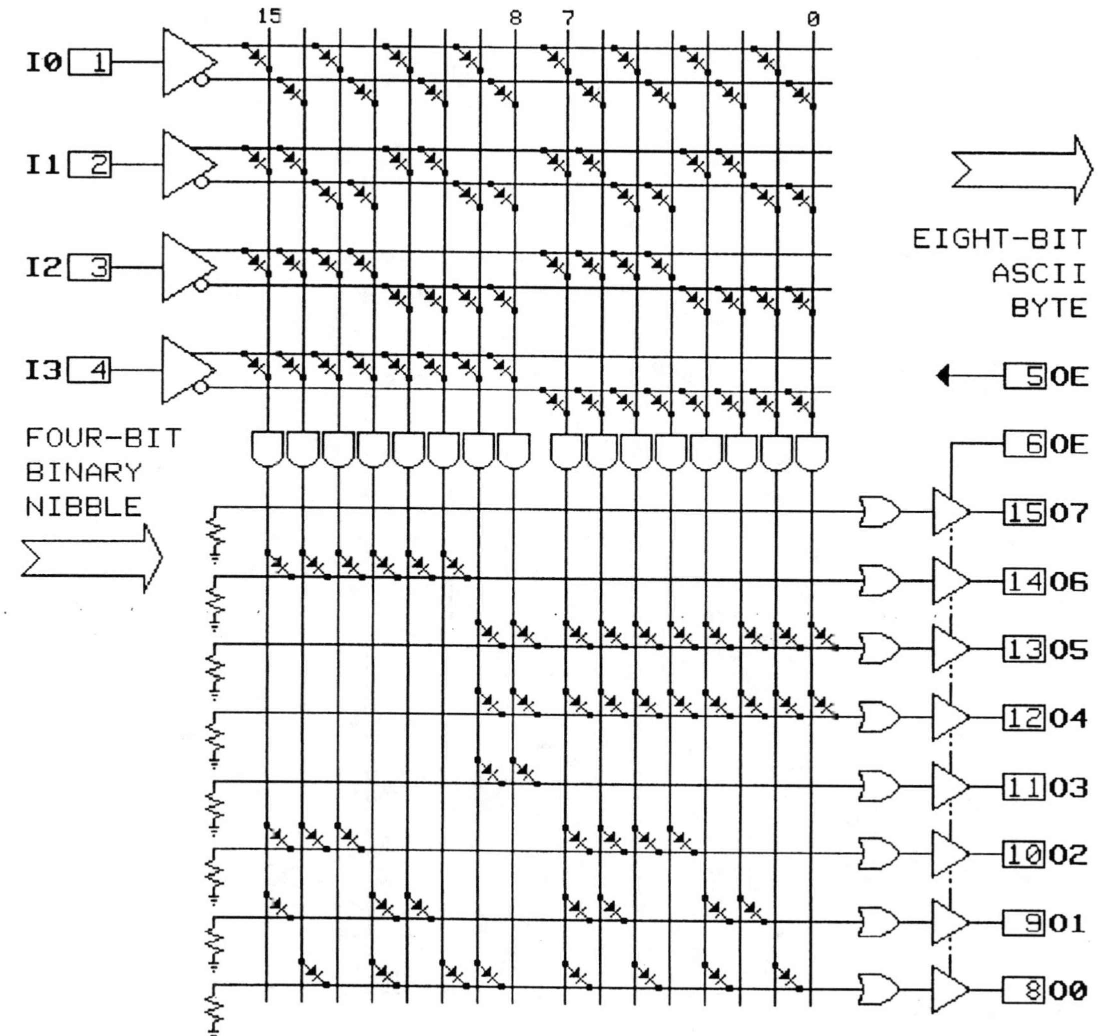

FIGURE 1.28 ***ROM-based binary to ASCII code converter.***

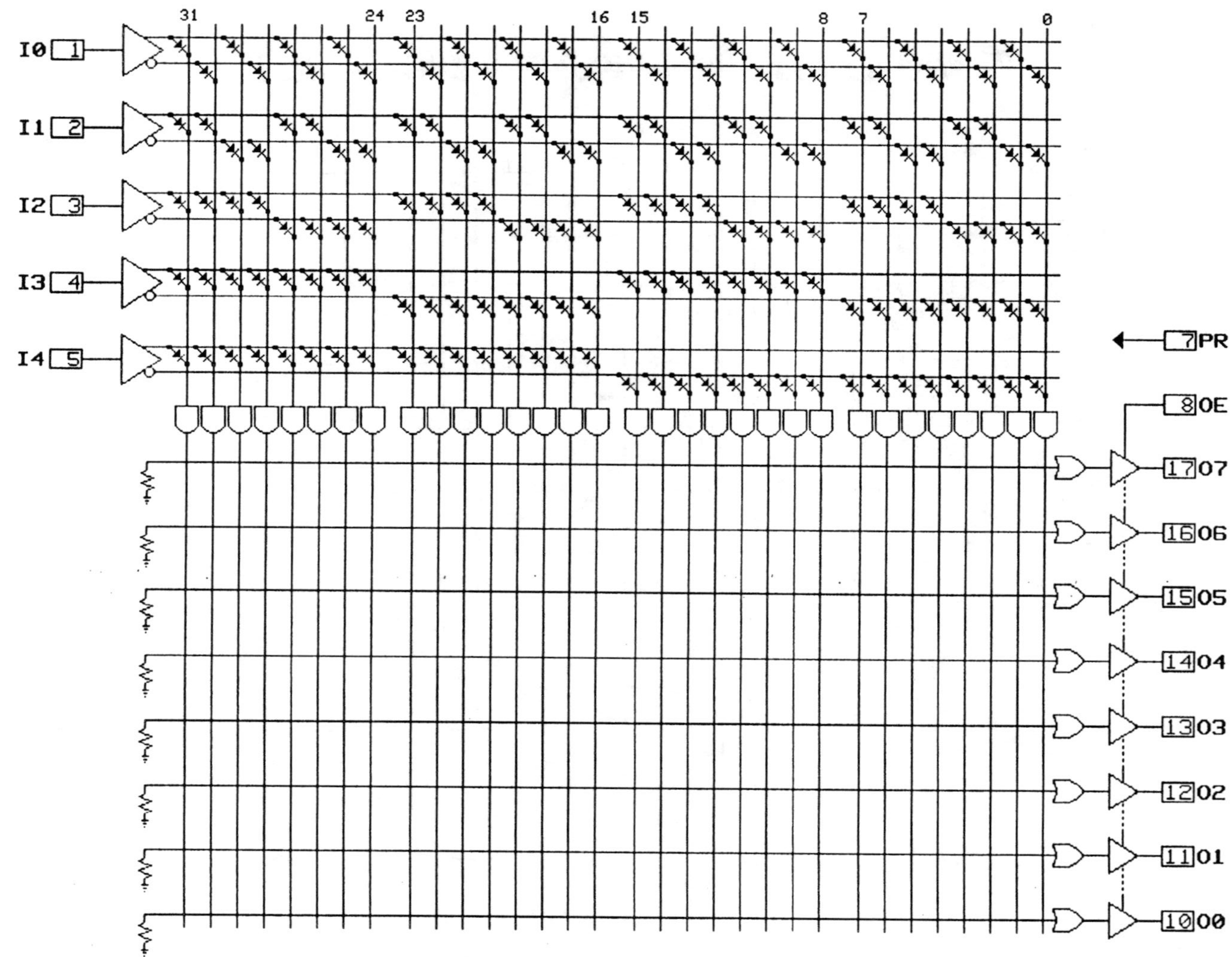

FIGURE 1.29 32 x 8 PROM schematic diagram.

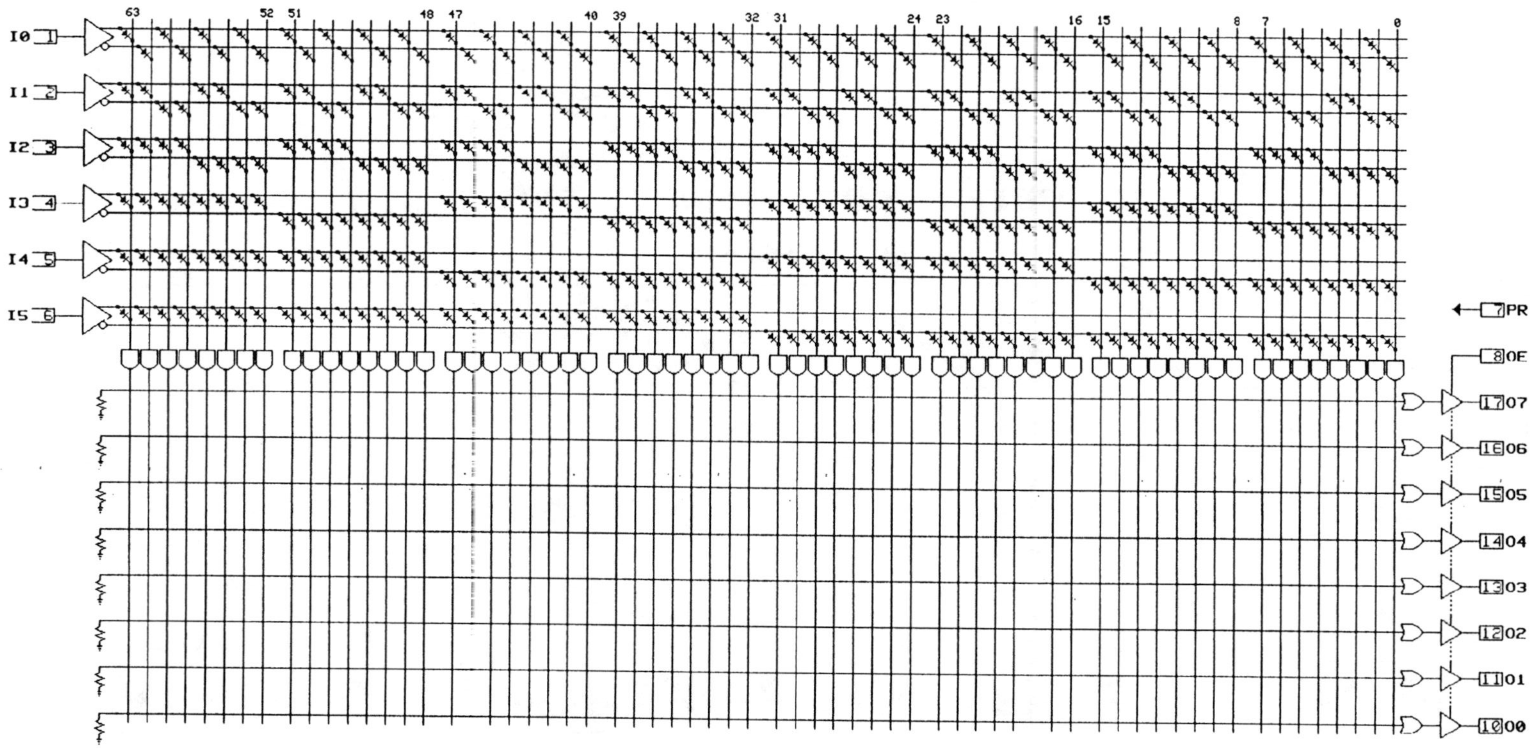

FIGURE 1.30 64 x 8 PROM schematic diagram.

Designation	Architecture	No. of Inputs	Total Bits
2716	2,048 x 8	11	16,384
2732	4,096 x 8	12	32,768
2764	8,192 x 8	13	65,536
27128	16,384 x 8	14	131,072
27256	32,768 x 8	15	262,144
27512	65,536 x 8	16	524,288

TABLE 1.6 Example PROM architectures.

$$W = AB\overline{C} + A\overline{B}C + \overline{A}\,\overline{B}\,\overline{C}$$

$$X = A\overline{B} + \overline{A}C$$

$$Y = AB + \overline{A}\,\overline{B} + AC$$

$$Z = A\overline{B}\,\overline{C} + BC$$

EXPRESSION SET 1.28 Practical Example 1.4, Logic Functions

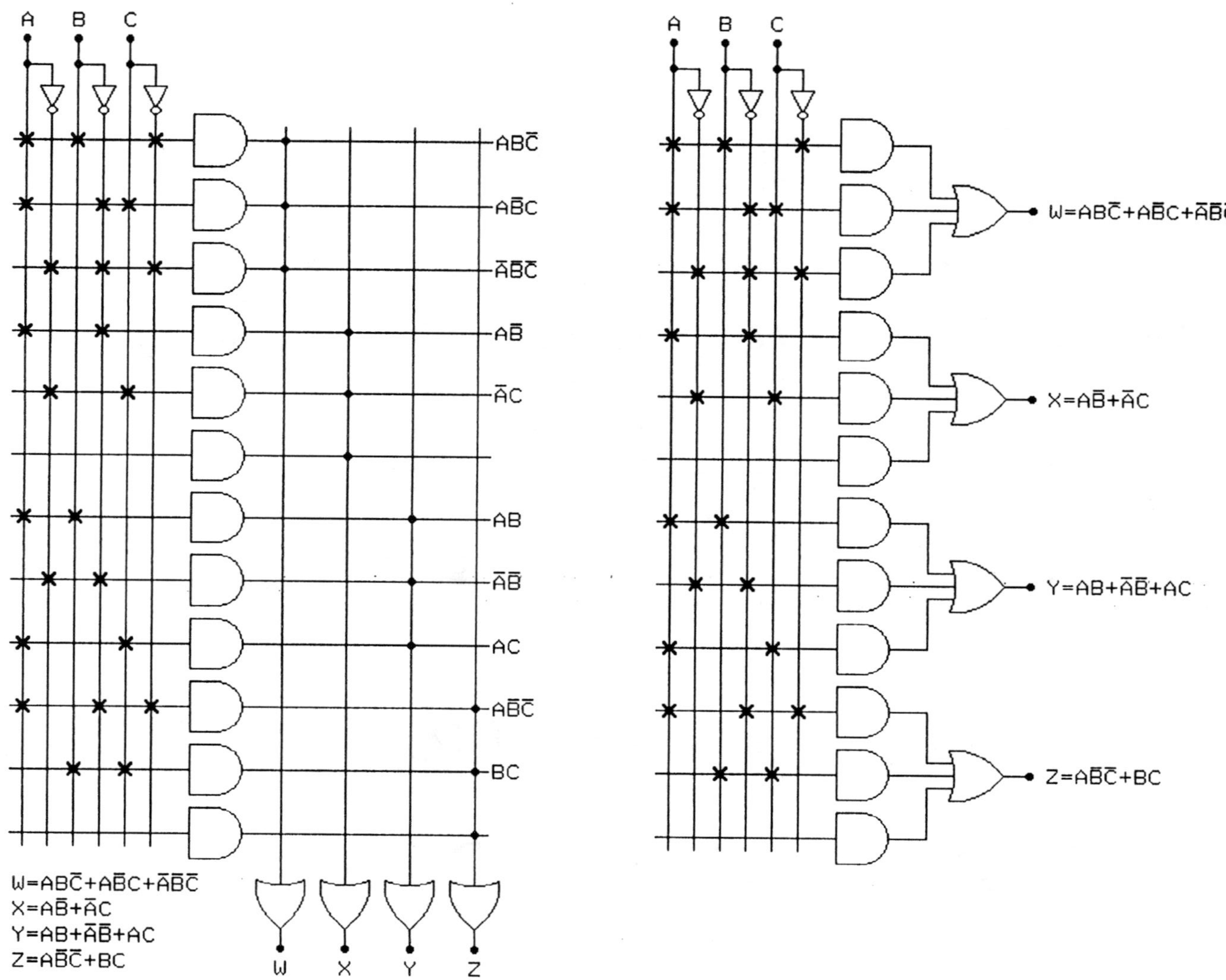

FIGURE 1.31 Practical Example 1.4, solutions.

$$W = \overline{A}\,\overline{C} + A\overline{C}\,\overline{D} + A\overline{B}D$$

$$X = A\overline{C} + BD + AB + \overline{A}\,\overline{B}C\overline{D}$$

$$Y = \overline{C}\,\overline{D} + \overline{A}B + B\overline{C} + A\overline{B}D$$

$$Z = \overline{C}D + AB\overline{C} + \overline{A}B\overline{D} + A\overline{B}C\overline{D}$$

EXPRESSION SET 1.29 Practical Exercise 1.4, Logic Functions

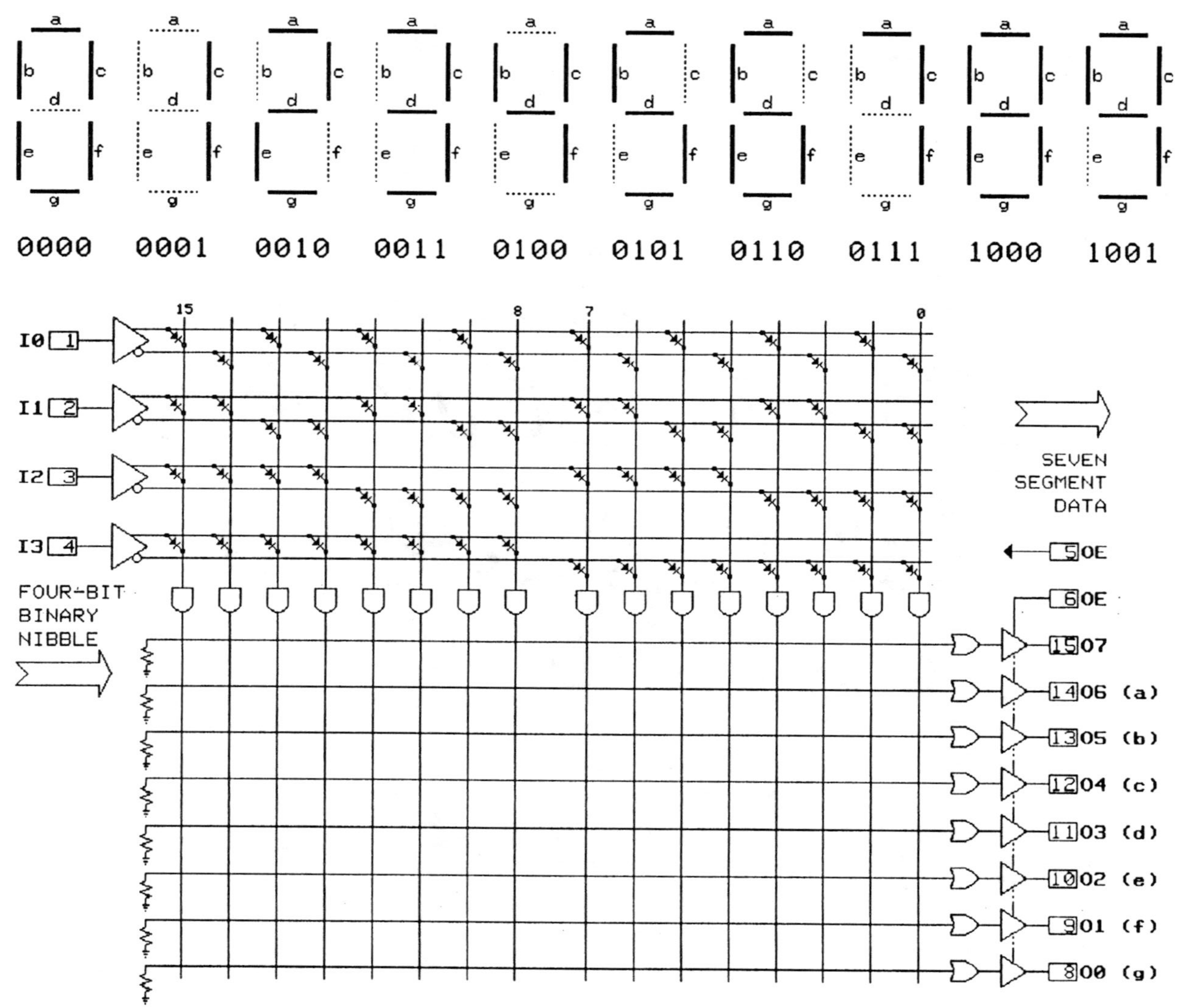

FIGURE 1.32 BCD to seven-segment code definitions and PROM.

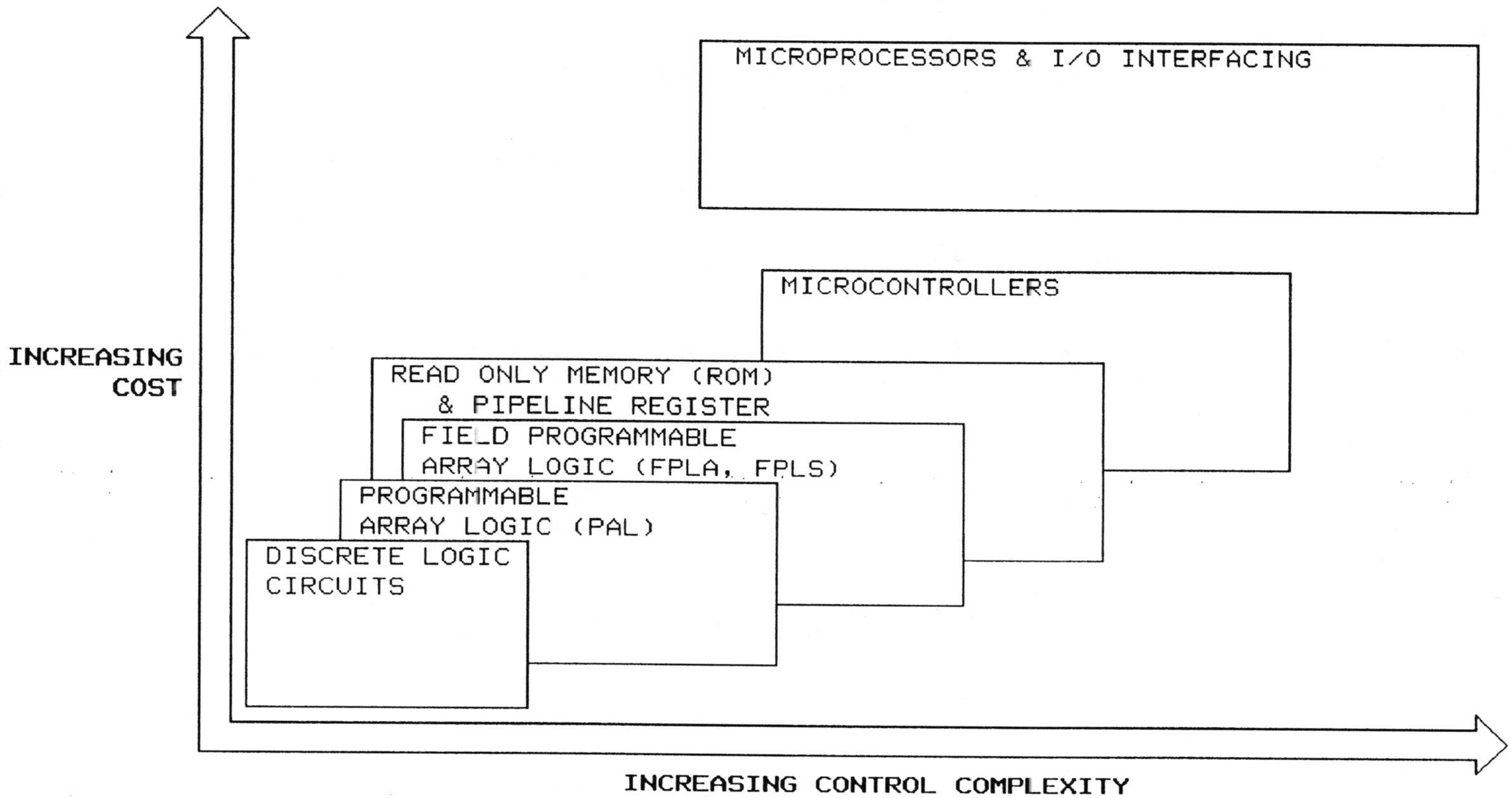

FIGURE 1.33 Spectrum of digital design complexity.

$$W = A\overline{B} + A\overline{B}\,\overline{D} + A\overline{B}D + ACD$$

$$X = A\overline{B} + CD + \overline{A}C + \overline{A}B\overline{C}\,\overline{D}$$

$$Y = \overline{B}\,\overline{D} + \overline{A}C + \overline{B}C + ACD$$

$$Z = \overline{B}\,\overline{D} + A\overline{B}C + \overline{A}C\overline{D} + \overline{A}B\overline{C}\,\overline{D}$$

EXPRESSION SET 1.30 Problems 24 - 26

$$R = (A\&B\&!D)\#(B\&!C)\#(!A\&B)\#(!C\&!D)$$

$$S = (!A\&!B\&!D)\#(A\&B)\#(B\&D)\#(A\&!C)$$

$$T = (A\&!B\&C\&!D)\#(!A\&B\&!D)\#(A\&B\&!C)\#(!C\&D)$$

$$U = (A\&!B\&C)\#(!A\&B\&C)\#(!A\&!B)\#(!B\&!C)$$

EXPRESSION SET 1.31 Problems 27 - 29

$$L = f(ABC) = \sum(0,2,4,5,7)$$

$$M = f(ABCD) = \sum(1,2,5,6,7,8,10,11,13,15)$$

$$N = f(ABC) = \sum(2,3,4,5,6)$$

$$P = f(ABCDE) = \sum(1,2,3,4,9,10,11,12,21,22,23,24)$$

EXPRESSION SET 1.32 Problems 30 - 31

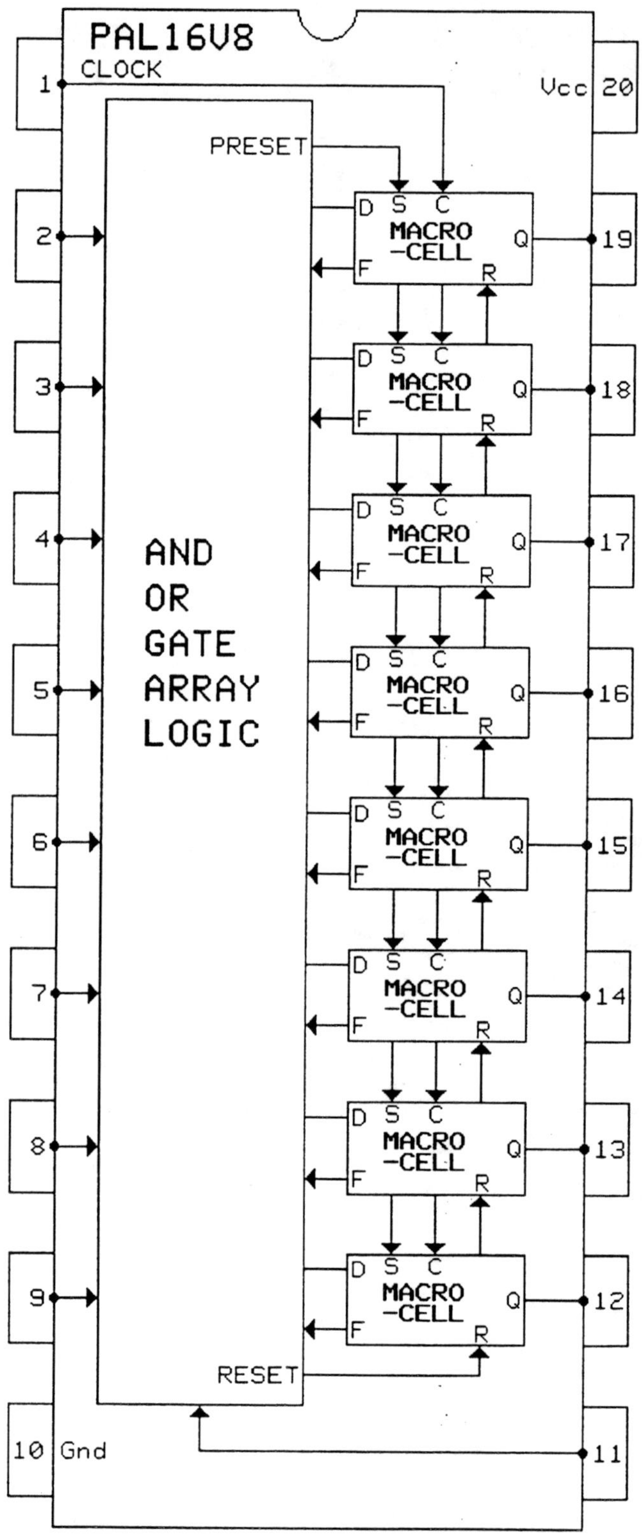

FIGURE 1.21 PAL16V8 device diagram

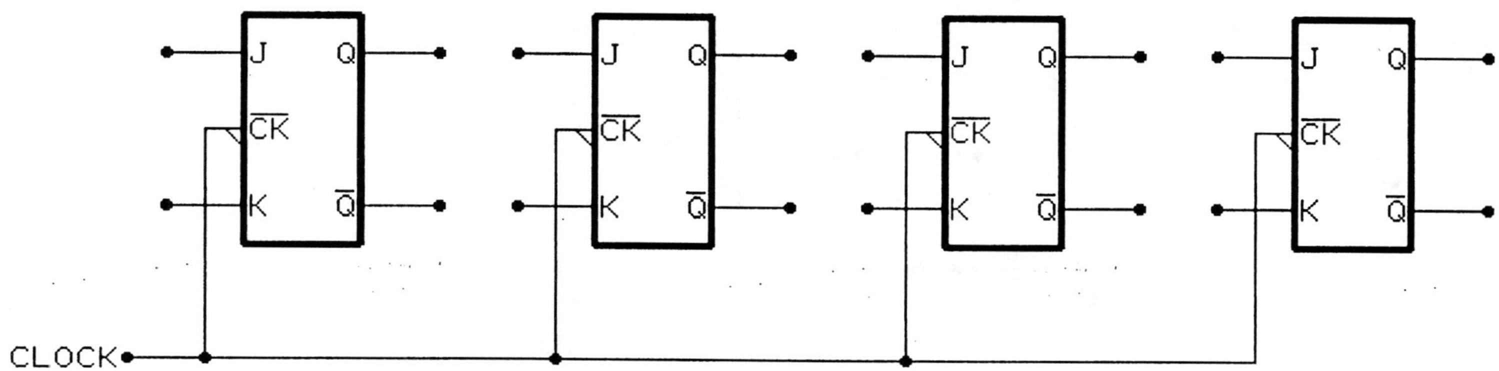

FIGURE 2.1 Synchronous binary register using JK flip-flops.

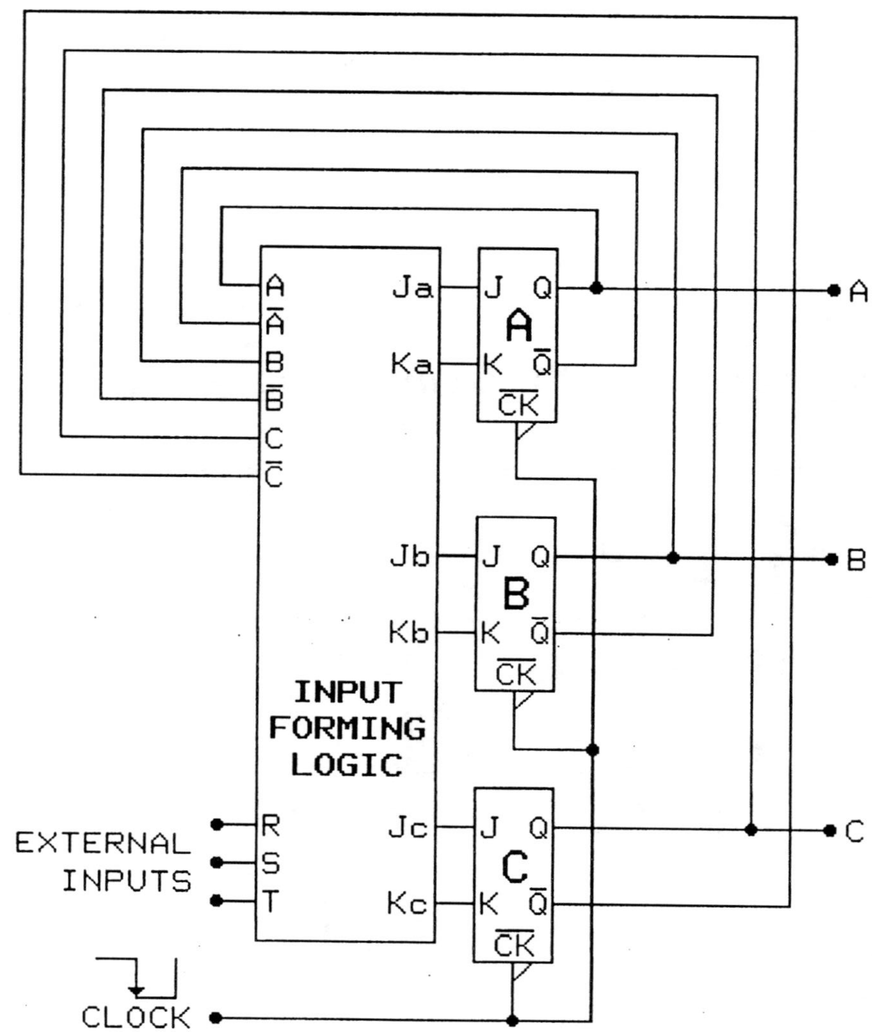

FIGURE 2.2 Synchronous binary counter.

PRESENT STATE Q_t	NEXT STATE Q_{t+1}	POSITIVE D-TYPE D_t	POSITIVE JK-TYPE J_t	K_t	POSITIVE SR TYPE S_t	R_t
0	0	0	0	*	0	*
0	1	1	1	*	1	0
1	0	0	*	1	0	1
1	1	1	*	0	*	0

* = Don't Care State

FIGURE 2.3 Flip-flop excitation tables.

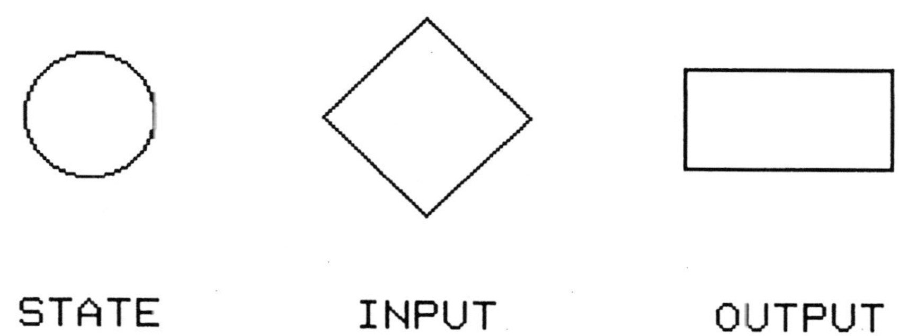

FIGURE 2.4 State diagram components.

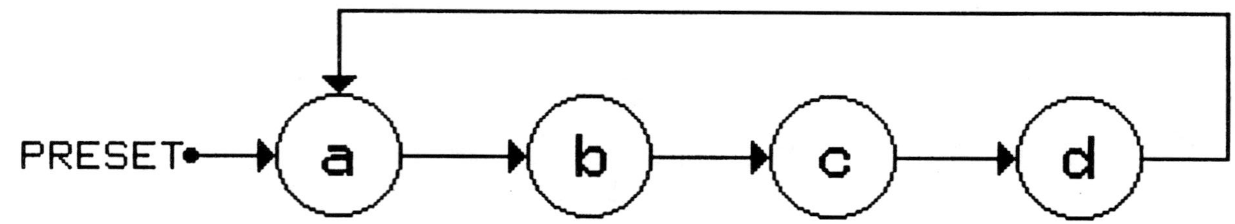

FIGURE 2.5 State diagram of a variable four-count sequence.

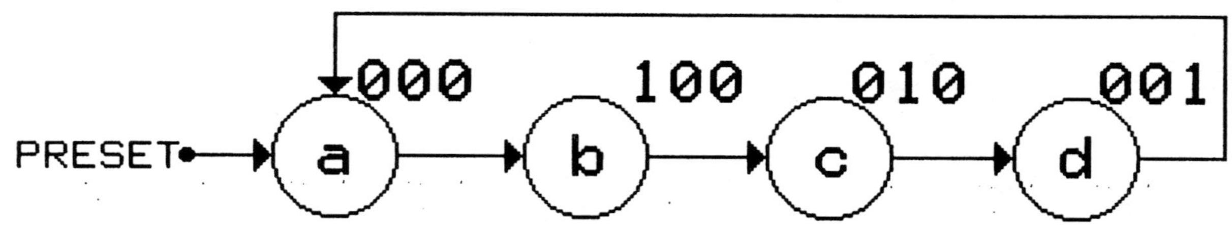

FIGURE 2.6 State diagram of a four-count sequence, first form.

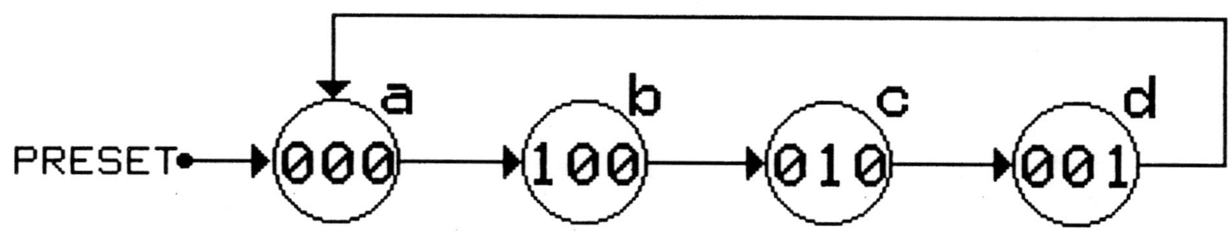

FIGURE 2.7 State diagram of a four-count sequence, second form.

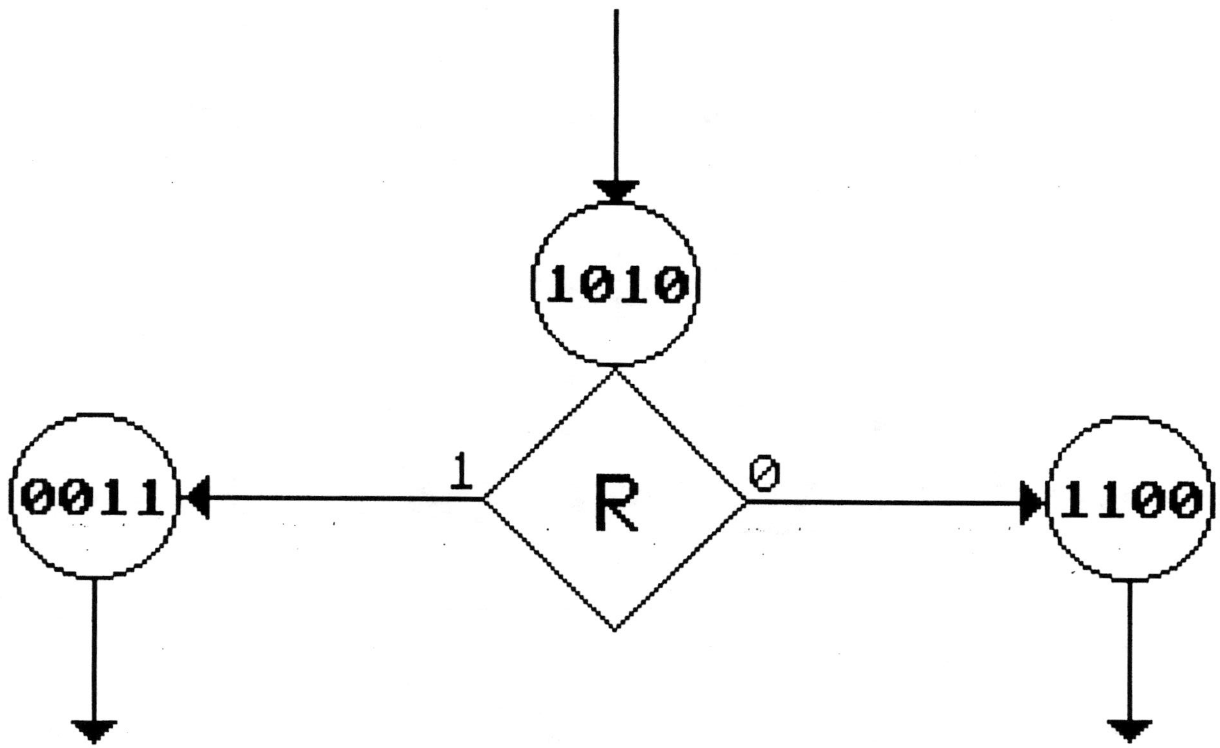

FIGURE 2.8 State diagram input box example.

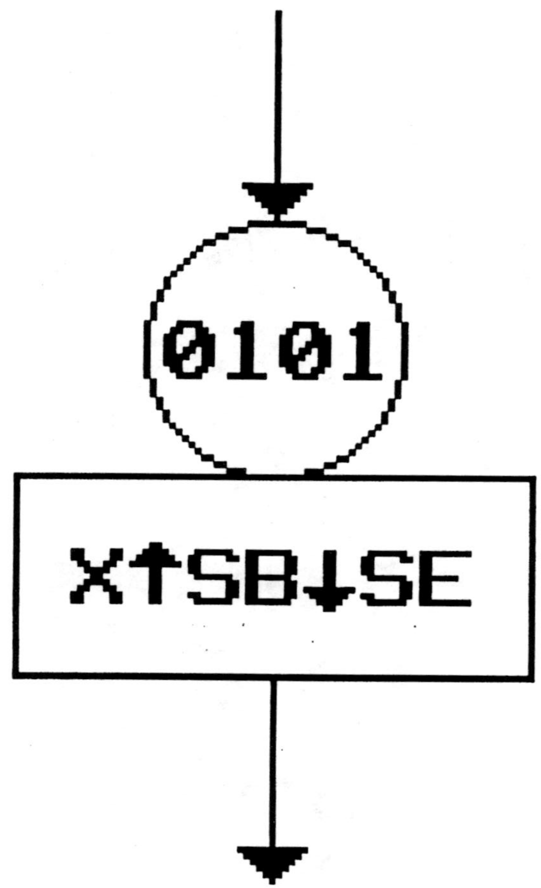

FIGURE 2.9 State diagram output box example.

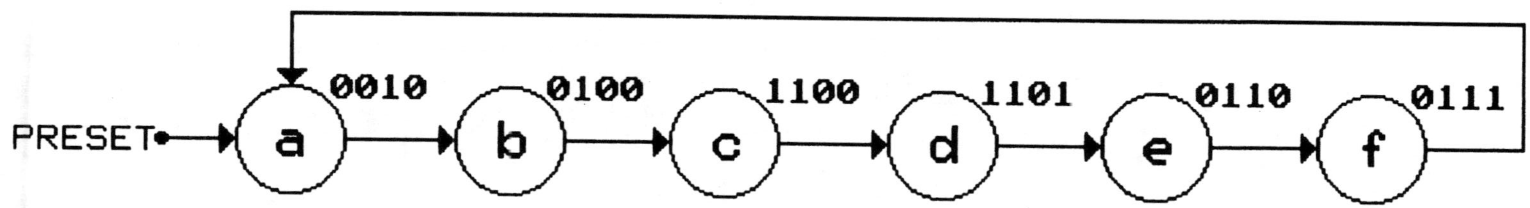

FIGURE 2.10 Practical example 2.1, solution.

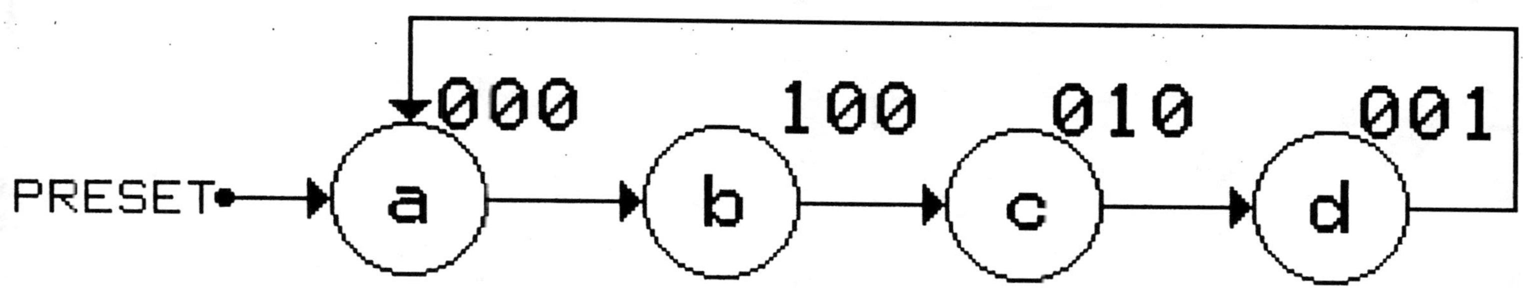

FIGURE 2.11 State diagram of example counter.

PRESENT STATE ABC	NEXT STATE ABC	Input Forming Logic D_A	D_B	D_C
000	100	1	0	0
001	000	0	0	0
010	001	0	0	1
011	***	*	*	*
100	010	0	1	0
101	***	*	*	*
110	***	*	*	*
111	***	*	*	*

D FF Excitation Q_t	Q_{t+1}	D_t
0	0	0
0	1	1
1	0	0
1	1	1

* = Don't Care

FIGURE 2.12 Example state table for D-type counter.

PRESENT STATE ABC	NEXT STATE ABC	Input Forming Logic D_A	D_B	D_C
000	100	1	0	0

FIGURE 2.13 Example state table entry.

PRESENT STATE ABC	NEXT STATE ABC	J_A	K_A	J_B	K_B	J_C	K_C
		Input Forming Logic					
000	100	1	*	0	*	0	*
001	000	0	*	0	*	*	1
010	001	0	*	*	1	1	*
011	***	*	*	*	*	*	*
100	010	*	1	1	*	0	*
101	***	*	*	*	*	*	*
110	***	*	*	*	*	*	*
111	***	*	*	*	*	*	*

JK FF Excitation

Q_t	Q_{t+1}	J_t	K_t
0	0	0	*
0	1	1	*
1	0	*	1
1	1	*	0

* = Don't Care

FIGURE 2.14 Example state table for JK-type counter.

PRESENT STATE ABC	NEXT STATE ABC	Input Forming Logic					
		S_A	R_A	S_B	R_B	S_C	R_C
000	100	1	0	0	*	0	*
001	000	0	*	0	*	0	1
010	001	0	*	0	1	1	0
011	***	*	*	*	*	*	*
100	010	0	1	1	0	0	*
101	***	*	*	*	*	*	*
110	***	*	*	*	*	*	*
111	***	*	*	*	*	*	*

SR FF Excitation			
Q_t	Q_{t+1}	S_t	R_t
0	0	0	*
0	1	1	0
1	0	0	1
1	1	*	0

* = Don't Care

FIGURE 2.15 Example state table for SR-type counter.

PRESENT STATE ABC	NEXT STATE ABC	IFL D_A	D_B	D_C
000	100	1	0	0
001	000	0	0	0
010	001	0	0	1
011	***	*	*	*
100	010	0	1	0
101	***	*	*	*
110	***	*	*	*
111	***	*	*	*

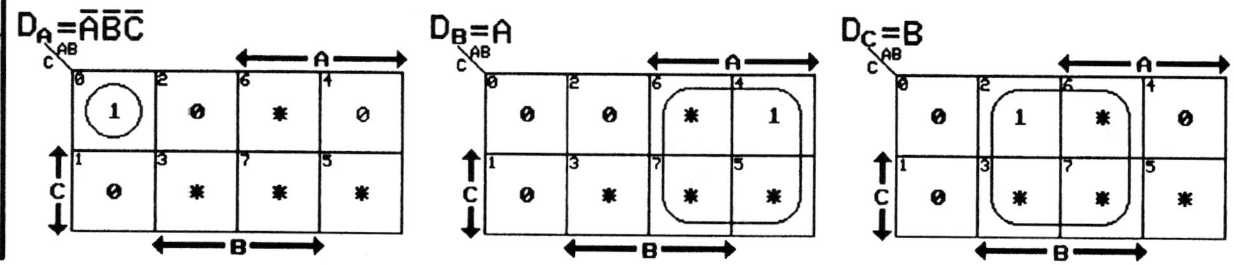

FIGURE 2.16 Extraction of SOP expressions for the D-type counter.

PRESENT STATE ABC	NEXT STATE ABC	INPUT FORMING LOGIC J_AK_A	J_BK_B	J_CK_C
000	100	1*	0*	0*
001	000	0*	0*	*1
010	001	0*	*1	1*
011	***	**	**	**
100	010	*1	1*	0*
101	***	**	**	**
110	***	**	**	**
111	***	**	**	**

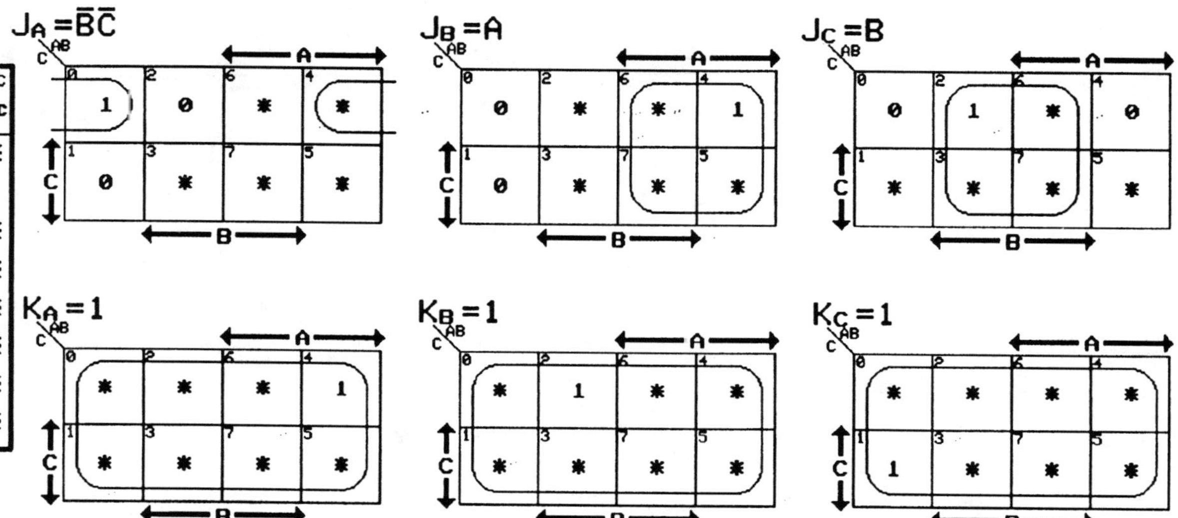

FIGURE 2.17 Extraction of SOP expressions for the JK-type counter.

PRESENT STATE ABC	NEXT STATE ABC	INPUT FORMING LOGIC S_AR_A	S_BR_B	S_CR_C
000	100	10	0*	0*
001	000	0*	0*	01
010	001	0*	01	10
011	***	**	**	**
100	010	01	10	0*
101	***	**	**	**
110	***	**	**	**
111	***	**	**	**

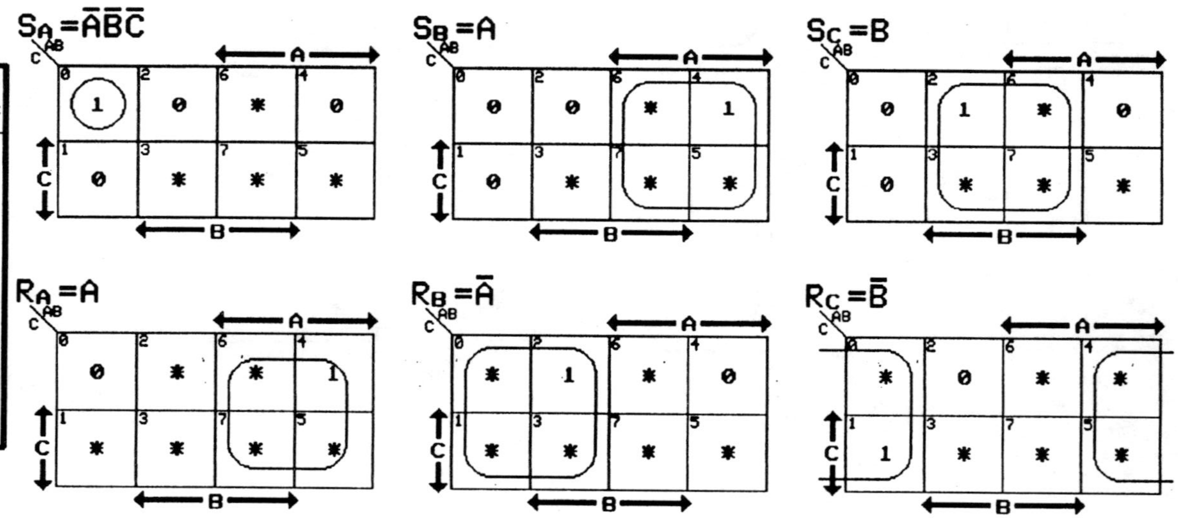

FIGURE 2.18 Extraction of SOP Expressions for the SR-type counter.

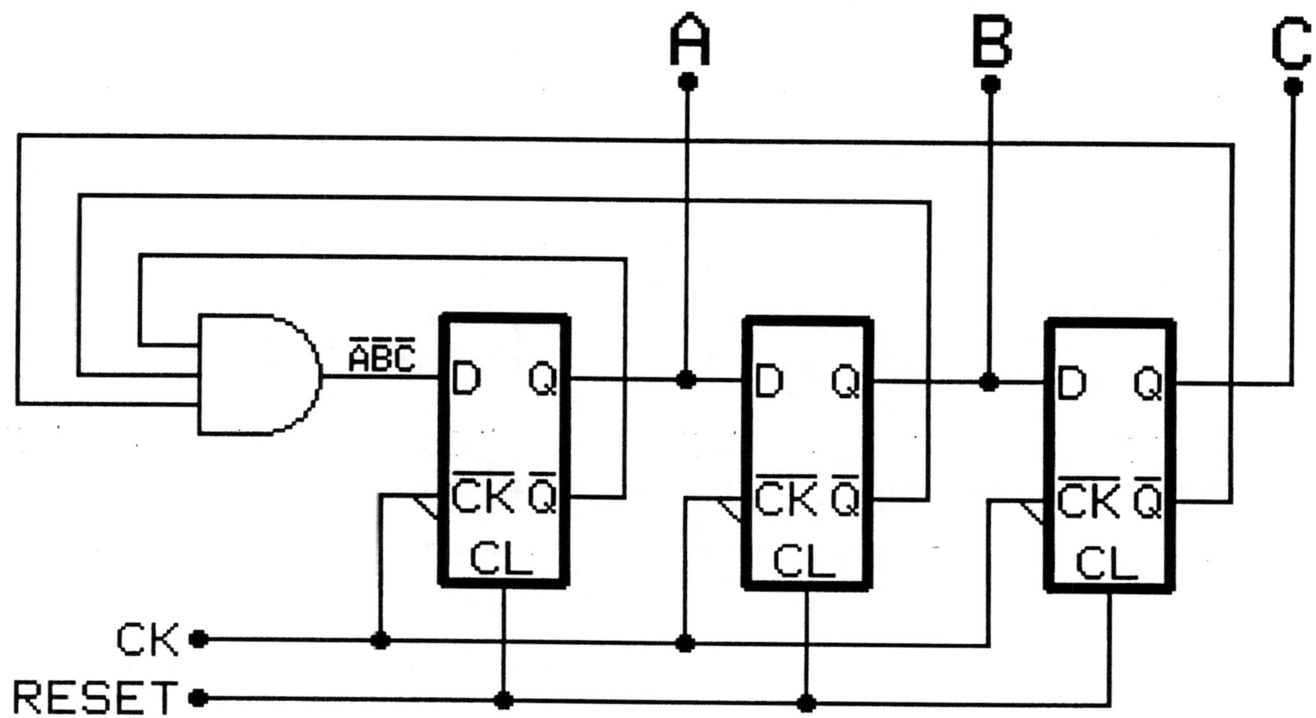

FIGURE 2.19 D-type synchronous binary counter example.

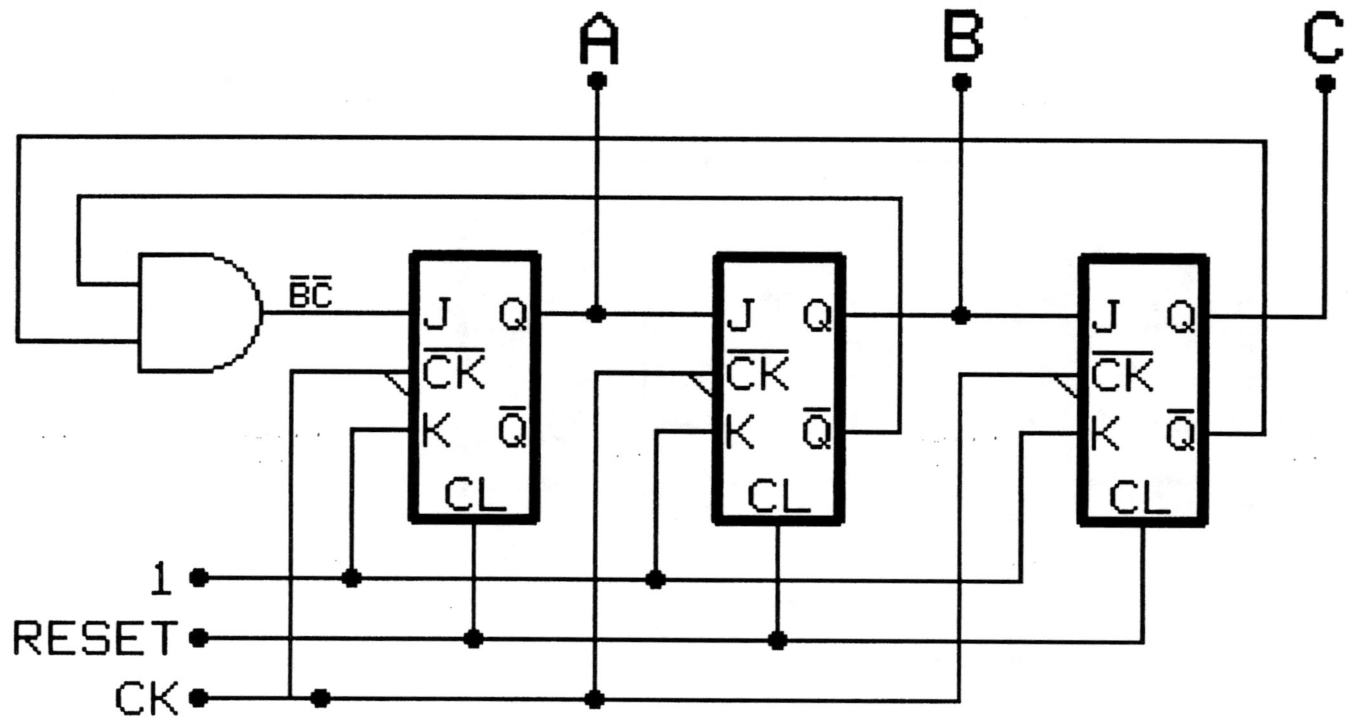

FIGURE 2.20 JK-type synchronous binary counter example.

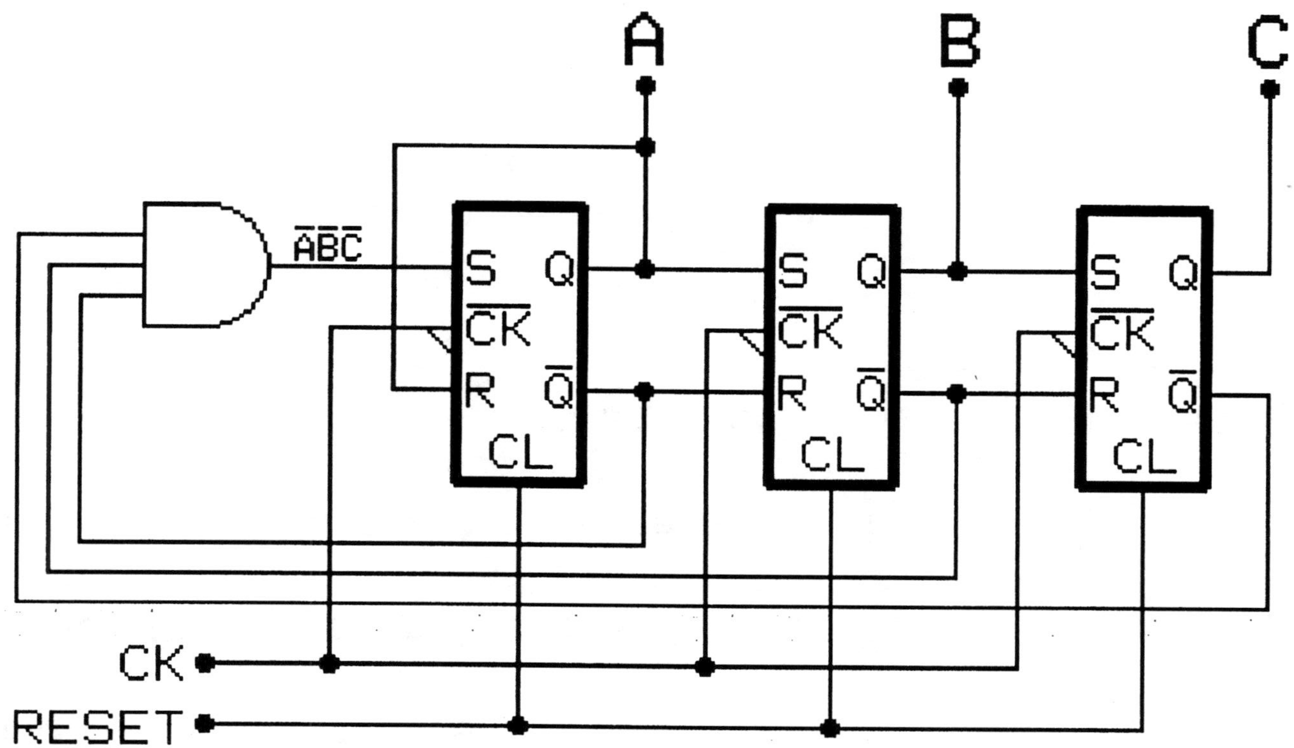

FIGURE 2.21 SR-type synchronous binary counter example.

Design Method Summary

The method used here for designing synchronous counters has five steps:

- Define the count sequence.
- Document the sequence as a state diagram.
- Generate a state table from the state diagram information.
- Use Karnaugh maps to extract the product terms for each flip-flop control input.
- Draw the final schematic diagram.

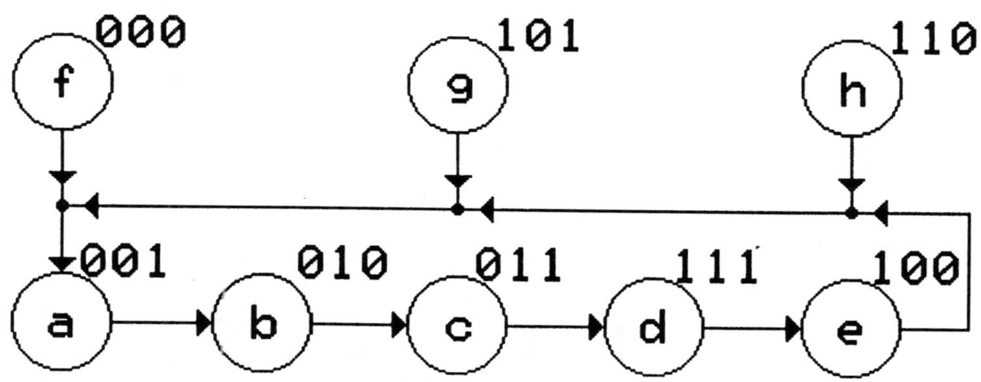

FIGURE 2.22 Practical example 2.2, state diagram.

PRESENT STATE ABC	NEXT STATE ABC	Input Forming Logic D_A	D_B	D_C
000	001	0	0	1
001	010	0	1	0
010	011	0	1	1
011	111	1	1	1
100	001	0	0	1
101	001	0	0	1
110	001	0	0	1
111	100	1	0	0

FIGURE 2.23 Practical example 2.2, state table.

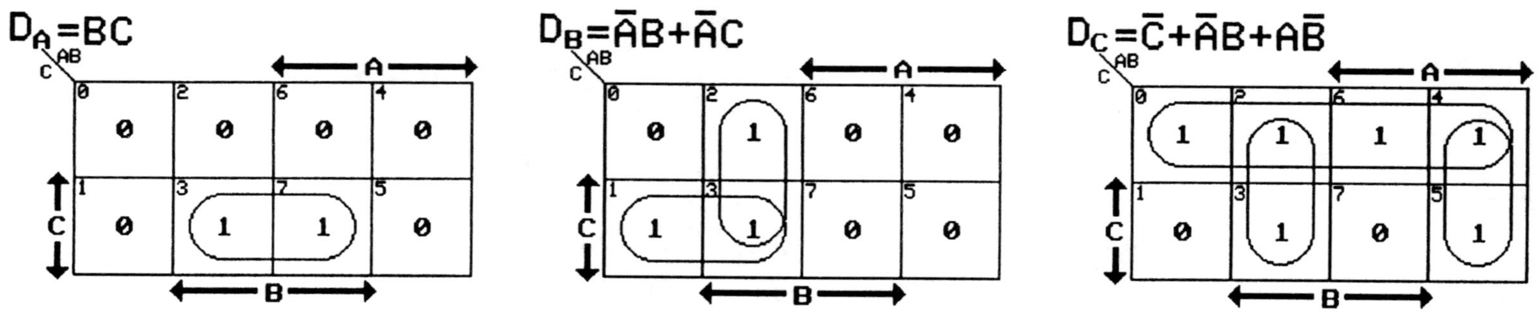

FIGURE 2.24 Practical example 2.2, Karnaugh map simplification.

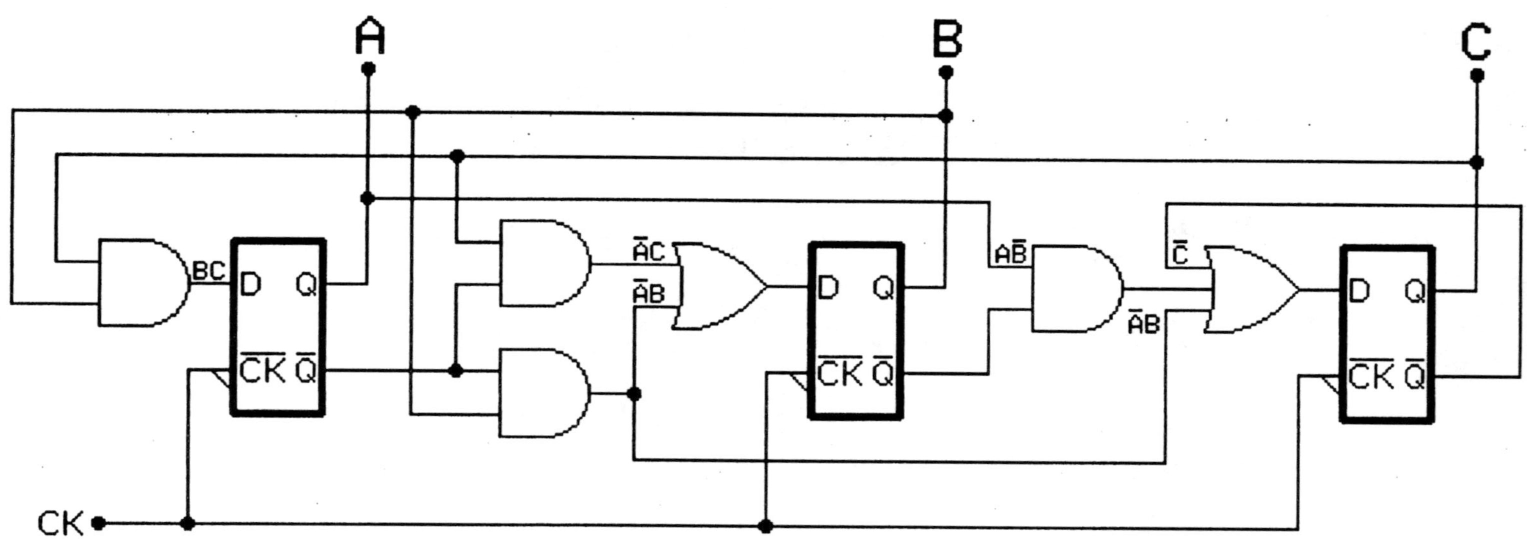

FIGURE 2.25 Practical example 2.2, schematic diagram.

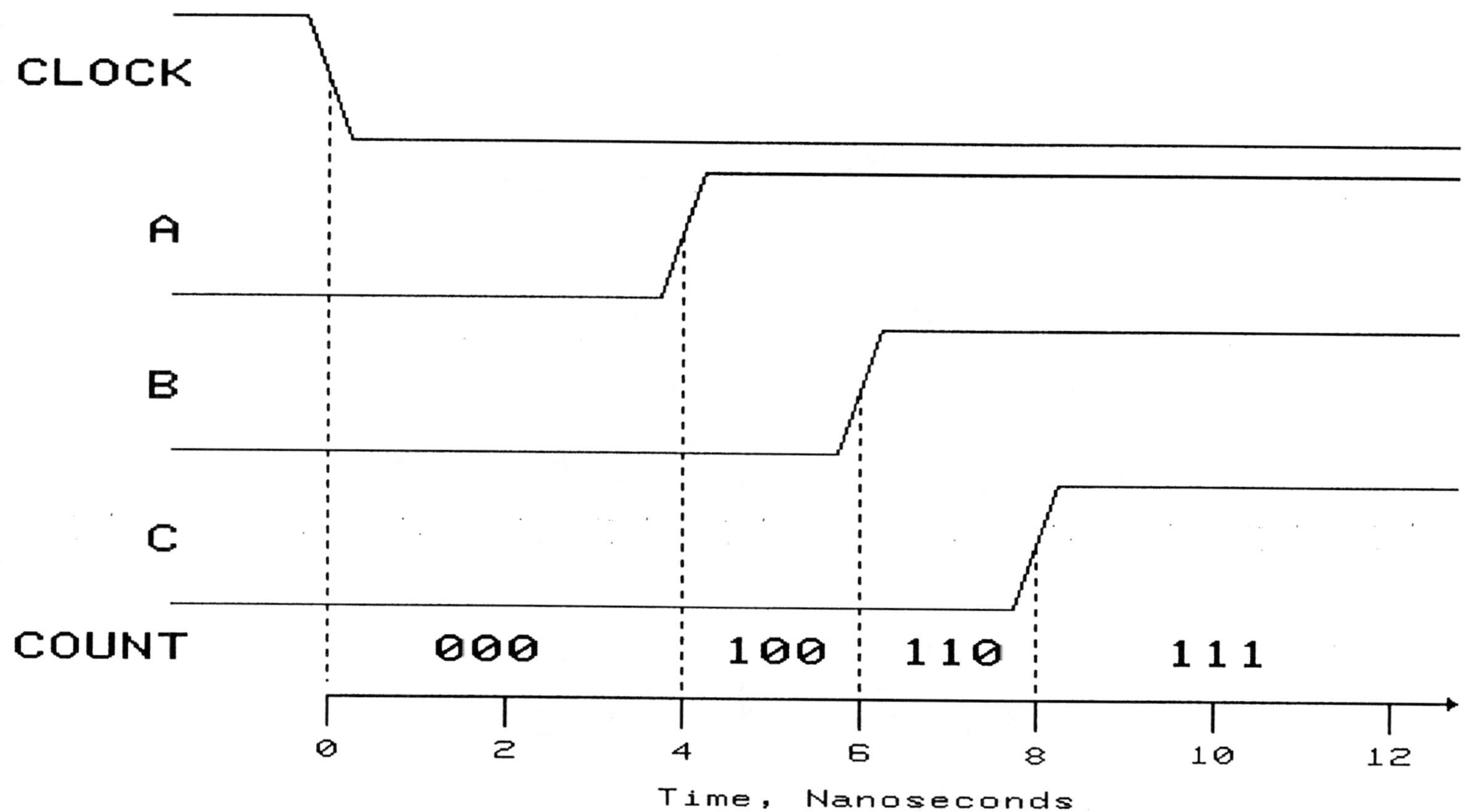

FIGURE 2.26 Glitches caused by unequal propagation delays.

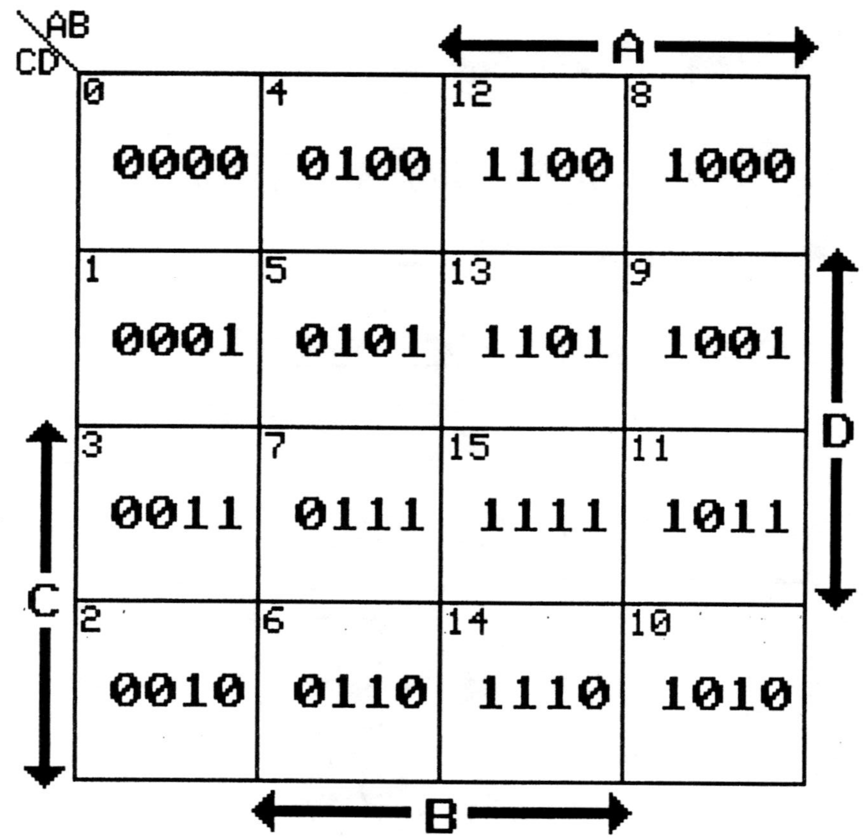

FIGURE 2.27 Karnaugh map.

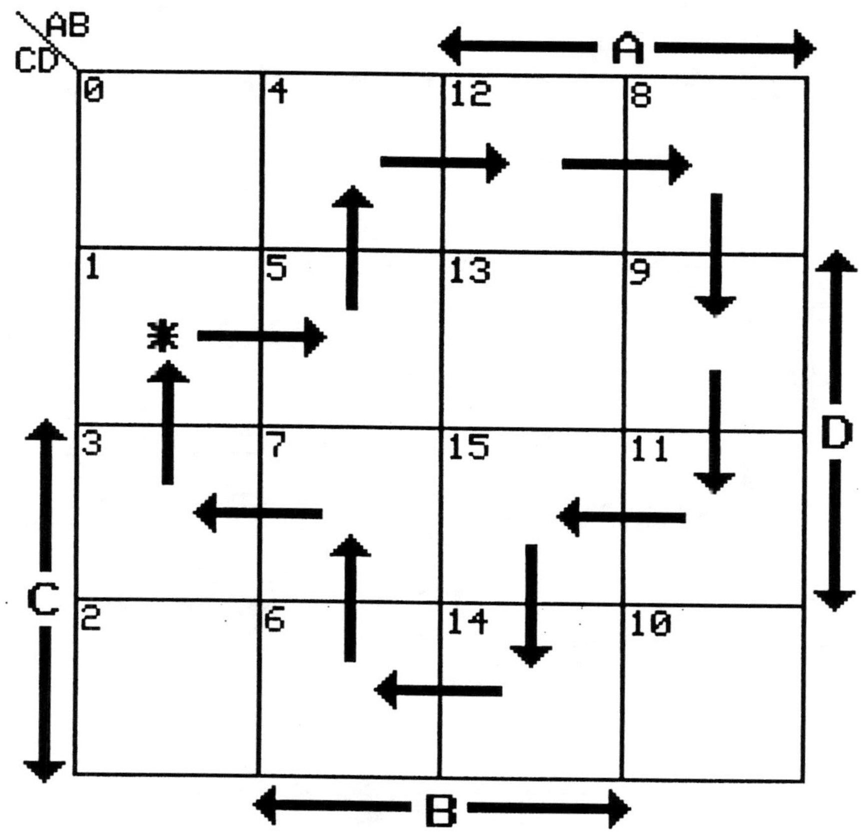

FIGURE 2.28 Next state map.

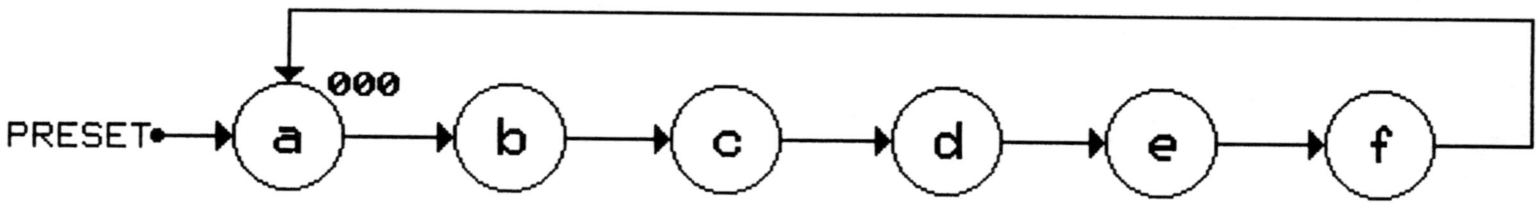

FIGURE 2.29 Practical example 2.3, counter definition.

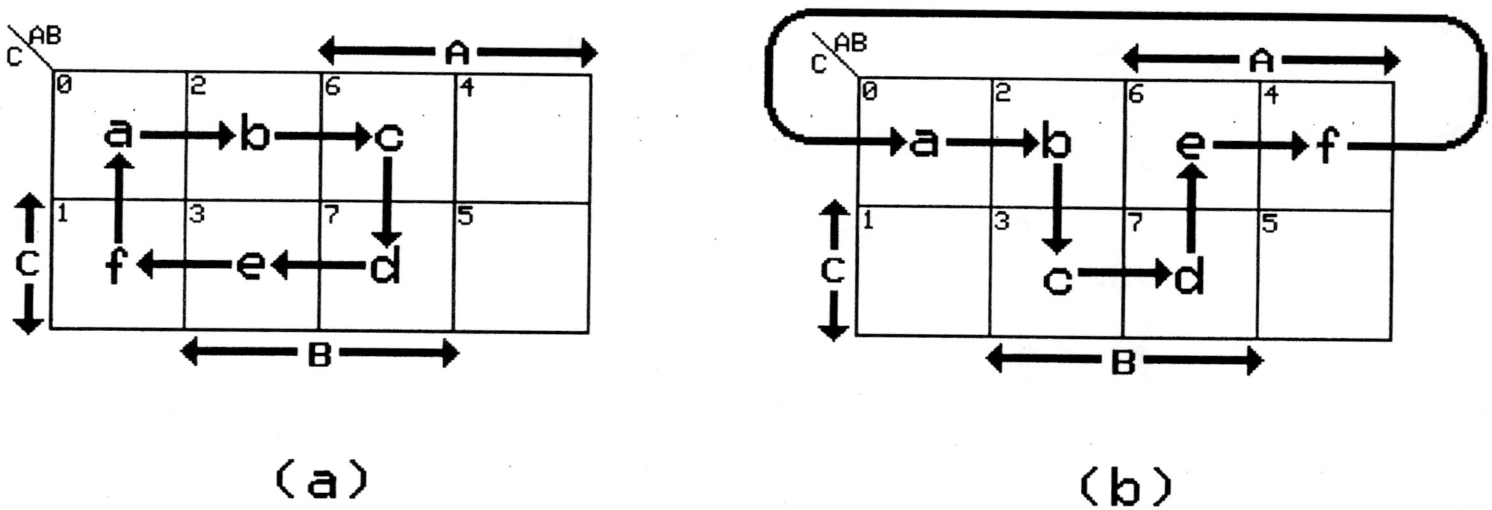

FIGURE 2.30 Practical example 2.3, two solutions.

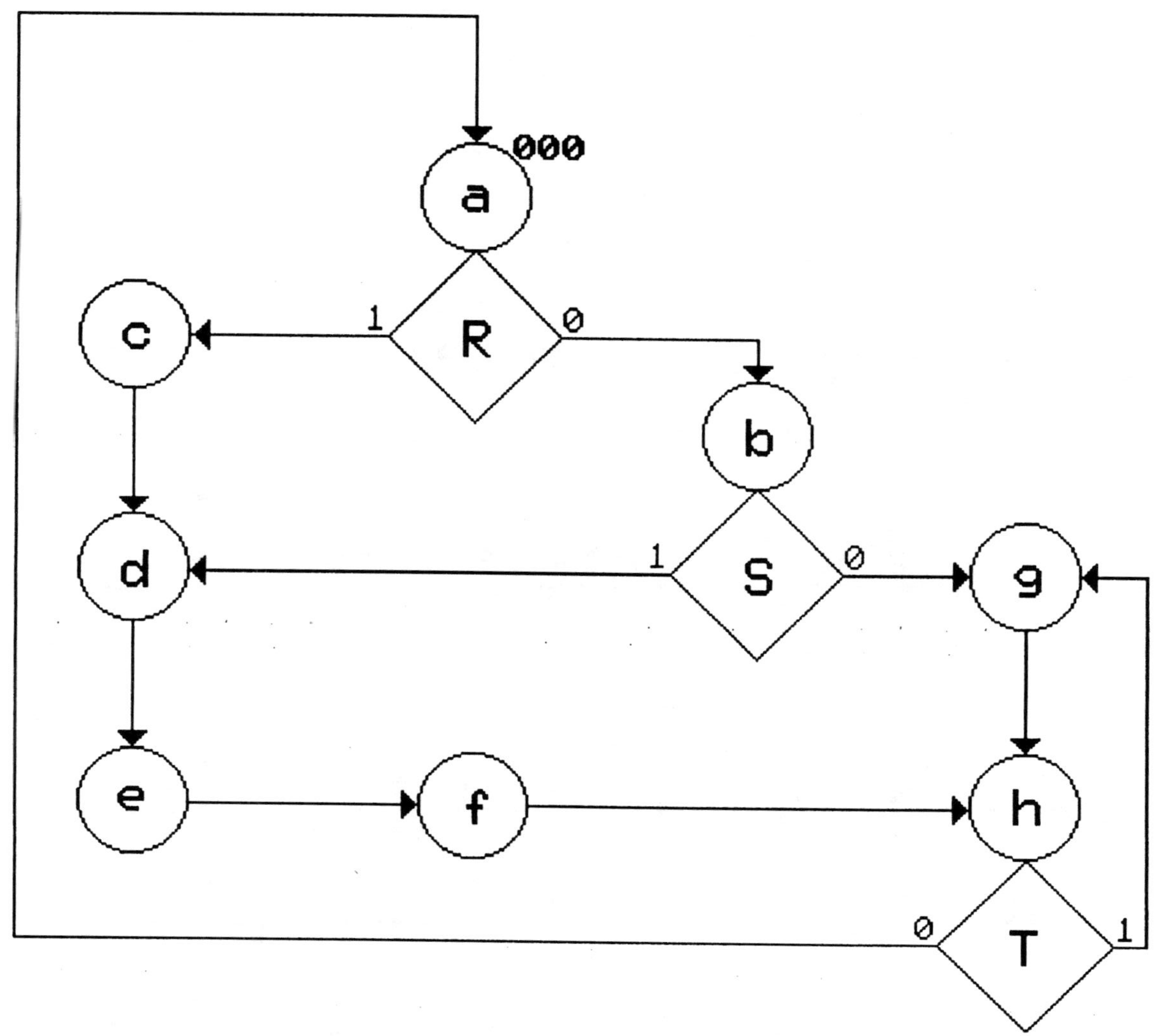

FIGURE 2.31 Practical example 2.4, state diagram.

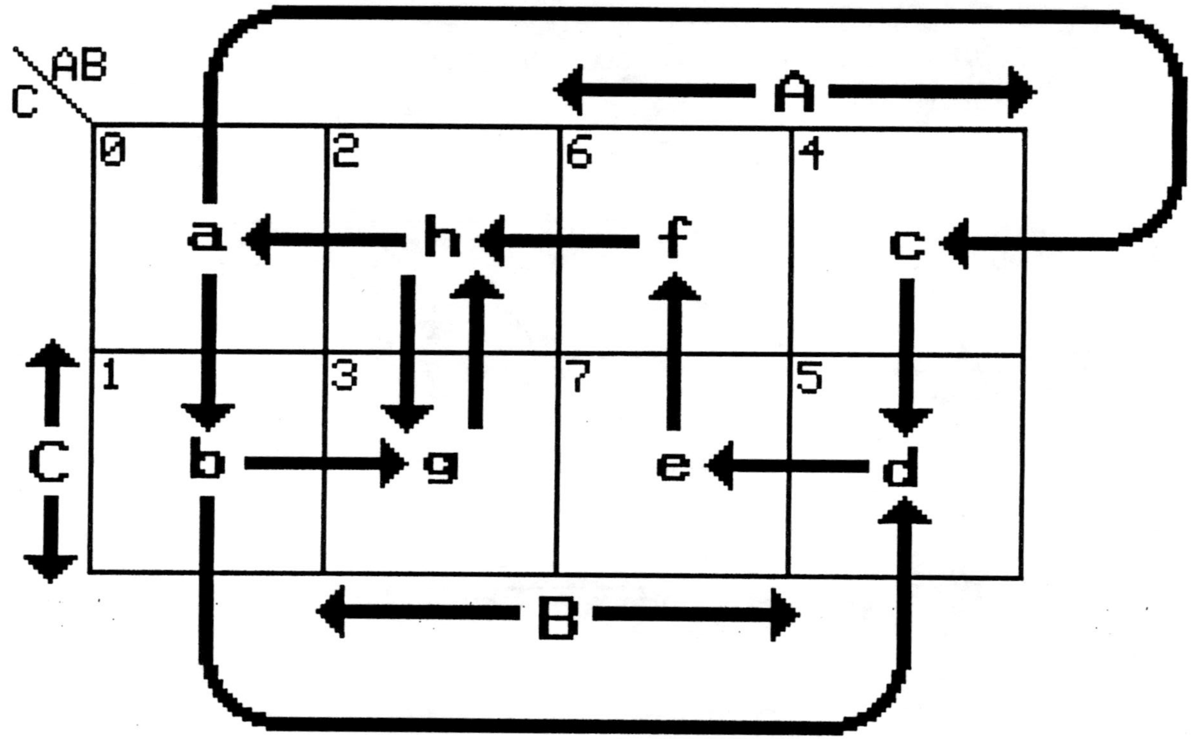

FIGURE 2.32 Practical example 2.4, next state map.

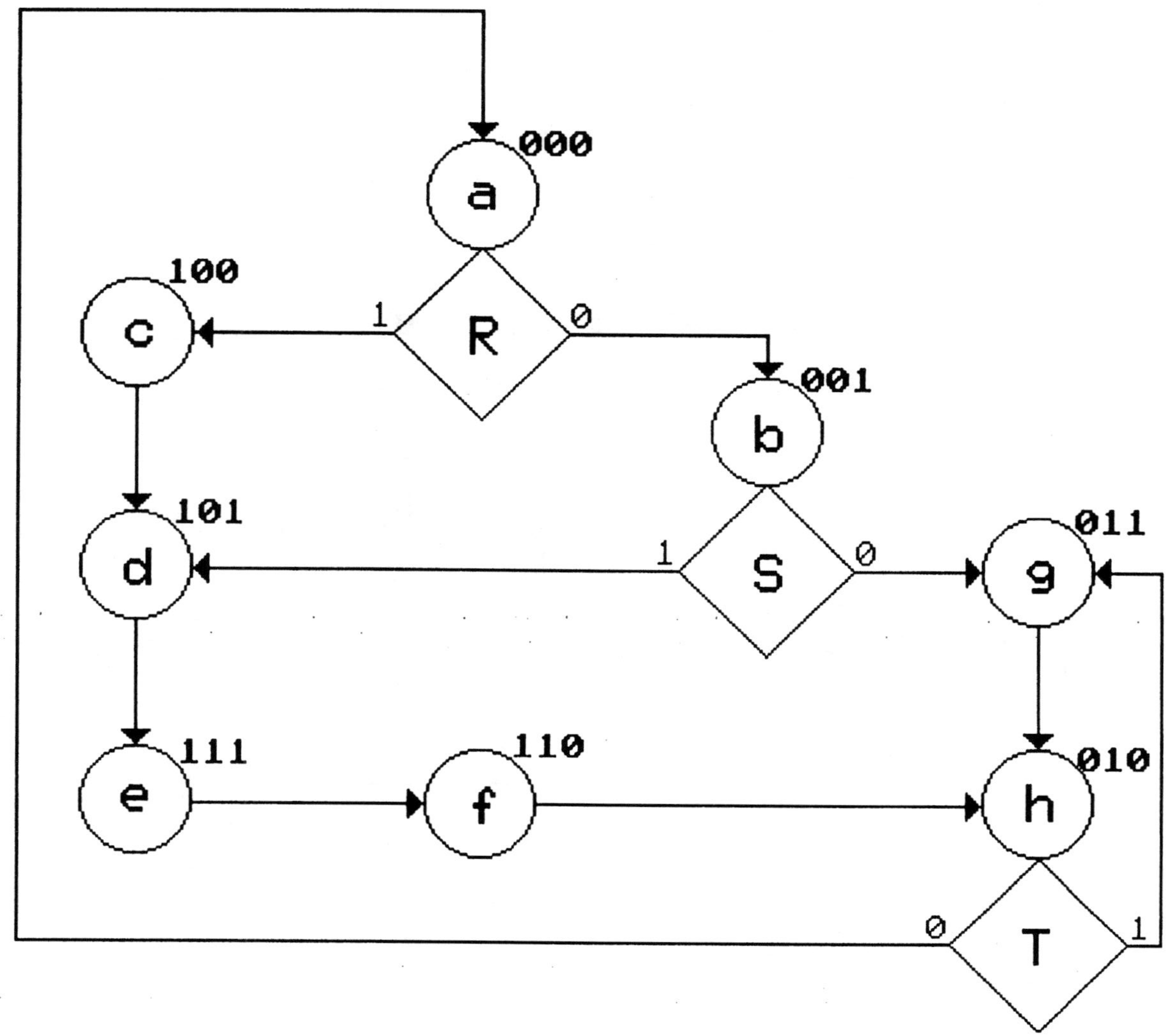

FIGURE 2.33 Practical example 2.4, solution.

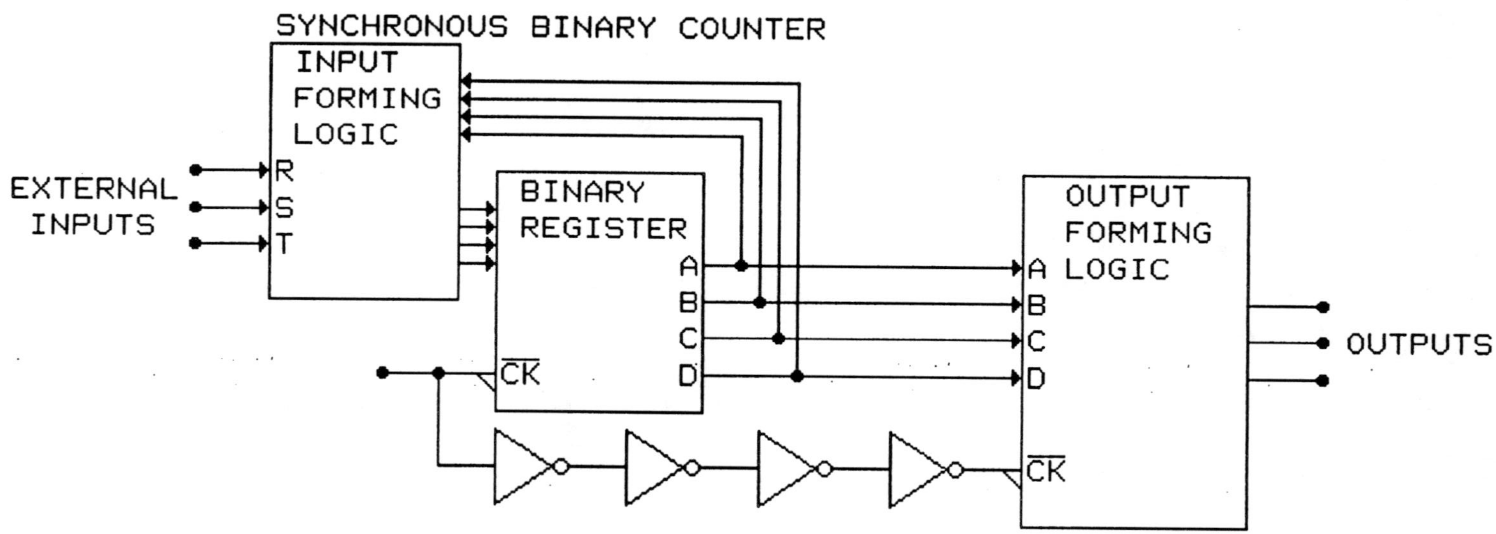

FIGURE 2.34 Delayed state timing example.

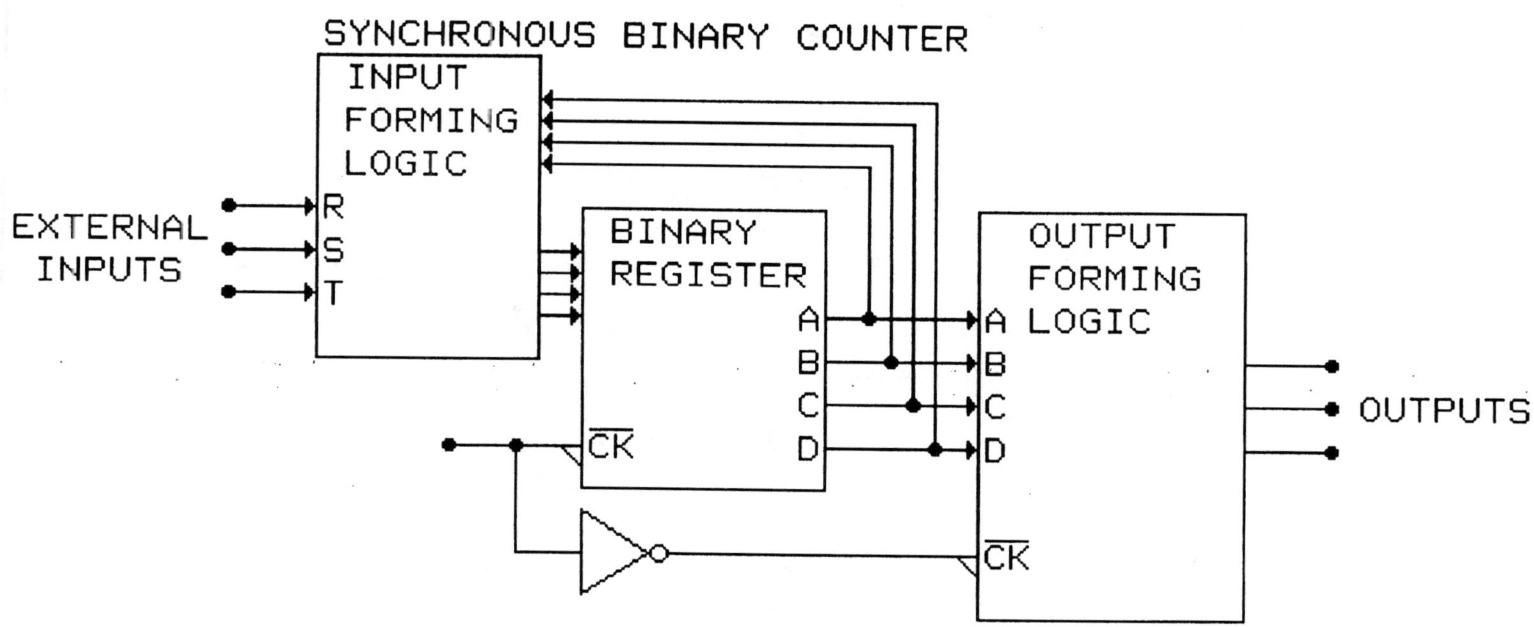

FIGURE 2.35 Alternate state timing.

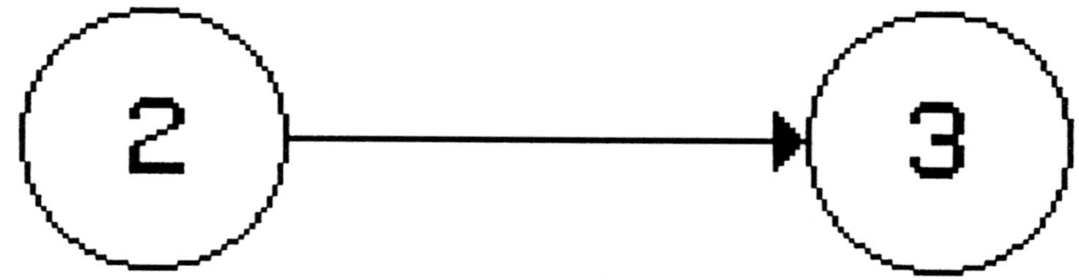

FIGURE 3.1 Unconditional numeric count sequence.

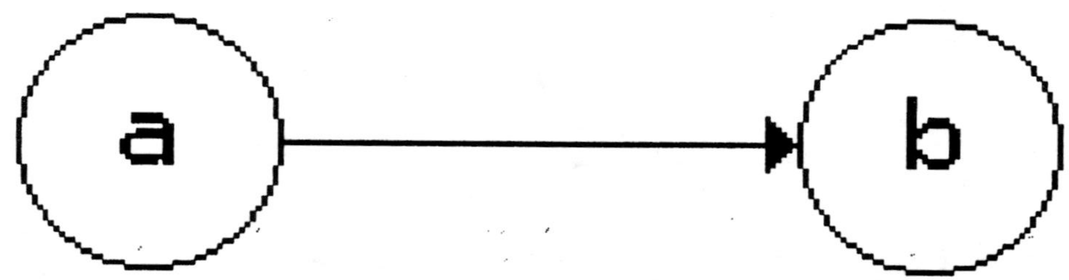

FIGURE 3.2 Unconditional variable count sequence.

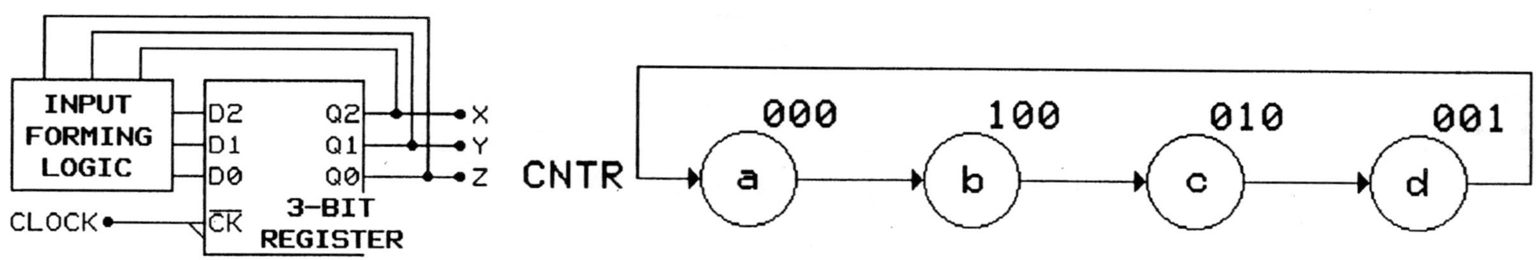

FIGURE 3.3 Practical Example 3.1, schematic and state diagram.

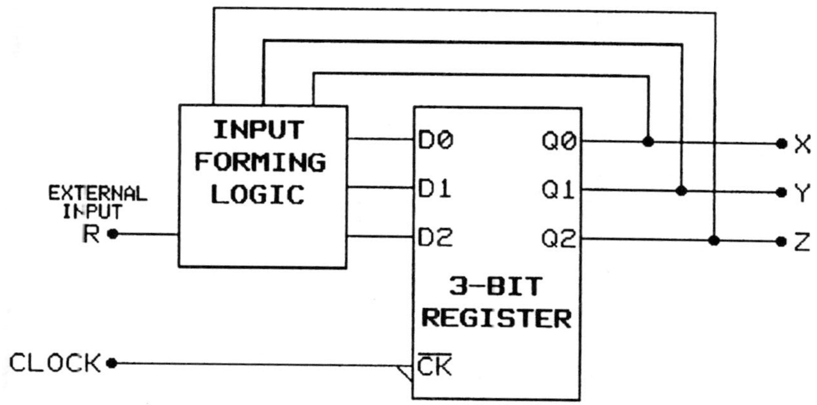

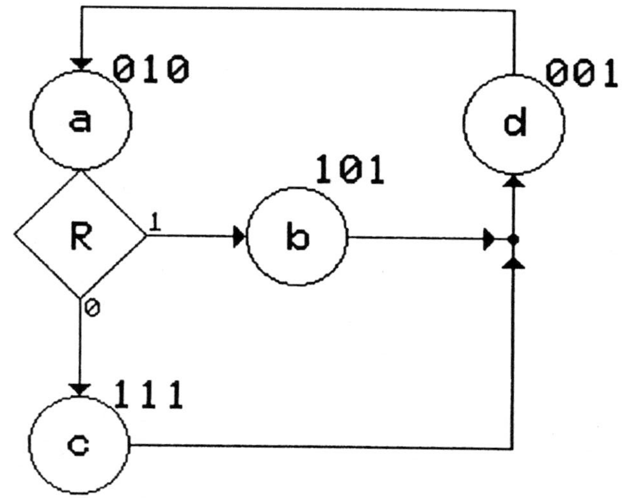

FIGURE 3.4 Practical Example 3.2, schematic and state diagram.

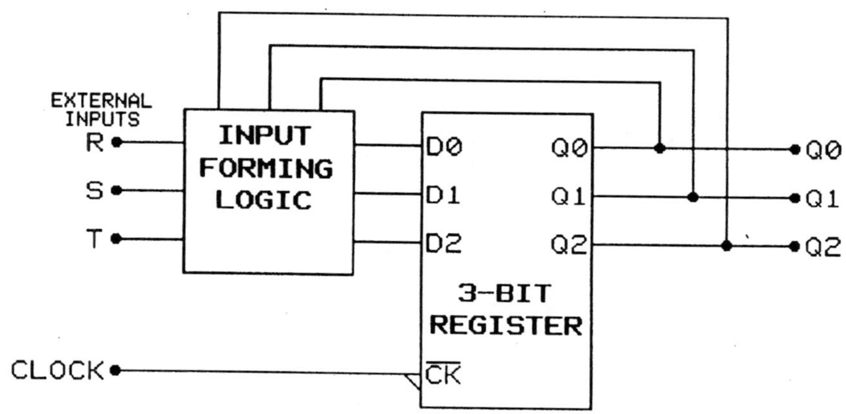

FIGURE 3.5 Practical Example 3.3, schematic diagram.

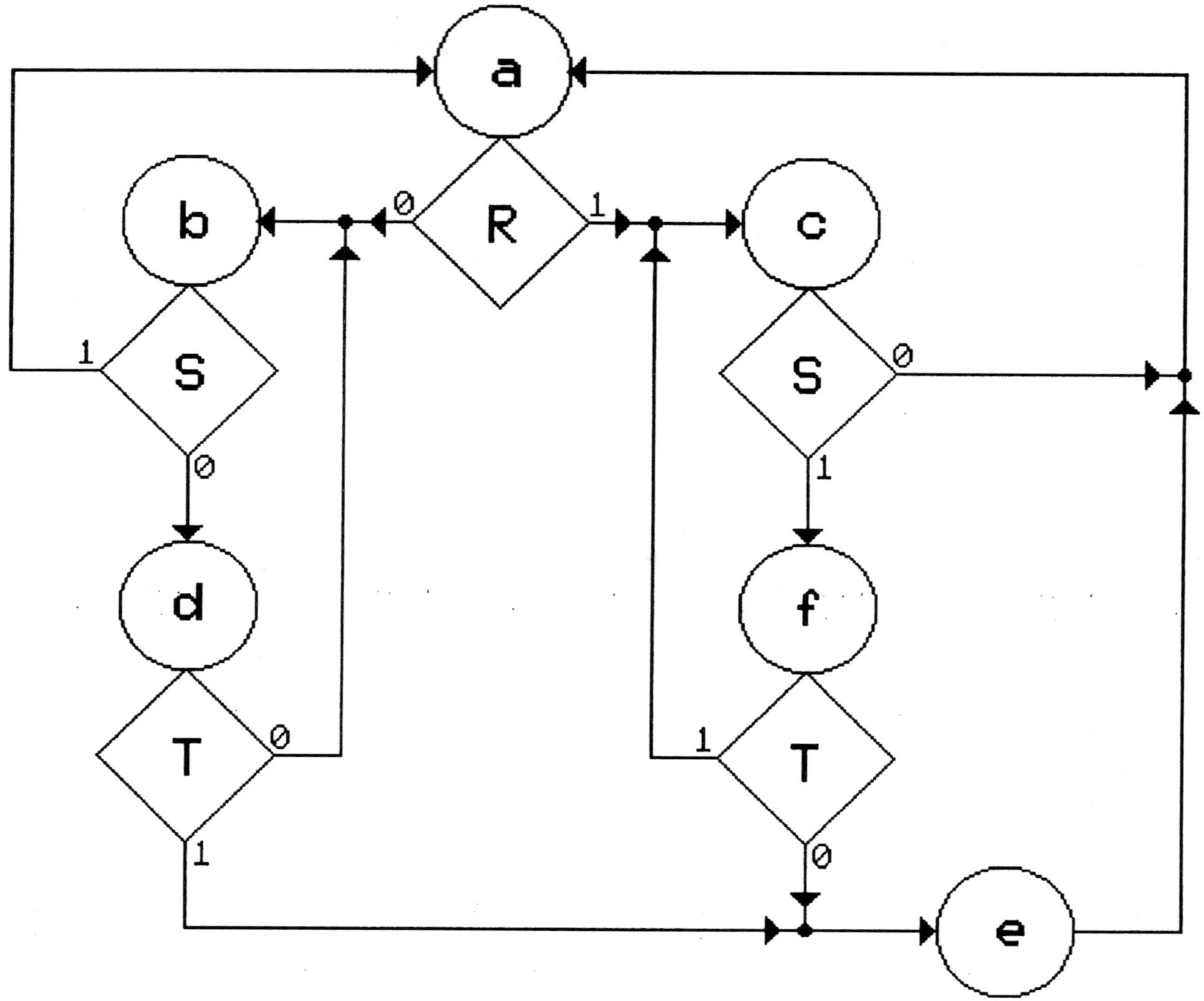

FIGURE 3.6 Practical Example 3.3, initial state diagram.

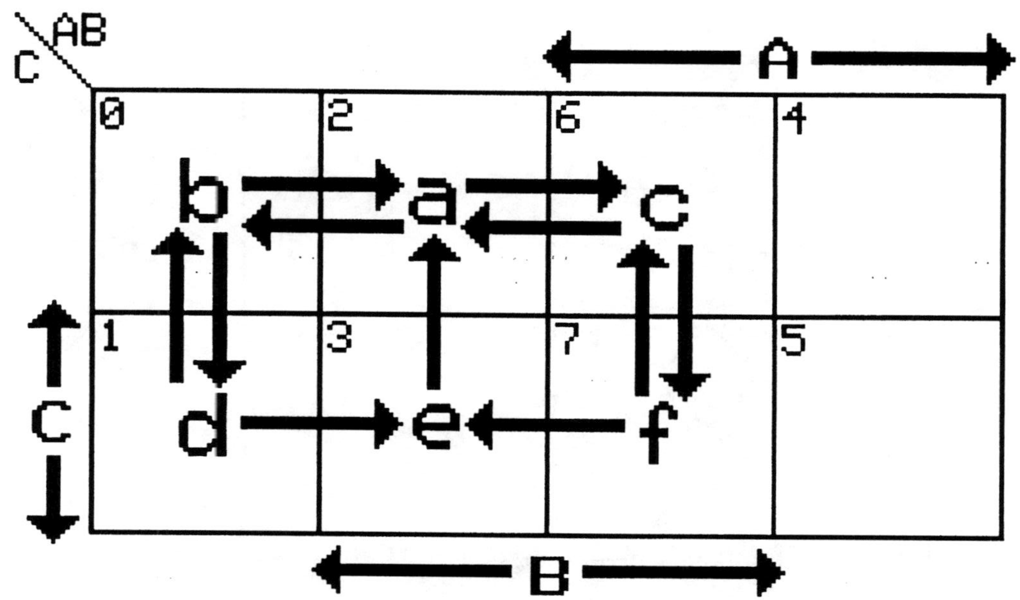

FIGURE 3.7 Practical Example 3.3, next state map.

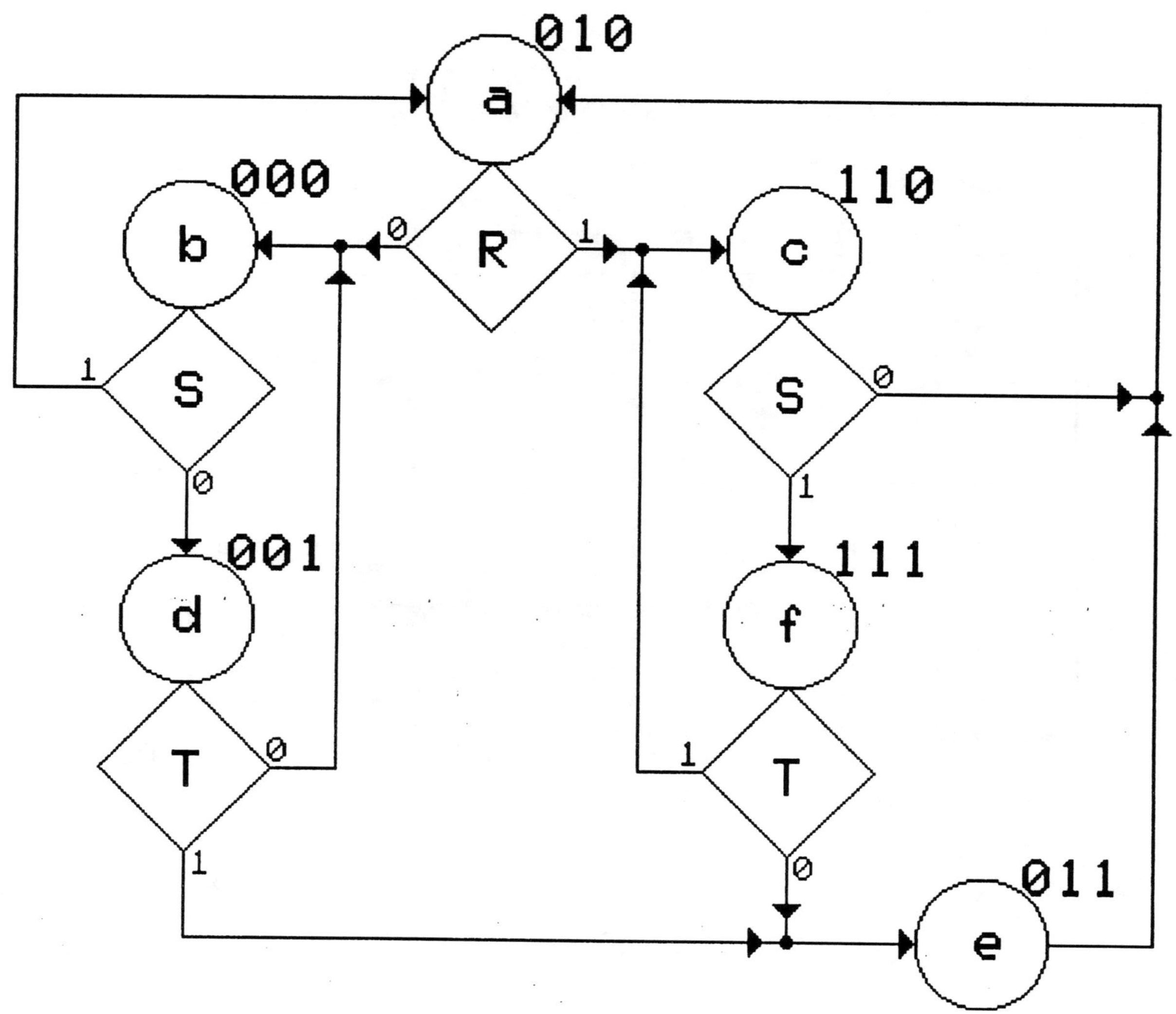

FIGURE 3.8 Practical Example 3.3, final state diagram.

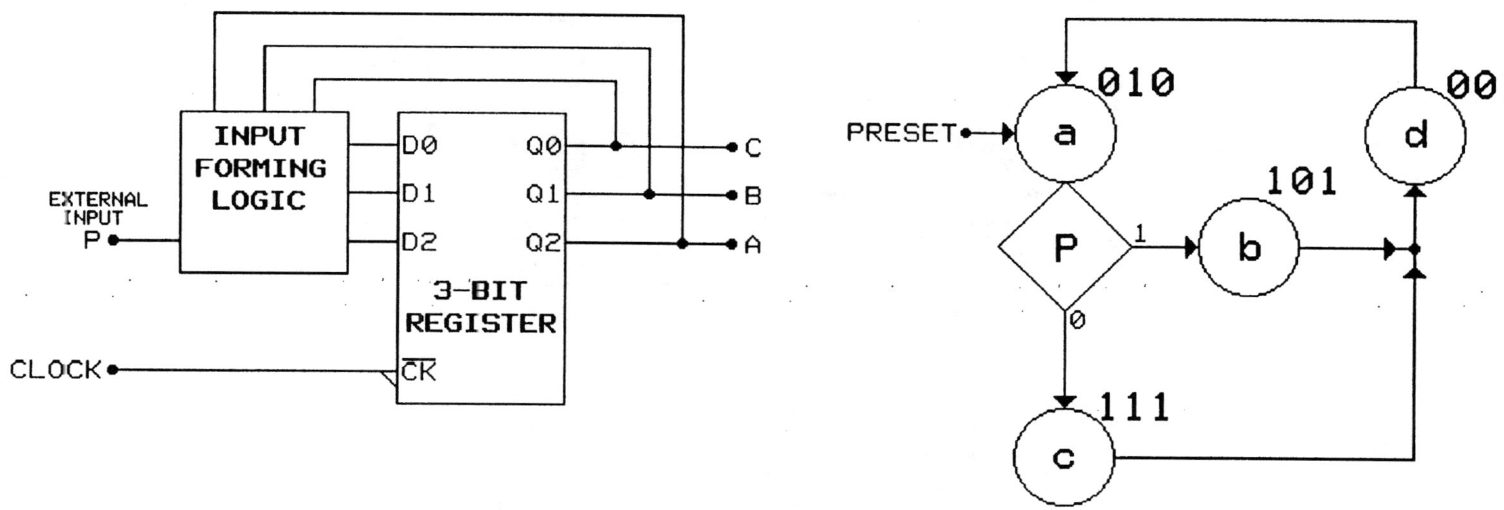

FIGURE 3.9 State machine with one decision.

	PRES ABC	INPUT P	NEXT ABC
Ø	ØØØ	*	***
1	ØØ1	*	Ø1Ø
2	Ø1Ø	Ø	111
		1	1Ø1
3	Ø11	*	***
4	1ØØ	*	***
5	1Ø1	*	ØØ1
6	11Ø	*	***
7	111	*	ØØ1

* = Don't Care

FIGURE 3.10 Single decision in a state table (State-010$_2$).

	PRES ABC	INPUT P	NEXT ABC	D_A	D_B	D_C
Ø	ØØØ	*	***	*	*	*
1	ØØ1	*	Ø1Ø	Ø	1	Ø
2	Ø1Ø	Ø	111			
		1	1Ø1			
3	Ø11	*	***	*	*	*
4	1ØØ	*	***	*	*	*
5	1Ø1	*	ØØ1	Ø	Ø	1
6	11Ø	*	***	*	*	*
7	111	*	ØØ1	Ø	Ø	1

* = Don't Care

FIGURE 3.11 Unconditional stimuli added to state table.

	PRES ABC	INPUT P	NEXT ABC	D_A	D_B	D_C
0	000	*	***	*	*	*
1	001	*	010	0	1	0
2	010	0	111	1	$\bar{P}$	1
		1	101			
3	011	*	***	*	*	*
4	100	*	***	*	*	*
5	101	*	001	0	0	1
6	110	*	***	*	*	*
7	111	*	001	0	0	1

* = Don't Care

FIGURE 3.12 Conditional stimulus added to state table.

	PRES ABC	INPUT P	NEXT ABC	J_A	K_A	J_B	K_B	J_C	K_C
0	000	*	***	*	*	*	*	*	*
1	001	*	010	0	*	1	*	*	1
2	010	0	111	1	*	*	P	1	*
		1	101						
3	011	*	***	*	*	*	*	*	*
4	100	*	***	*	*	*	*	*	*
5	101	*	001	*	1	0	*	*	0
6	110	*	***	*	*	*	*	*	*
7	111	*	001	*	1	*	1	*	0

* = Don't Care

JK FF Excitation

Q_t	Q_{t+1}	J_t	K_t
0	0	0	*
0	1	1	*
1	0	*	1
1	1	*	0

FIGURE 3.13 Conditional stimulus in a JK state table.

	PRES ABC	INPUT P	NEXT ABC	S_A	R_A	S_B	R_B	S_C	R_C
0	000	*	***	*	*	*	*	*	*
1	001	*	010	0	*	1	0	0	1
2	010	0	111	1	0	$\bar{P}$	P	1	0
		1	101						
3	011	*	***	*	*	*	*	*	*
4	100	*	***	*	*	*	*	*	*
5	101	*	001	0	1	0	*	*	0
6	110	*	***	*	*	*	*	*	*
7	111	*	001	0	1	0	1	*	0

* = Don't Care

SR FF Excitation

Q_t	Q_{t+1}	S_t	R_t
0	0	0	*
0	1	1	0
1	0	0	1
1	1	*	0

FIGURE 3.14 Conditional stimulus in a SR state table.

	D_A	D_B	D_C
0	*	*	*
1	0	1	0
2	1	$\overline{P}$	1
3	*	*	*
4	*	*	*
5	0	0	1
6	*	*	*
7	0	0	1

FROM STATE TABLE

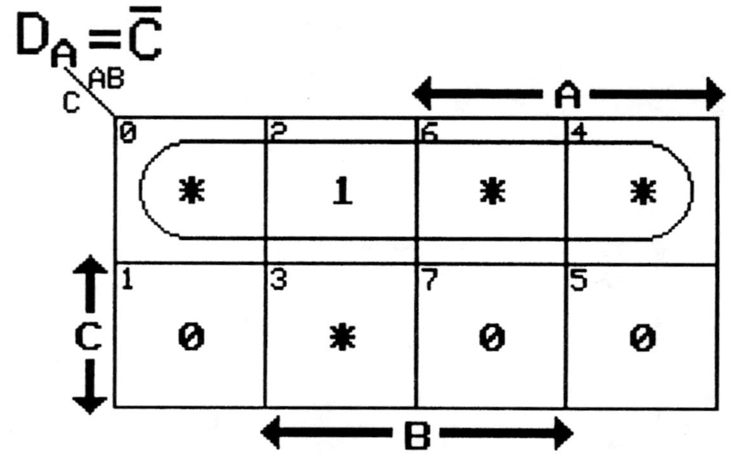

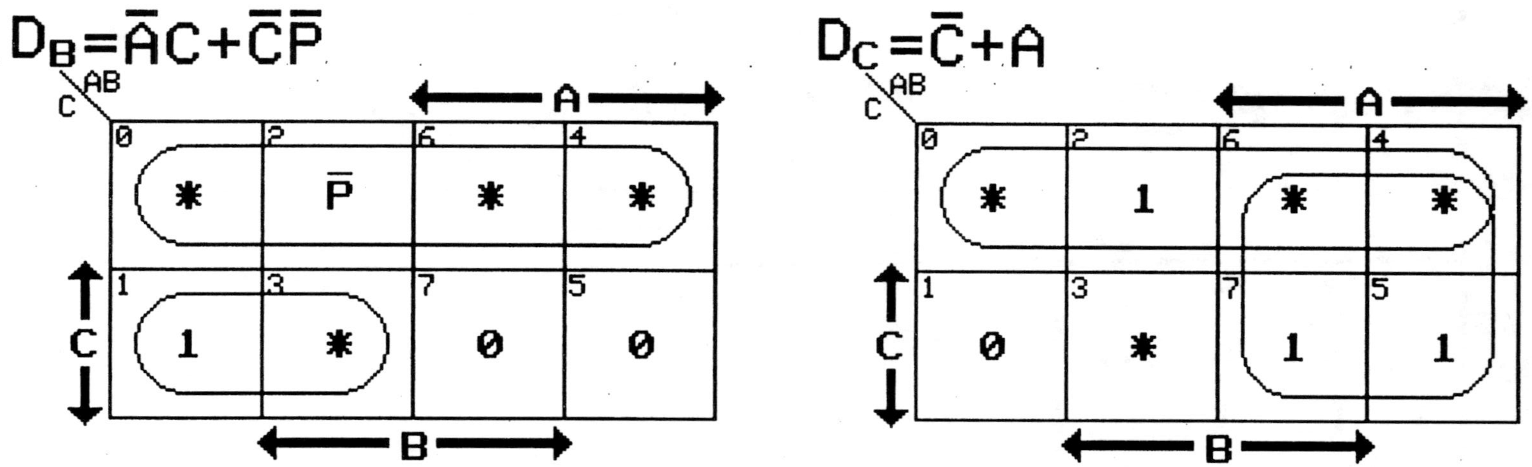

FIGURE 3.15 Karnaugh maps for D-type implementation of example.

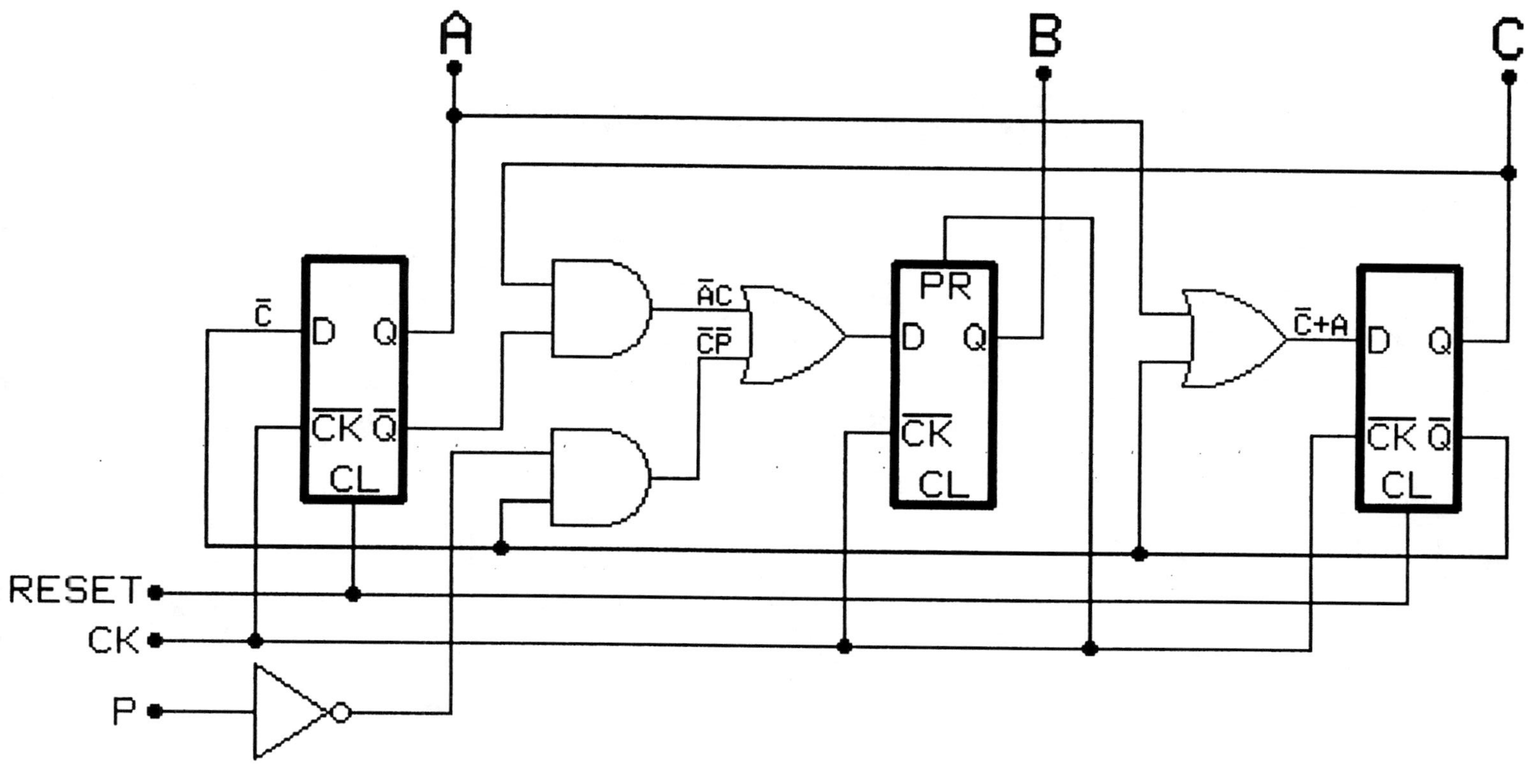

FIGURE 3.16 D-type implementation of controlled synchronous binary counter.

		J_A	K_A	J_B	K_B	J_C	K_C
0		*	*	*	*	*	*
1		0	*	1	*	*	1
2		1	*	*	P	1	*
3		*	*	*	*	*	*
4		*	*	*	*	*	*
5		*	1	0	*	*	0
6		*	*	*	*	*	*
7		*	1	*	1	*	0

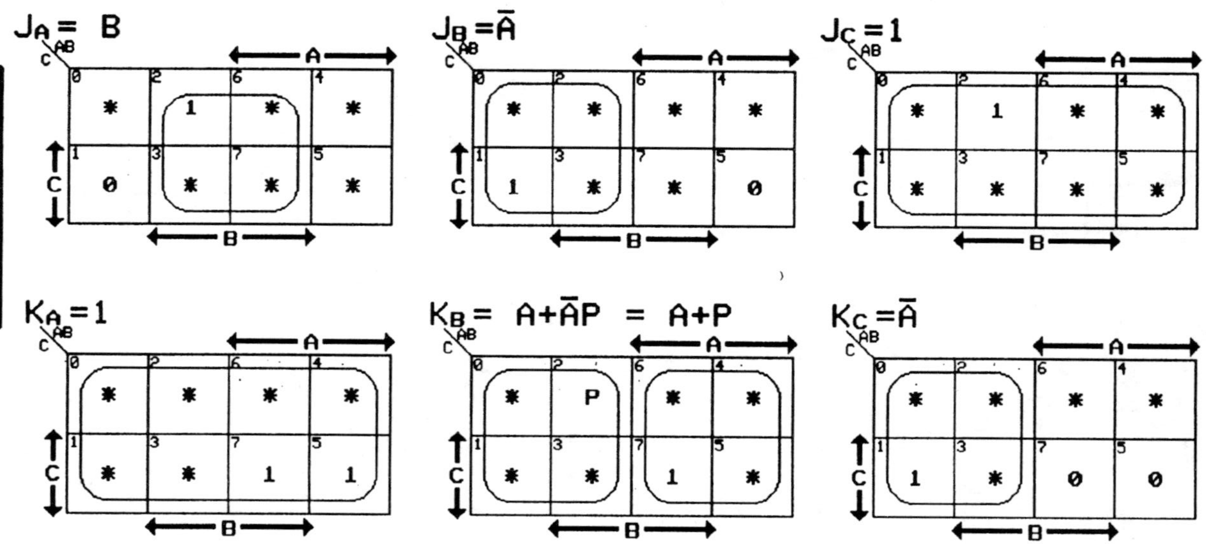

FIGURE 3.17 Karnaugh maps for JK-type implementation of example.

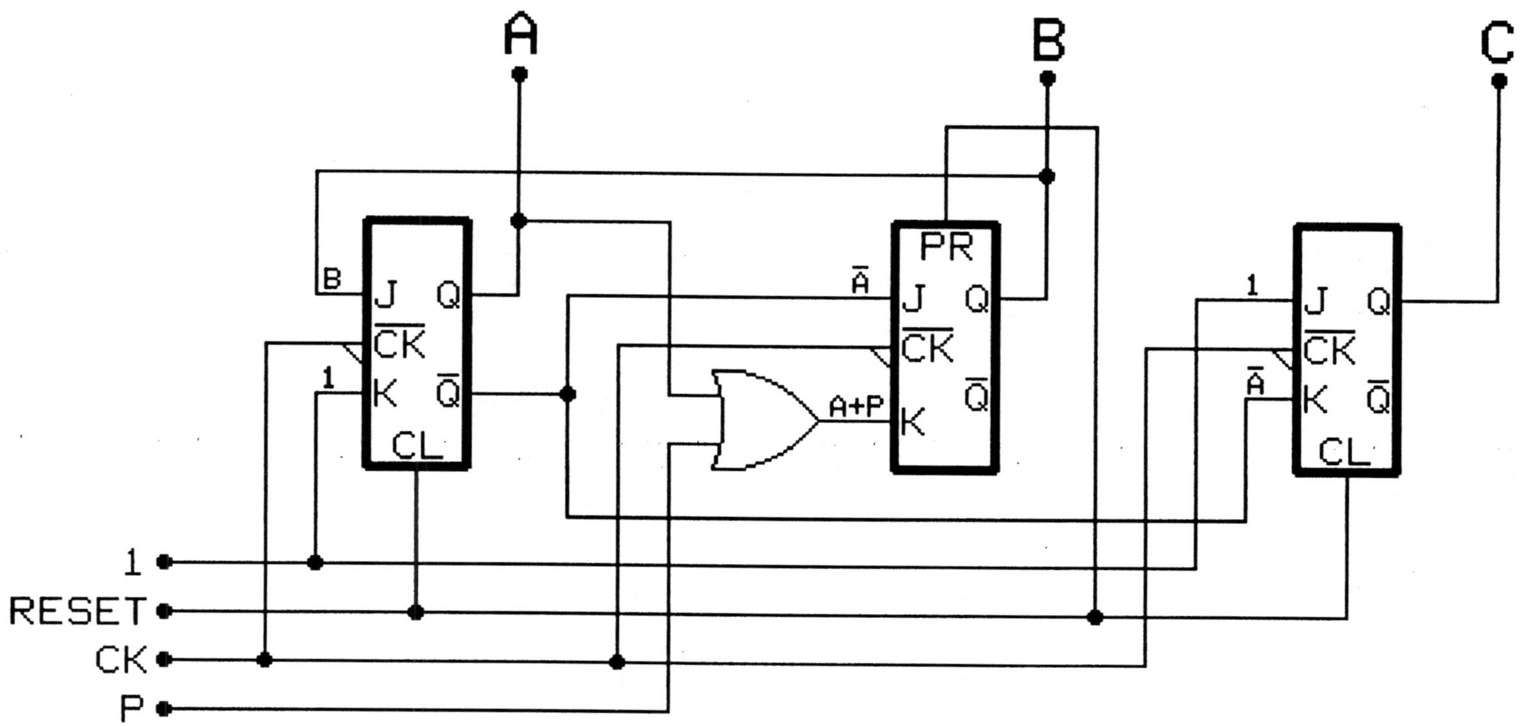

FIGURE 3.18 JK-type implementation of controlled synchronous binary counter.

	S_A	R_A	S_B	R_B	S_C	R_C
0	*	*	*	*	*	*
1	0	*	1	0	0	1
2	1	*	$\bar{P}$	P	1	*
3	*	*	*	*	*	*
4	*	*	*	*	*	*
5	0	1	0	*	*	0
6	*	*	*	*	*	*
7	0	1	0	1	*	0

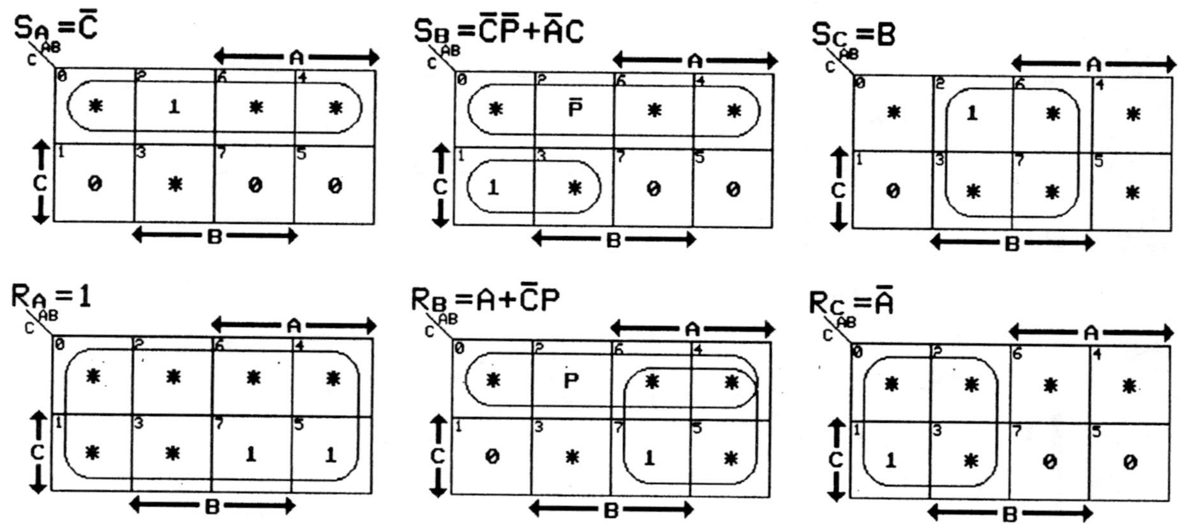

FIGURE 3.19 *Karnaugh maps for SR-type implementation of example.*

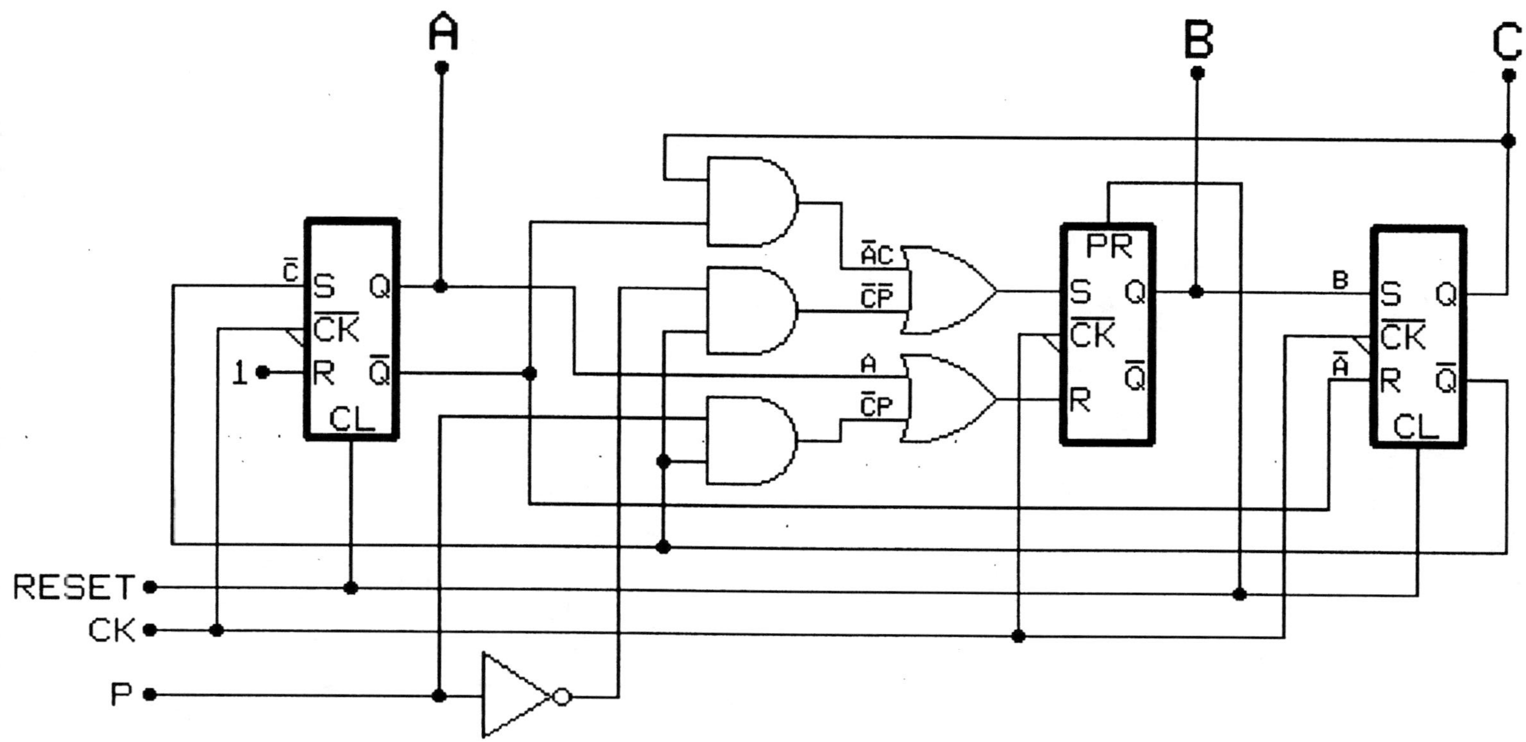

FIGURE 3.20 SR-type implementation of controlled synchronous binary counter

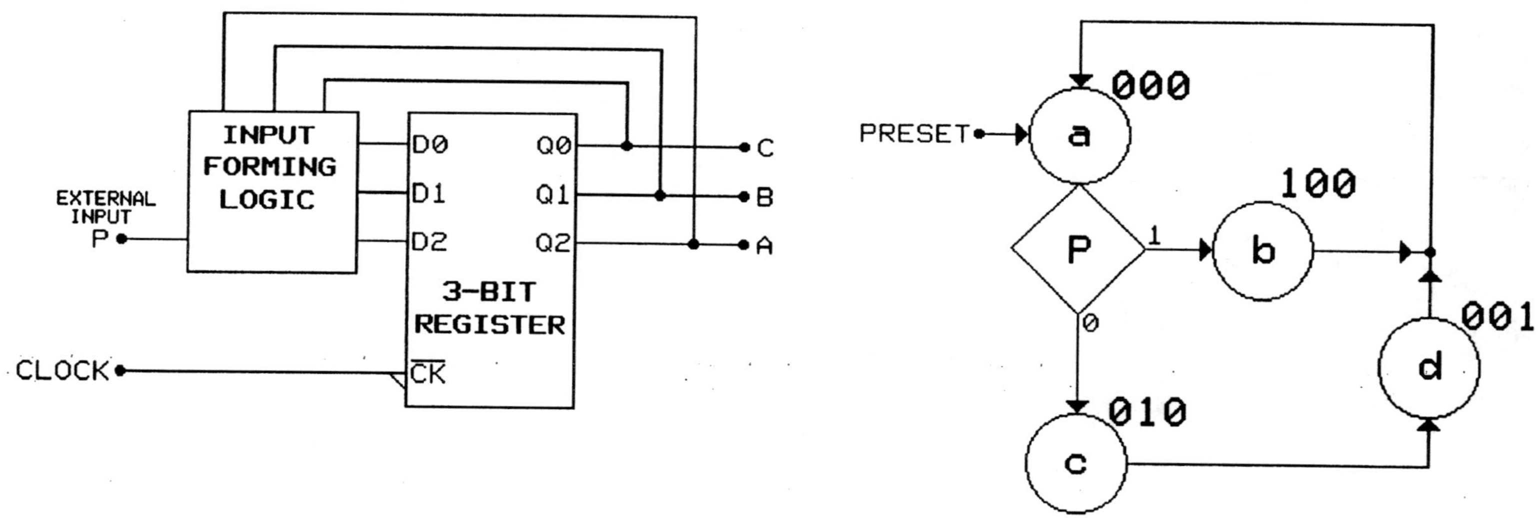

FIGURE 3.21 Practical exercise 3.2, problem definition.

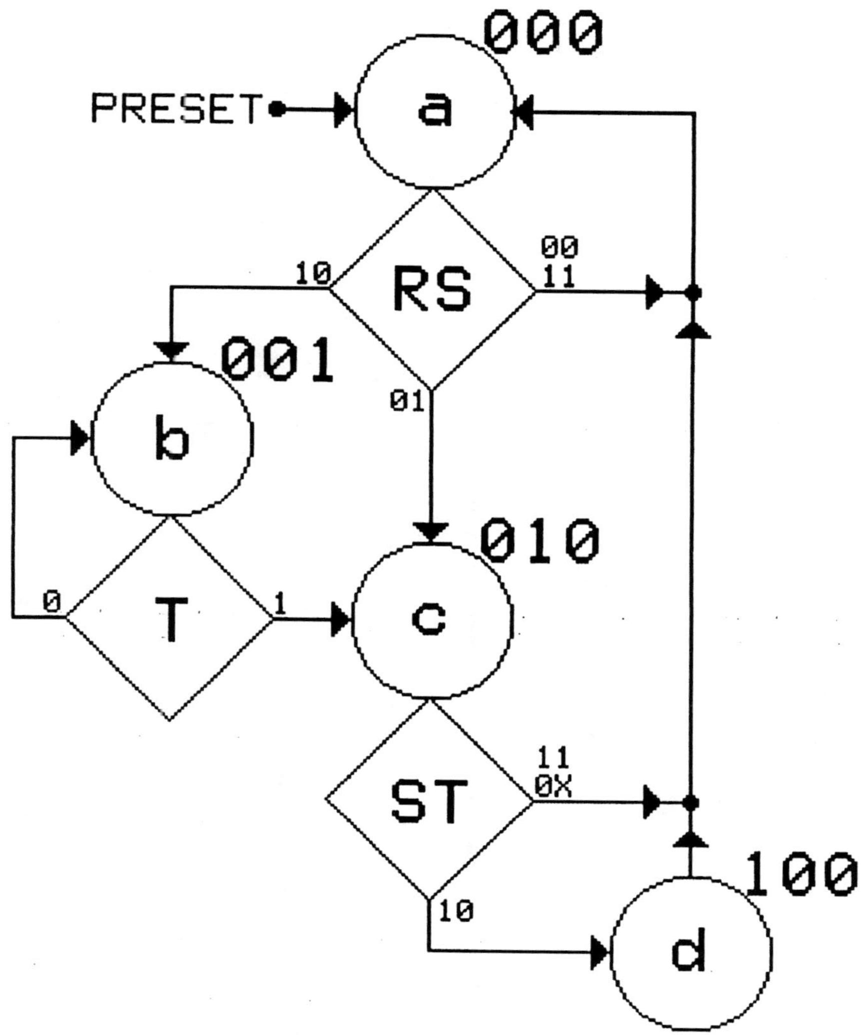

FIGURE 3.22 Practical Example 3.4, state diagram.

STATE	N	PRES ABC	INPUT RST	NEXT ABC	D_A	D_B	D_C
a	0	000	00*	000	0	$\bar{R}S$	$R\bar{S}$
			01*	010			
			10*	001			
			11*	000			
b	1	001	**0	001	0	T	$\bar{T}$
			**1	010			
c	2	010	*00	000	$S\bar{T}$	0	0
			*01	000			
			*10	100			
			*11	000			
	3	011	***	***	*	*	*
d	4	100	***	000	0	0	0
	5	101	***	***	*	*	*
	6	110	***	***	*	*	*
	7	111	***	***	*	*	*

* = Don't Care

FIGURE 3.23 Practical Example 3.4, state table.

STATE	N	PRES ABC	INPUT RST	NEXT ABC
a	Ø	ØØØ	ØØ*	ØØØ
			Ø1*	Ø1Ø
			1Ø*	ØØ1
			11*	ØØØ

FIGURE 3.24 Practical Example 3.4, State-0 present-input-next columns.

STATE	N	PRES ABC	INPUT RST	NEXT ABC	D_A	D_B	D_C
a	Ø	ØØØ	ØØ*	ØØØ	Ø	$\bar{R}S$	$R\bar{S}$
			Ø1*	Ø1Ø			
			1Ø*	ØØ1			
			11*	ØØØ			

FIGURE 3.25 Practical Example 3.4, state table, State-a.

STATE	N	PRES ABC	INPUT RST	NEXT ABC	D_A	D_B	D_C
b	1	001	**0	001	Ø	T	$\overline{T}$
			**1	010			

FIGURE 3.26 Practical Example 3.4, state table, State-b.

STATE	N	PRES ABC	INPUT RST	NEXT ABC	D_A	D_B	D_C
c	2	010	*00	000	$S\overline{T}$	Ø	Ø
			*01	000			
			*10	100			
			*11	000			

FIGURE 3.27 Practical Example 3.4, state table, State-c.

PRES ABC	D_A	D_B	D_C
000	0	$\bar{R}S$	$R\bar{S}$
001	0	T	$\bar{T}$
010	$S\bar{T}$	0	0
011	*	*	*
100	0	0	0
101	*	*	*
110	*	*	*
111	*	*	*

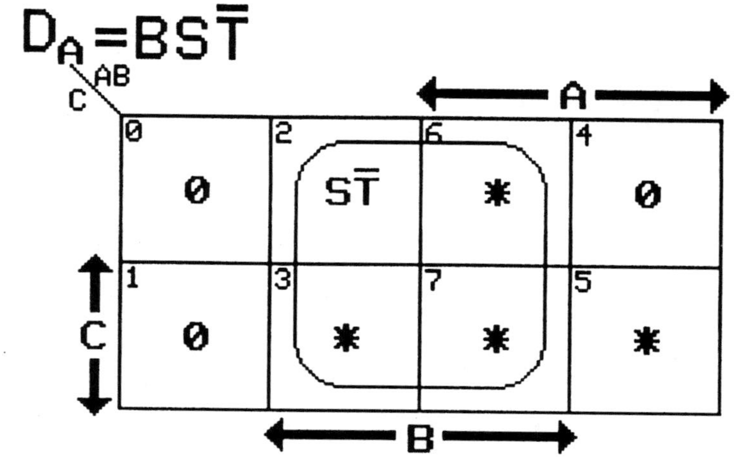

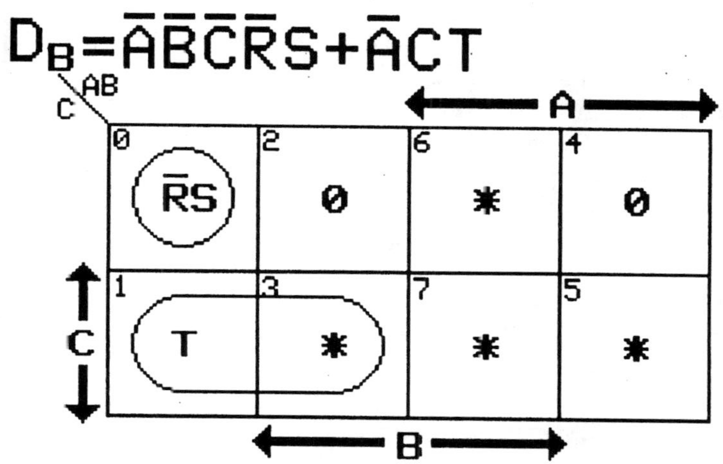

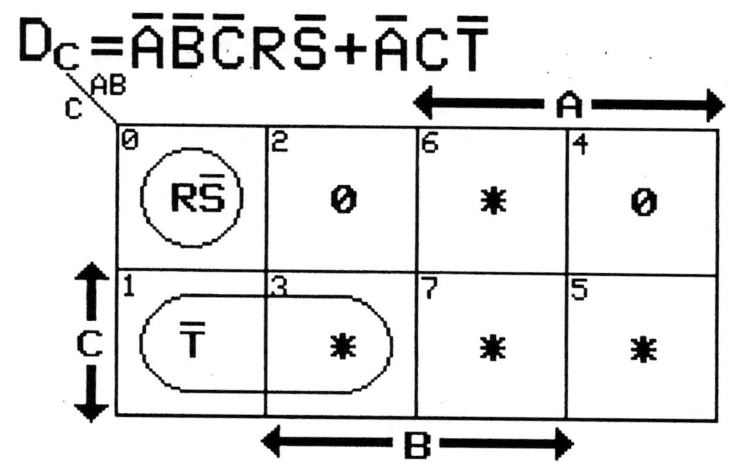

FIGURE 3.28 Practical Example 3.4, karnaugh maps.

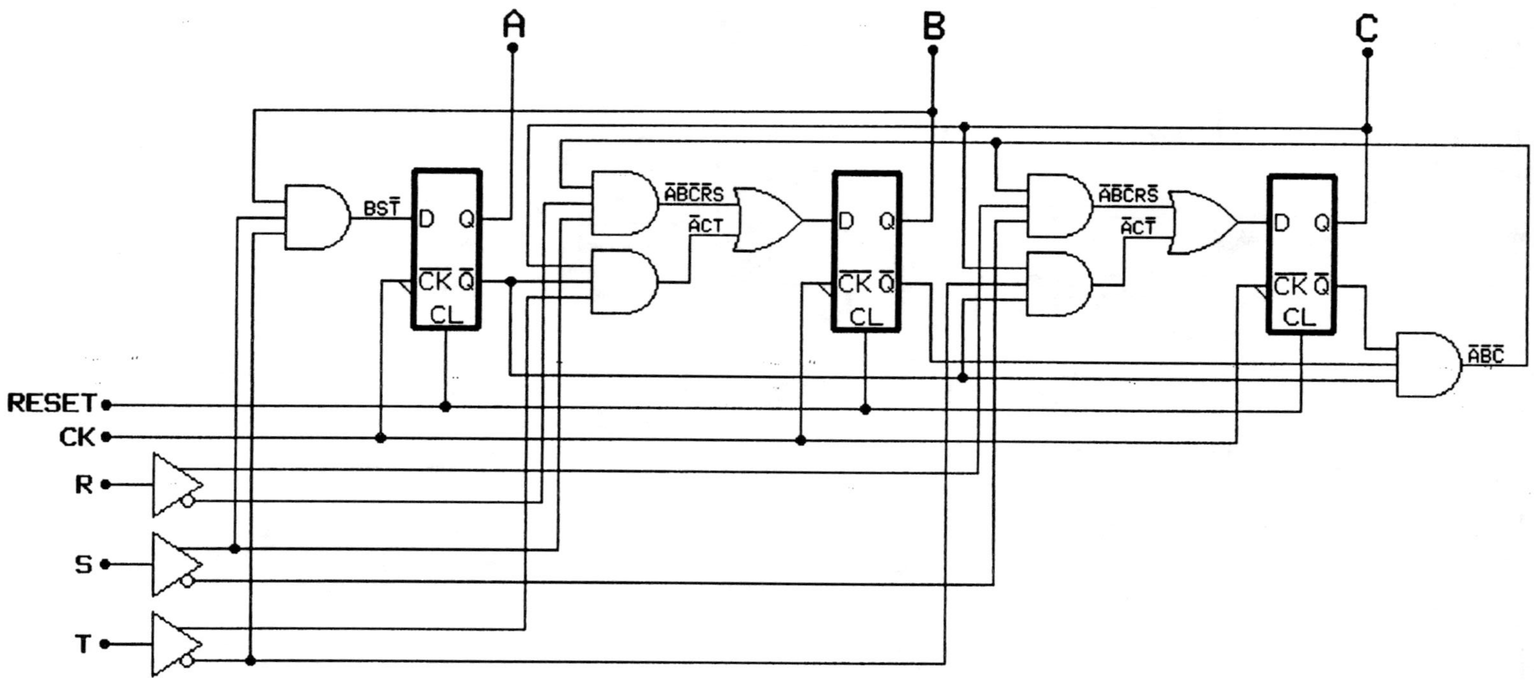

FIGURE 3.29 Practical Example 3.4, schematic diagram.

STATE	N	PRES ABC	INPUT RST	NEXT ABC	D_A	D_B	D_C
c	2	010	*00	000	$S\bar{T}$	0	0
			*01	000			
			*10	100			
			*11	000			

FIGURE 3.30 ***Practical Example 3.4, state table, State-c.***

STATE	N	PRES ABC	INPUT RST	NEXT ABC	D_A	D_B	D_C
c	2	010	*00	101	$\bar{S}\bar{T}+S\bar{T}+ST$ $=S+\bar{T}$	$\bar{S}T+S\bar{T}+ST$ $=S+T$	$\bar{S}\bar{T}+\bar{S}T$ $=\bar{S}$
			*01	011			
			*10	110			
			*11	110			

FIGURE 3.31 ***Next state value selection with complex circuitry.***

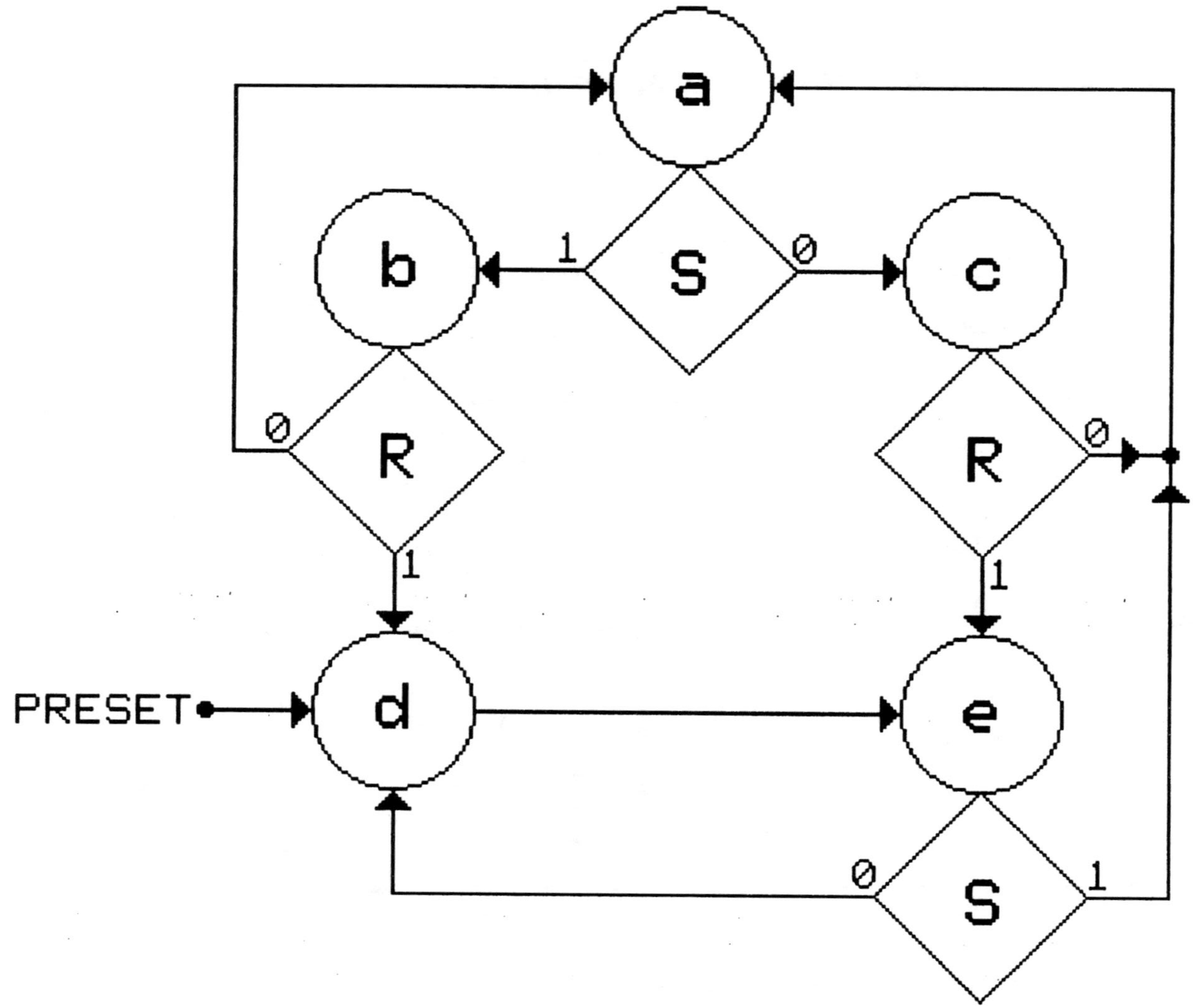

FIGURE 3.32 State diagram for questions 1 - 9.

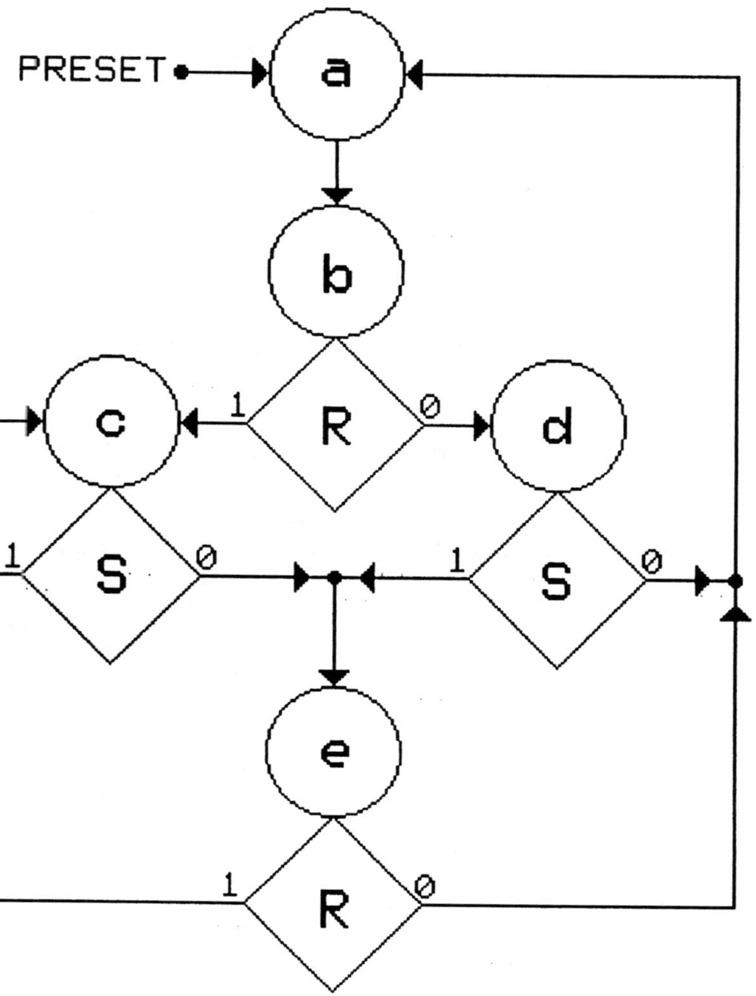

FIGURE 3.33 State diagram for questions 10 - 18.

```
SEQUENCE[A,B,C] {
        PRESENT a  NEXT b;
        PRESENT b  IF R NEXT c;  DEFAULT d;
        PRESENT c  IF S NEXT d;  DEFAULT a;
        PRESENT d  IF T NEXT a;  DEFAULT b;    }
```

FIGURE 3.34 State machine program for problems 19 - 27.

```
SEQUENCE[A,B,C,D] {
    PRESENT a  NEXT b;
    PRESENT b  NEXT c;
    PRESENT c  NEXT d;
    PRESENT d  NEXT e;
    PRESENT e  IF   (!R&!S)  NEXT f;  DEFAULT NEXT b;
    PRESENT f  IF   (!R#!S)  NEXT g;  DEFAULT NEXT f;
    PRESENT g  IF   ( R# S)  NEXT j;  DEFAULT NEXT h;
    PRESENT h  NEXT a;
    PRESENT j  IF (R&S)#(!R&!S) NEXT e;  DEFAULT NEXT a;  }
```

FIGURE 3.35 State machine program for problems 28 - 36.

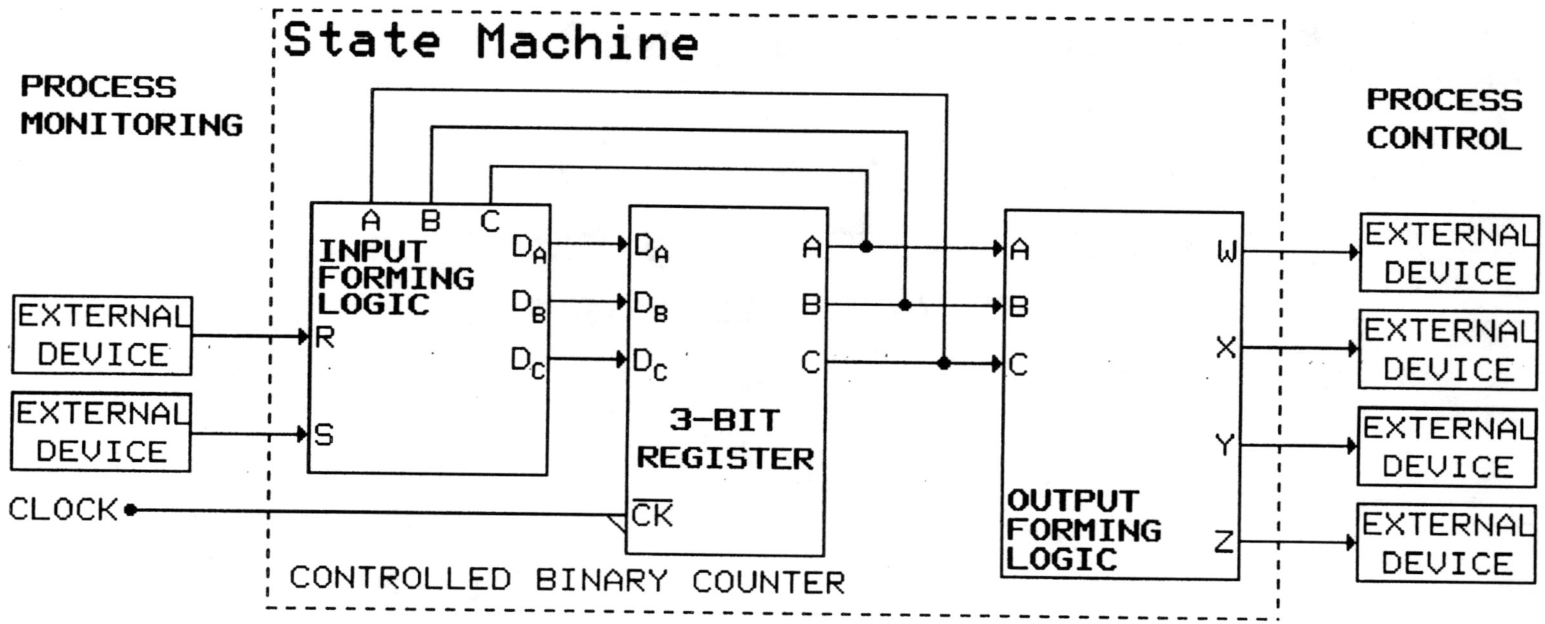

FIGURE 4.1 State machine overview.

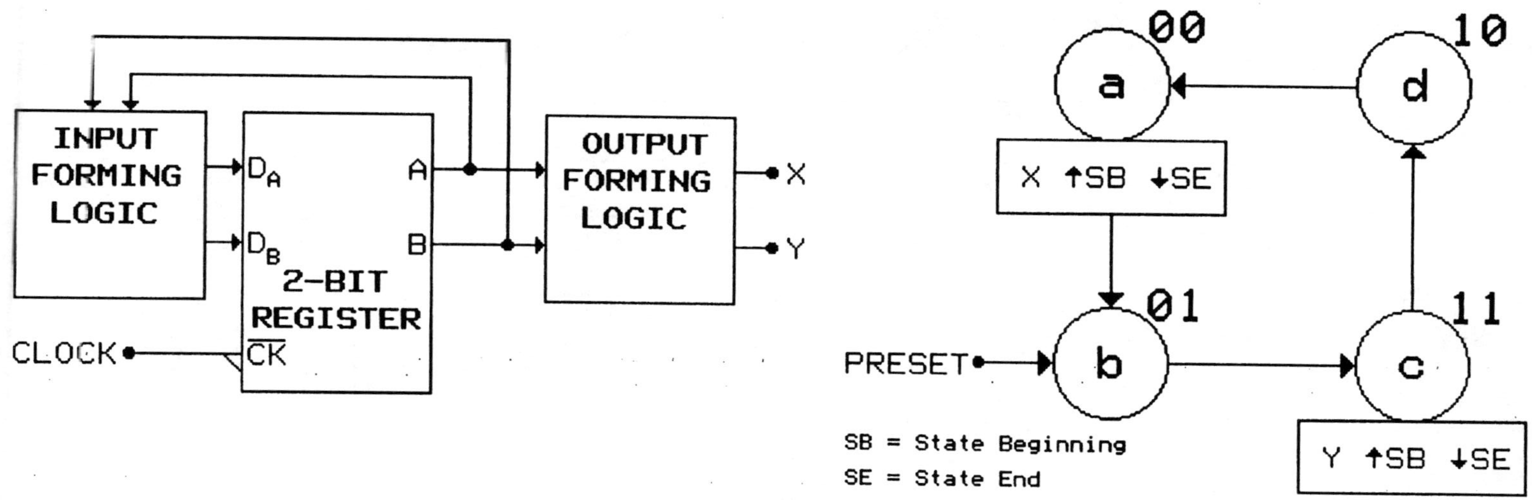

FIGURE 4.2 2-bit state machine with output.

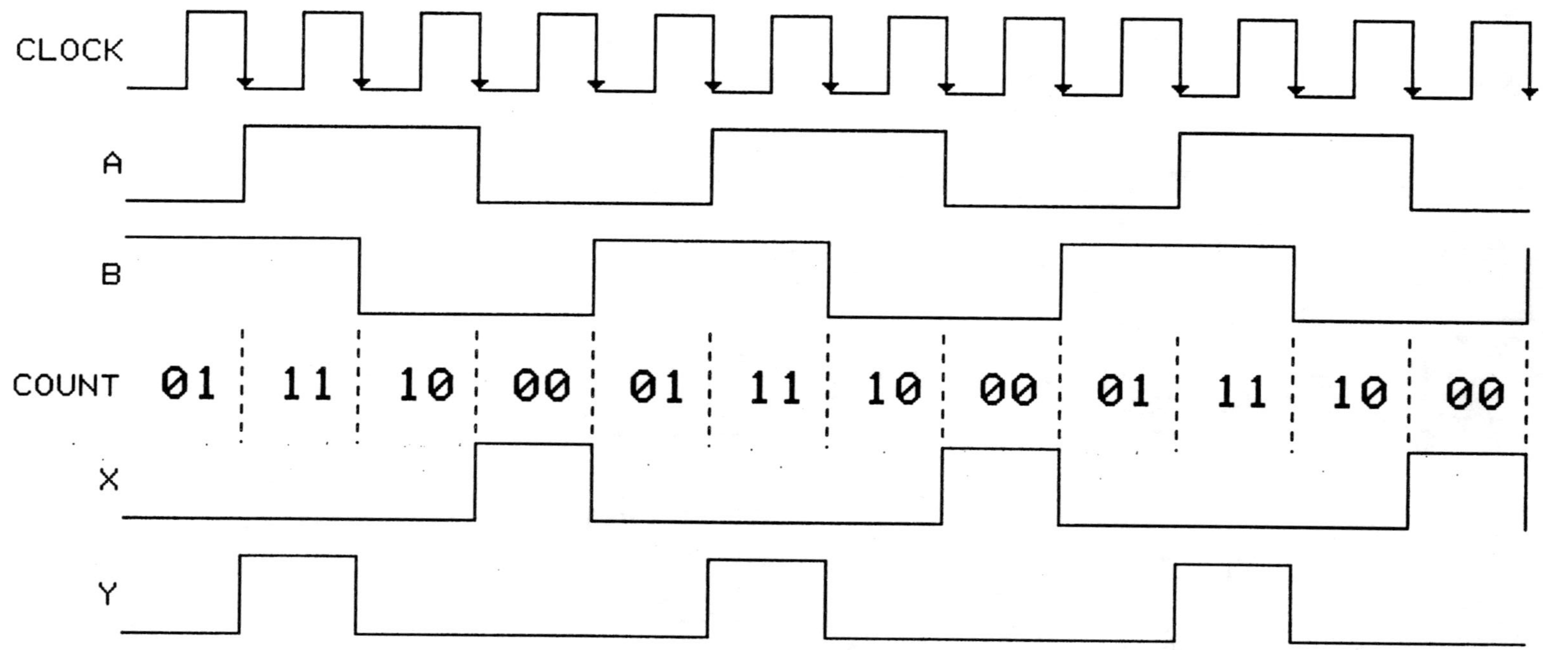

FIGURE 4.3 Example timing diagram.

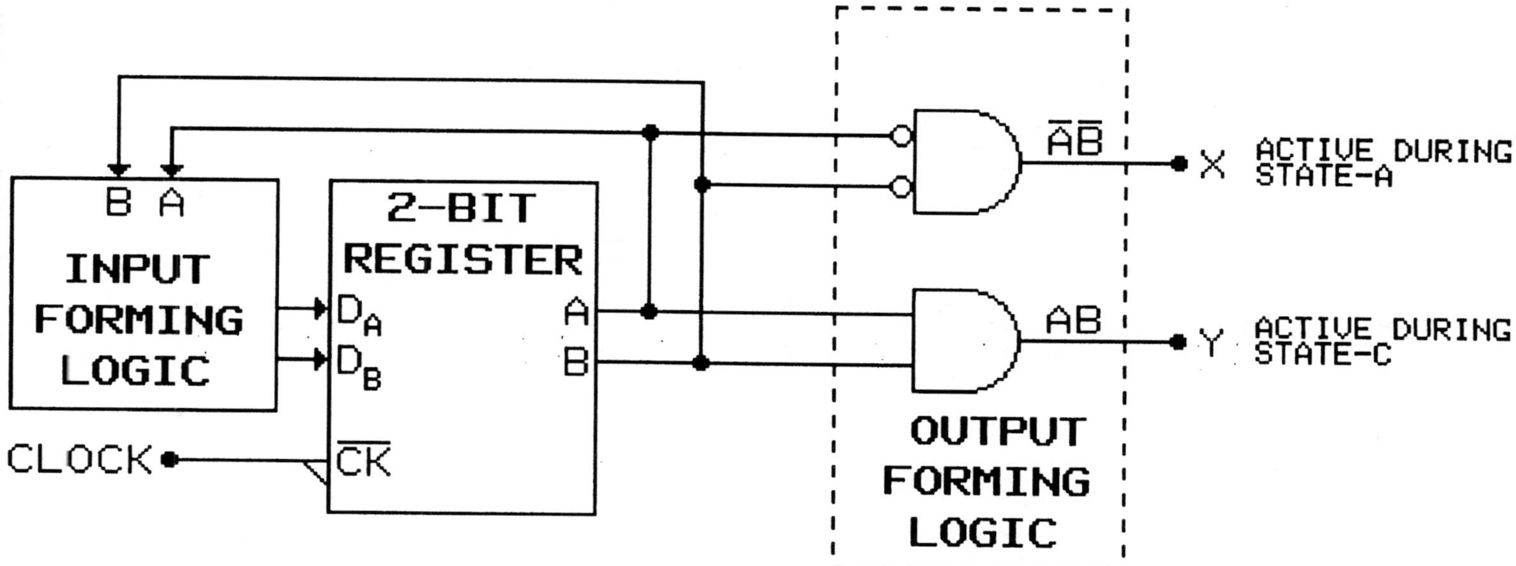

FIGURE 4.4 ***Simple output forming logic.***

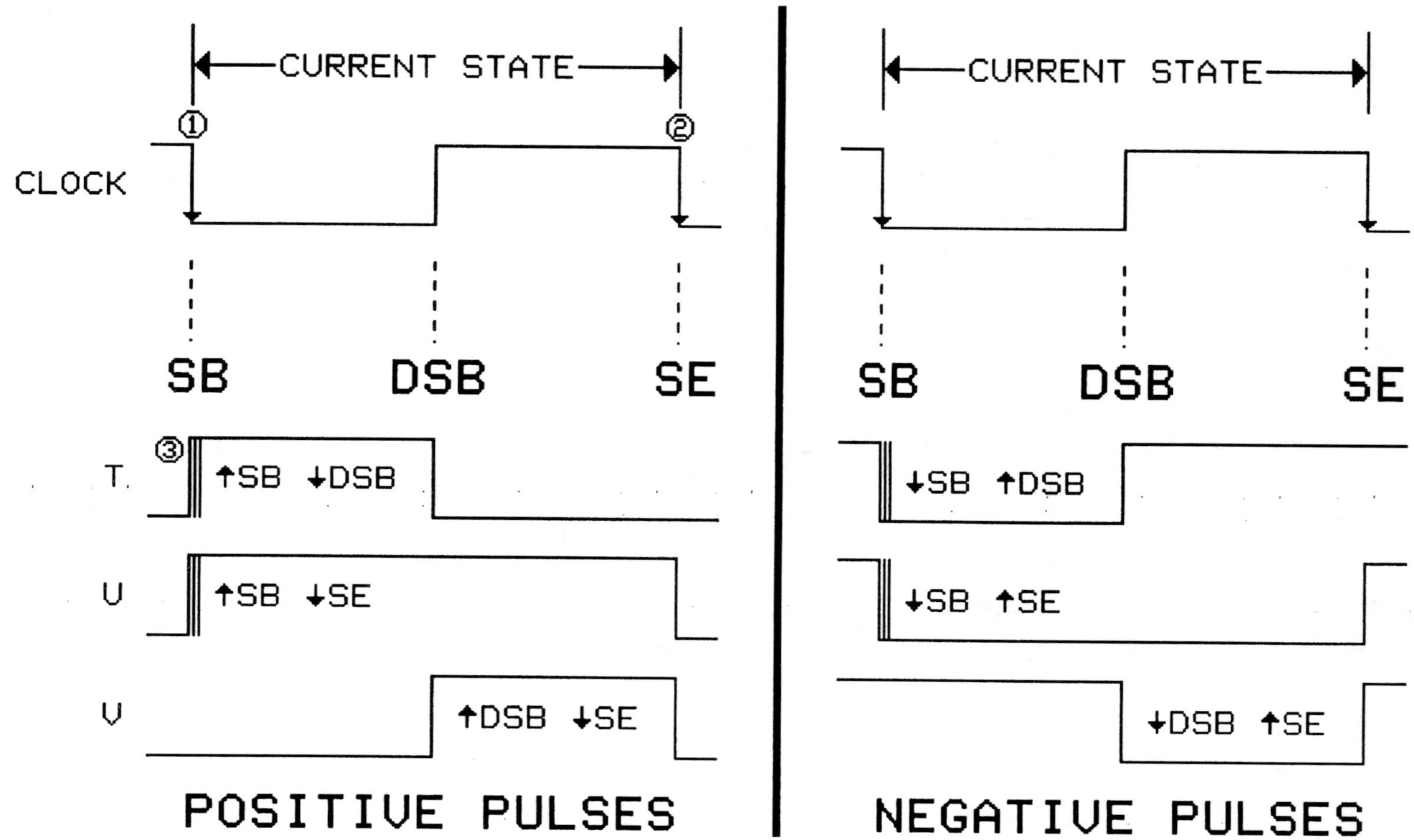

FIGURE 4.5 Non-delayed state machine output pulses.

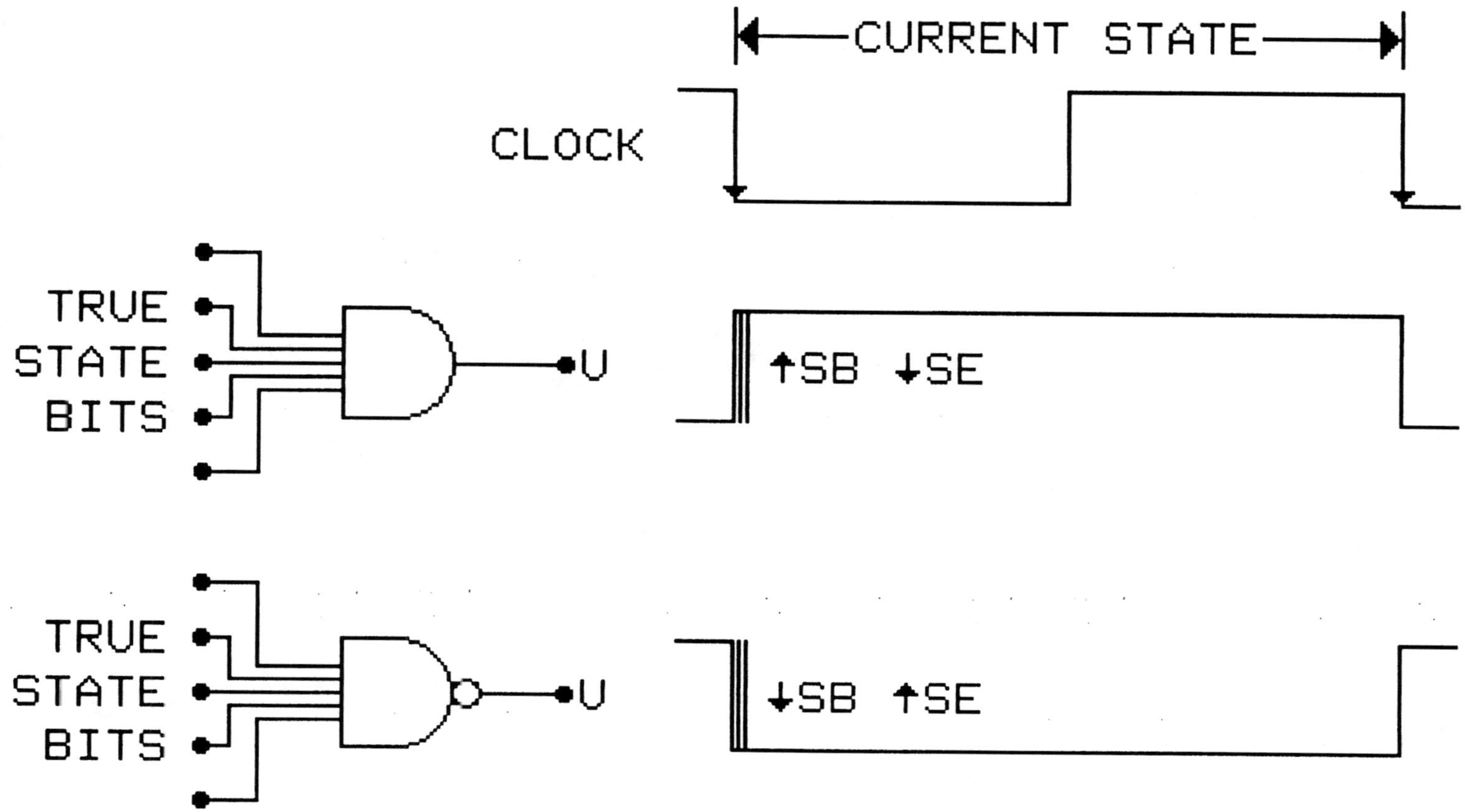

FIGURE 4.6 Unconditional SB to SE pulse generation.

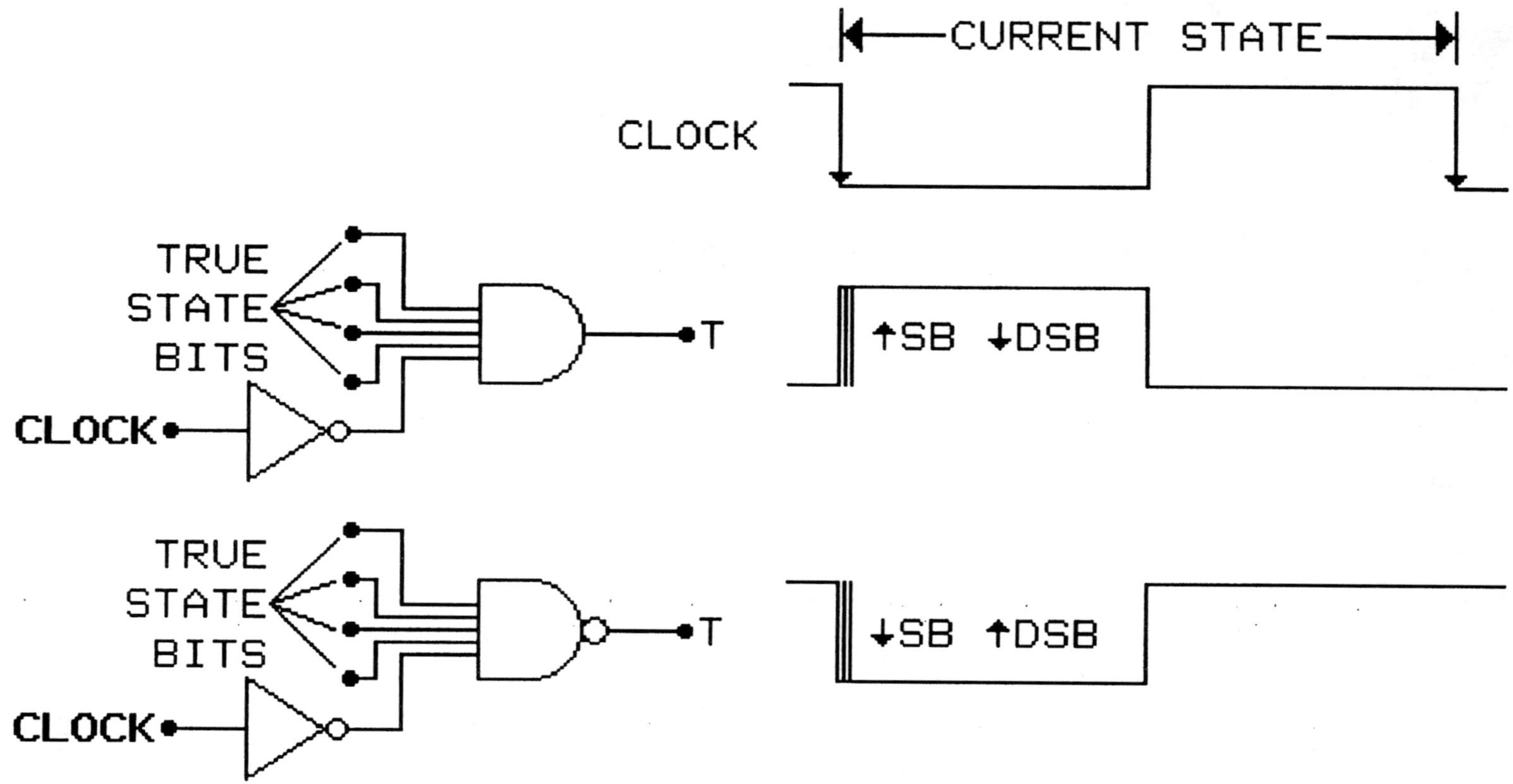

FIGURE 4.7 Unconditional SB to DSB pulse generation.

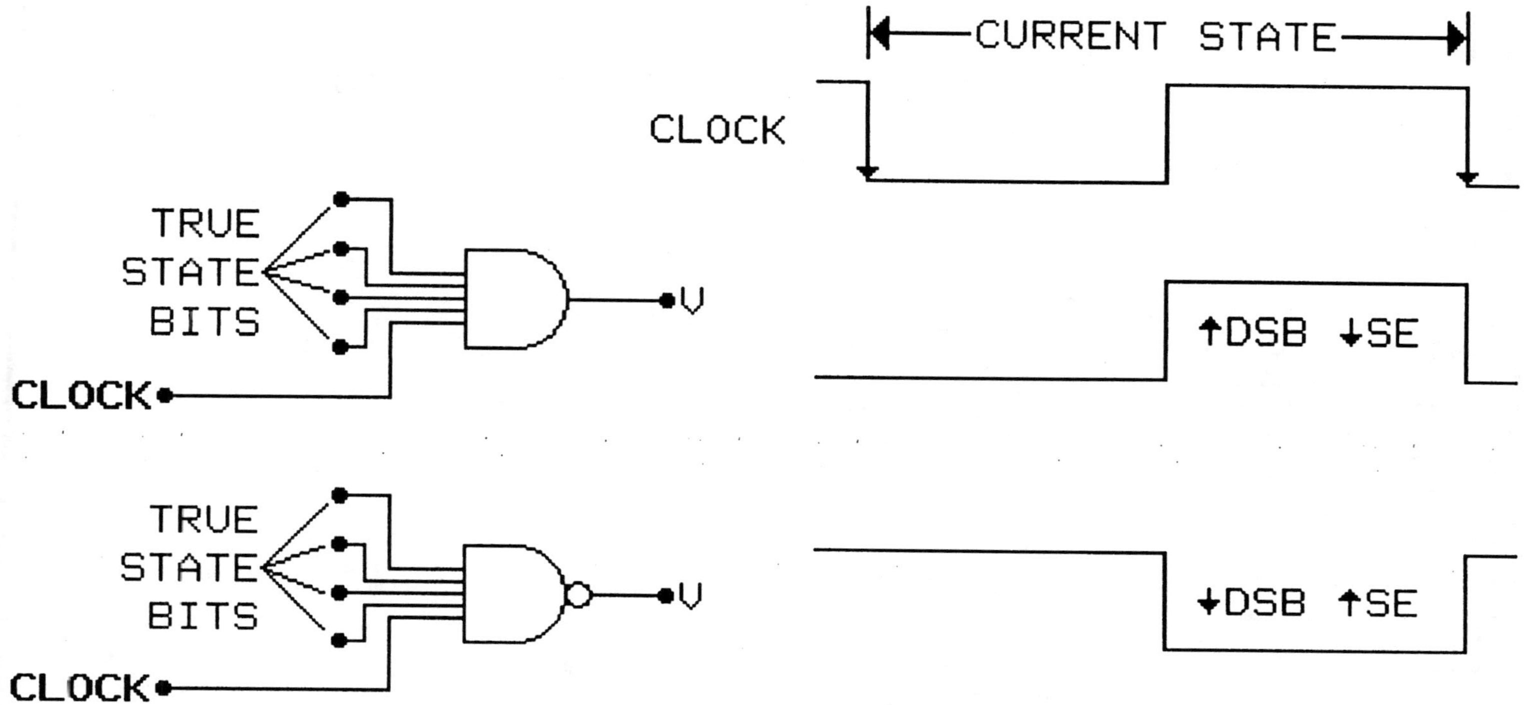

FIGURE 4.8 Unconditional DSB to SE pulse generation.

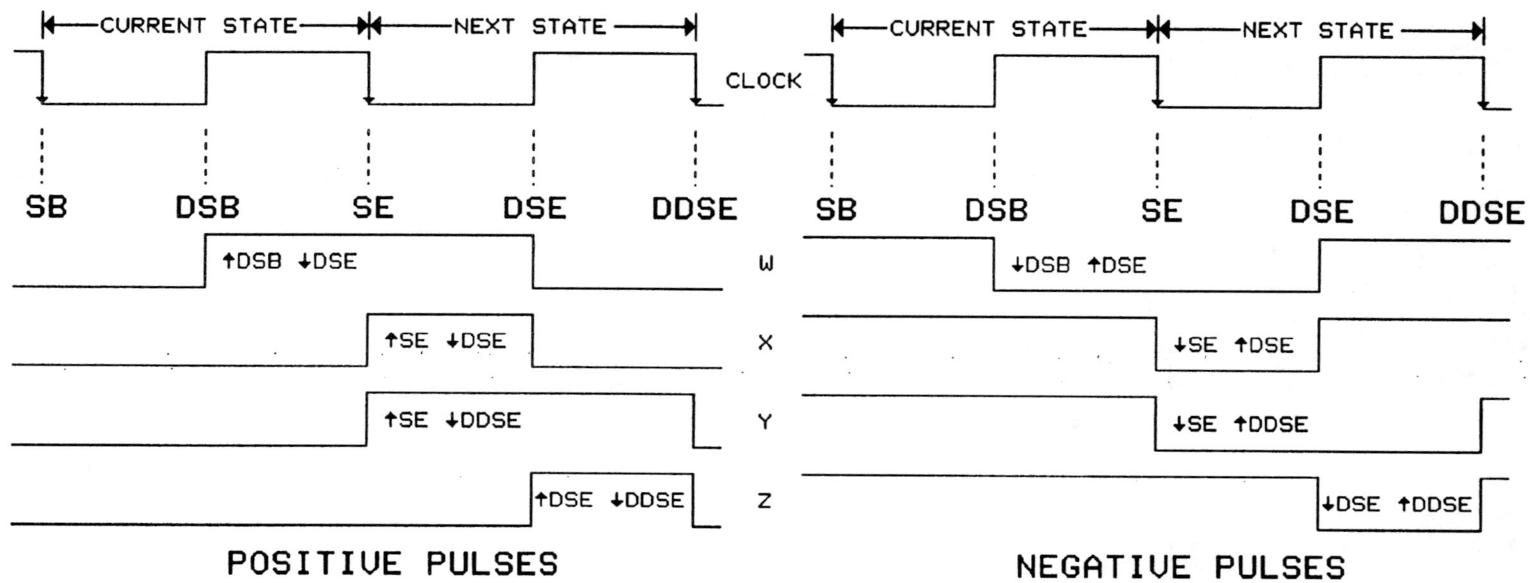

FIGURE 4.9 Delayed state machine output pulses.

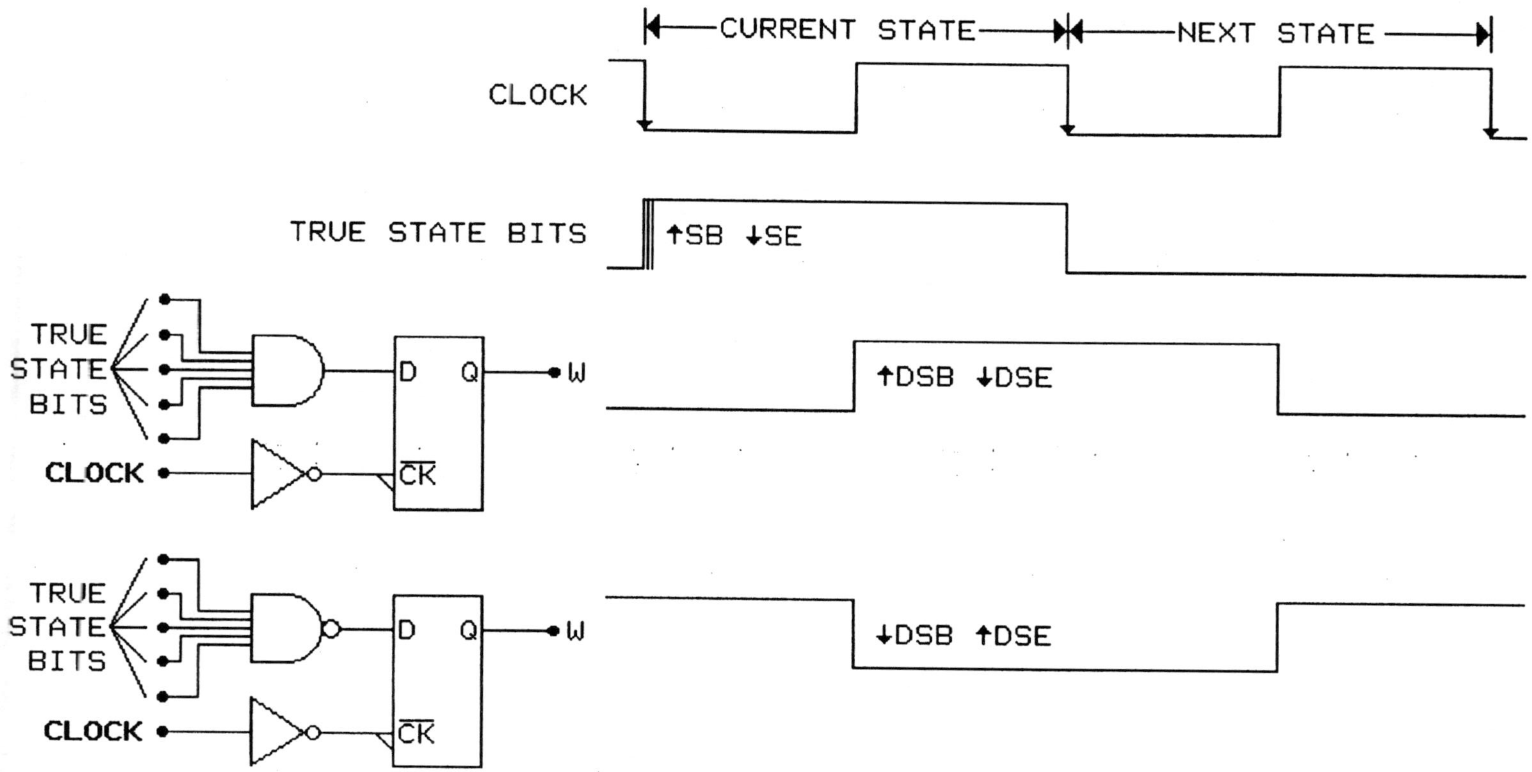

FIGURE 4.10 Unconditional DSB to DSE pulse generation.

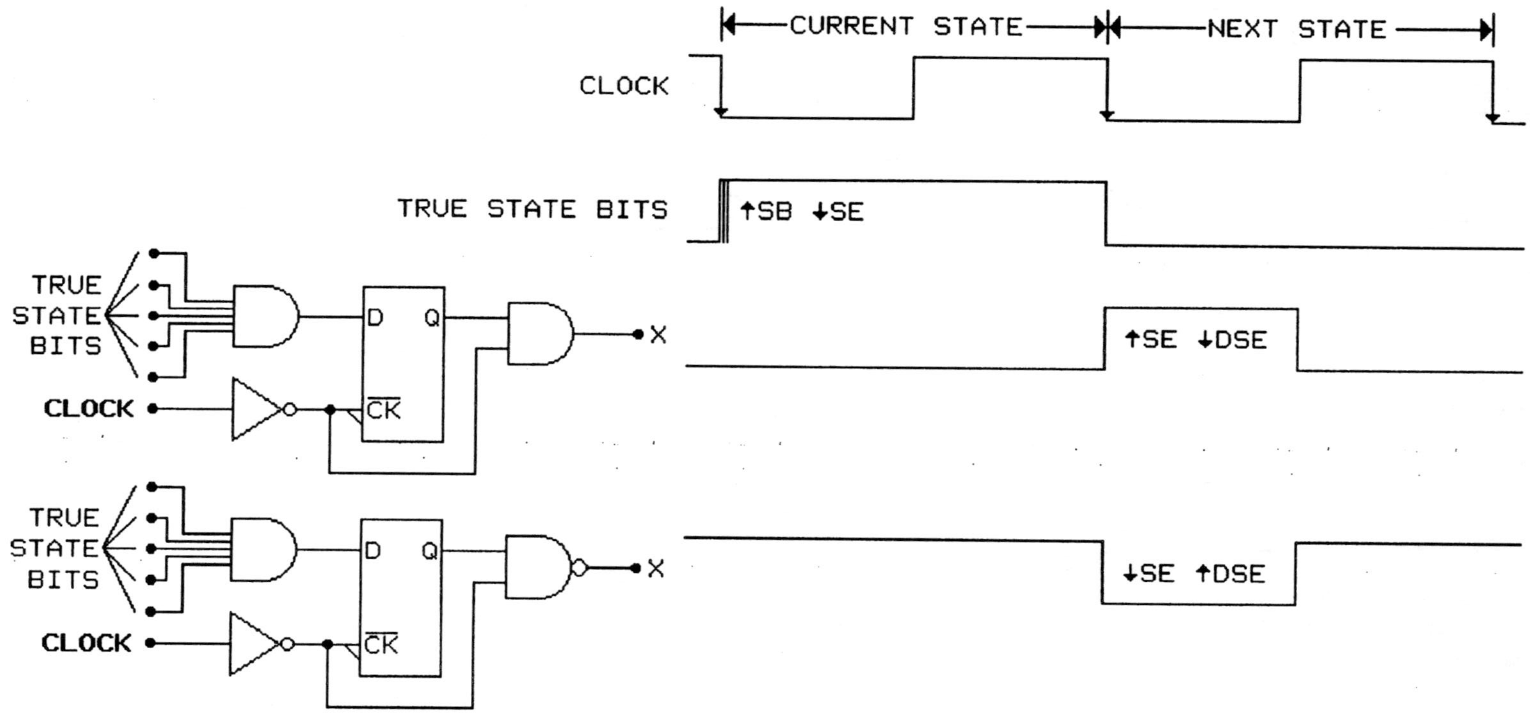

FIGURE 4.11 Unconditional SE to DSE pulse generation.

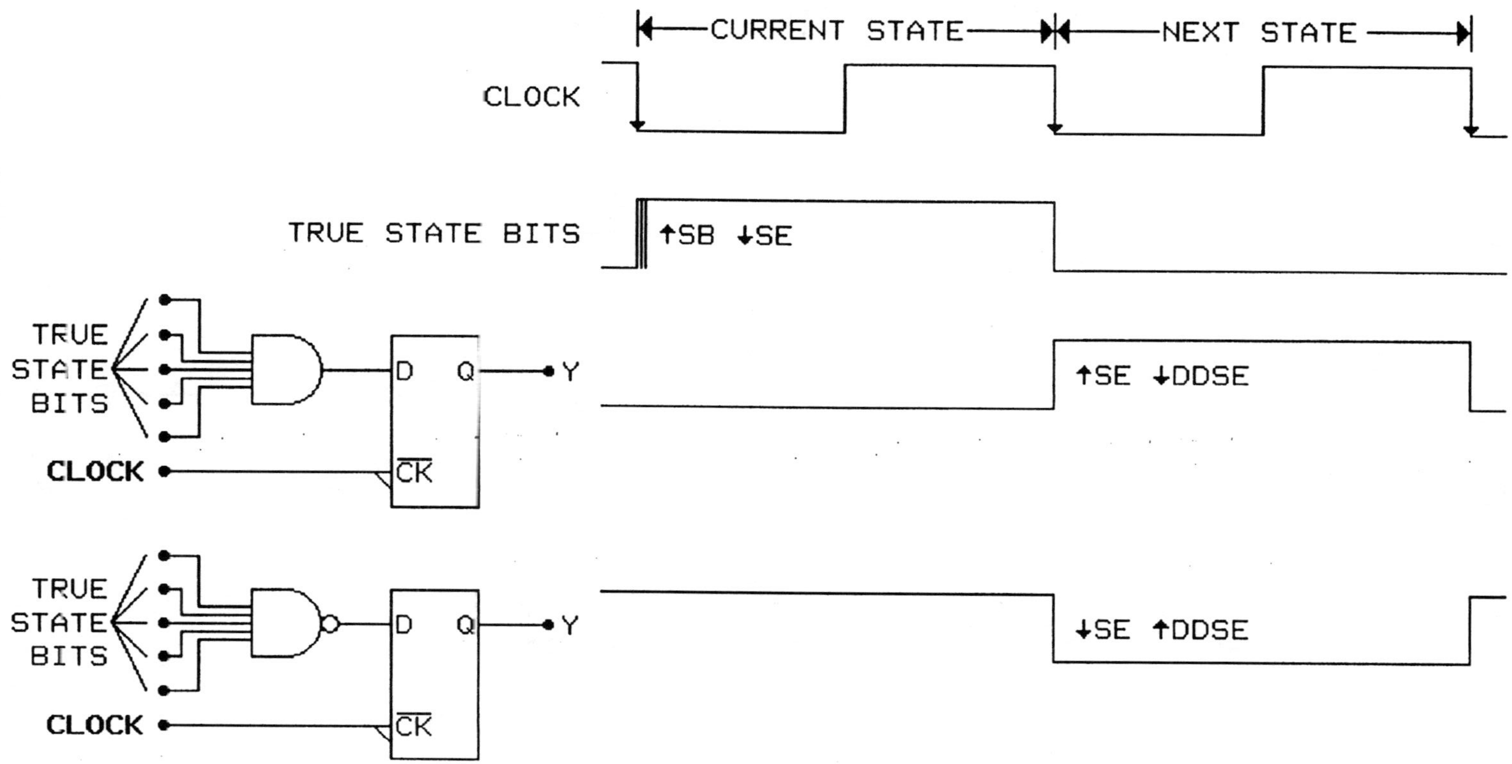

FIGURE 4.12 Unconditional SE to DDSE pulse generation.

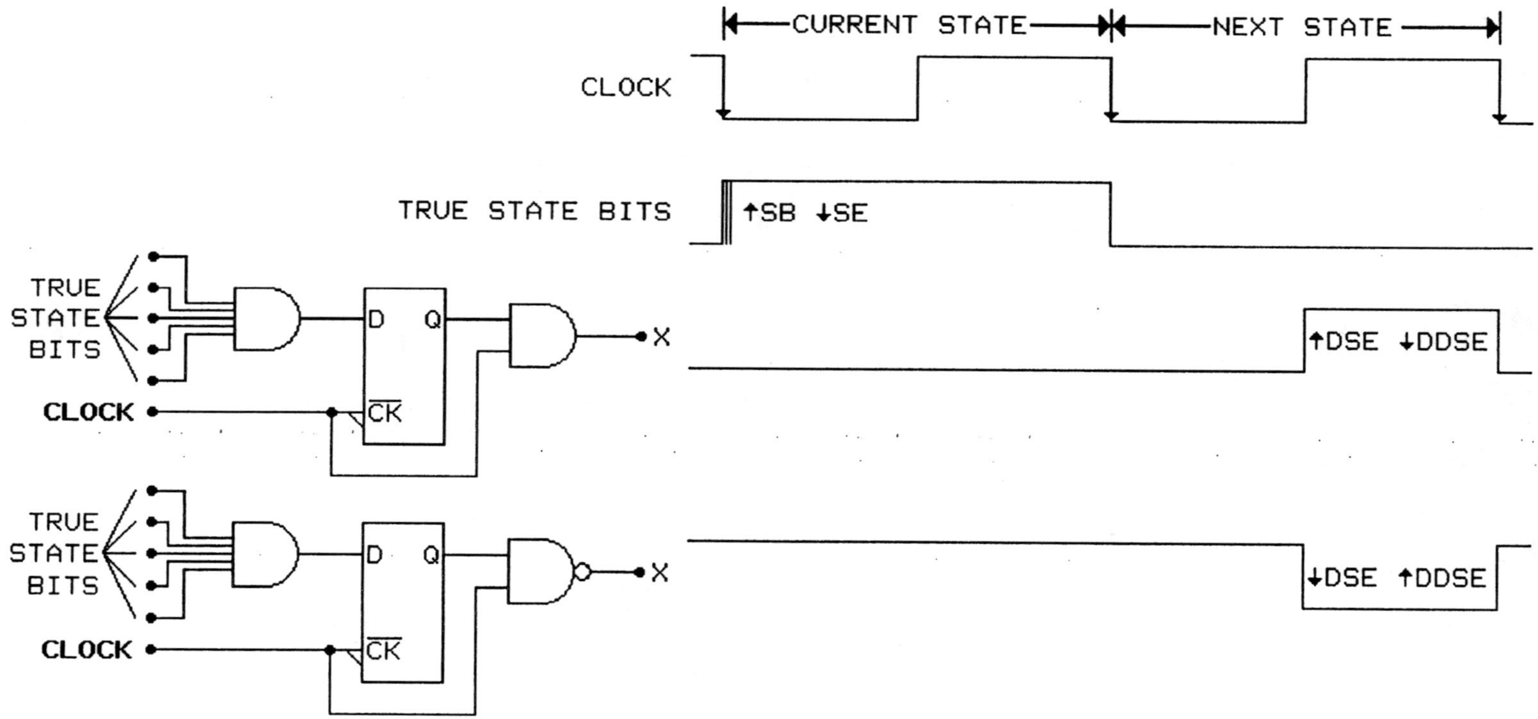

FIGURE 4.13 Unconditional DSE to DDSE pulse generation.

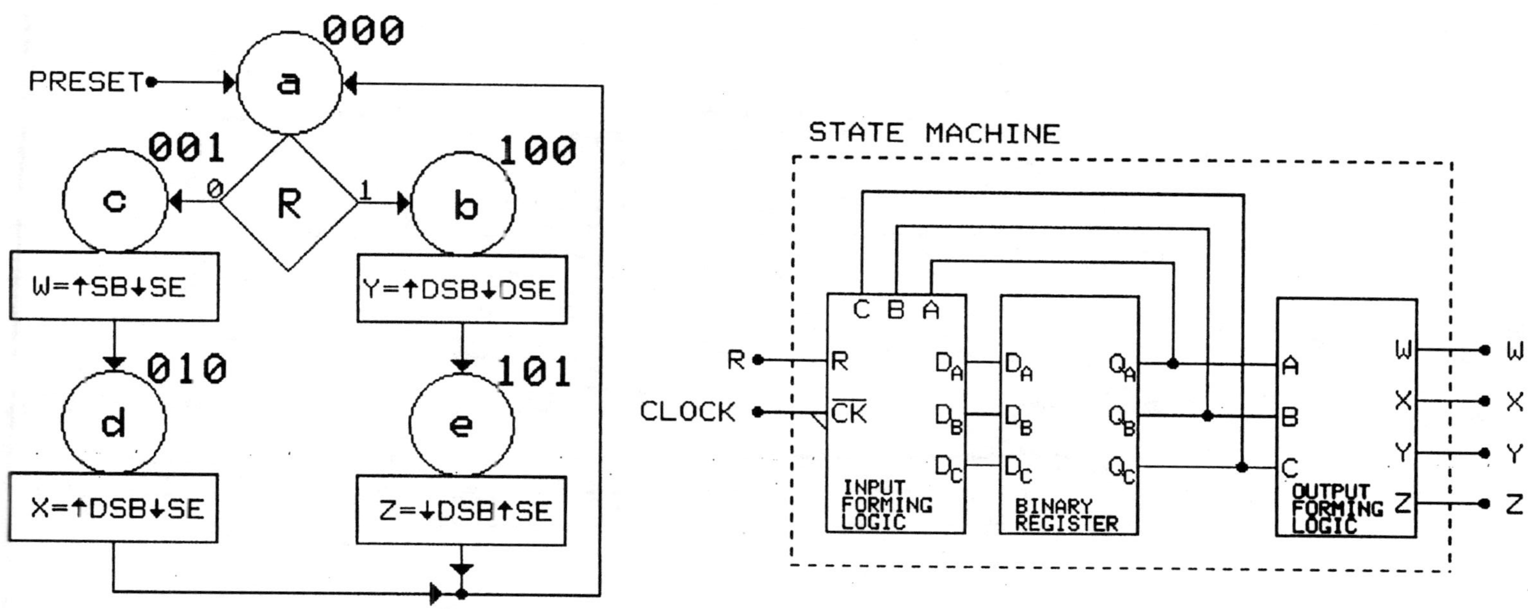

FIGURE 4.14 Practical Example 4.1, state diagram and block diagram.

	PRES ABC	INPUT R	NEXT ABC	D_A	D_B	D_C	W	X	Y	Z
a	000	0 1	001 100	R	0	$\overline{R}$				
c	001	*	010	0	1	0	↑SB↓SE			
d	010	*	000	0	0	0		↑DSB↓SE		
	011	*	***	*	*	*				
b	100	*	101	1	0	1			↑DSB↓DSE	
e	101	*	000	0	0	0				↓DSB↑SE
	110	*	***	*	*	*				
	111	*	***	*	*	*				

(W, X, Y, Z: OUTPUT FORMING LOGIC)

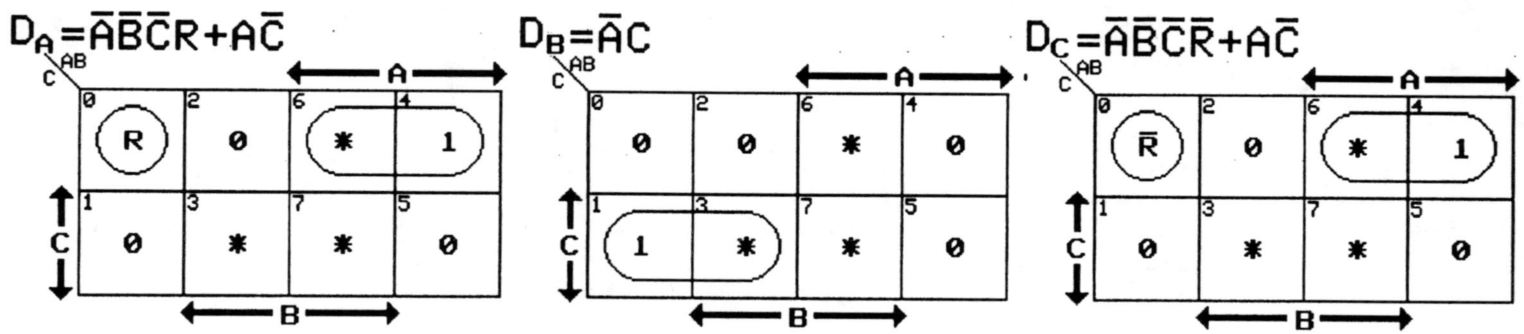

FIGURE 4.15 Practical Example 4.1, state table and karnaugh maps.

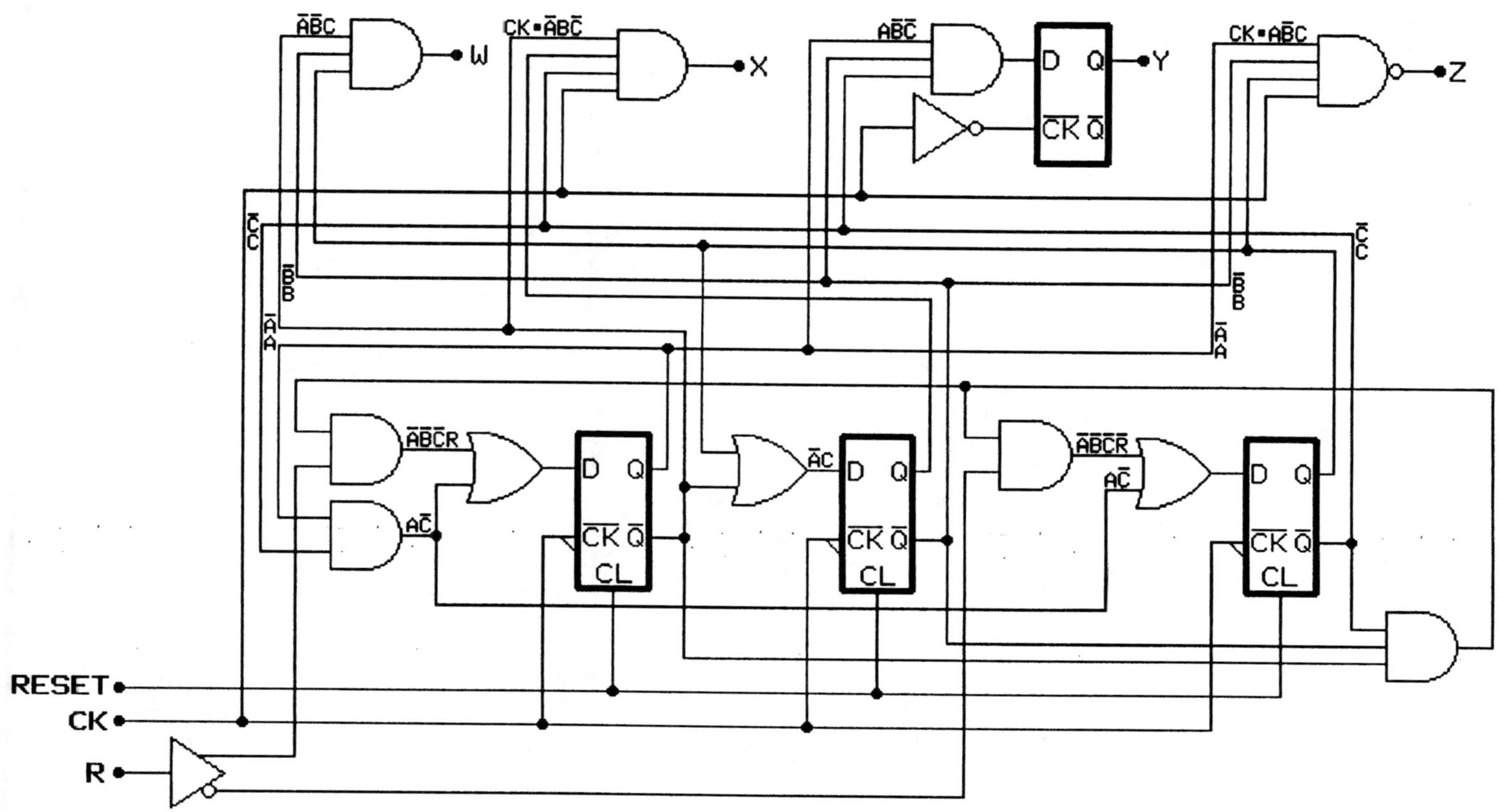

FIGURE 4.16 Practical Example 4.1, schematic diagram.

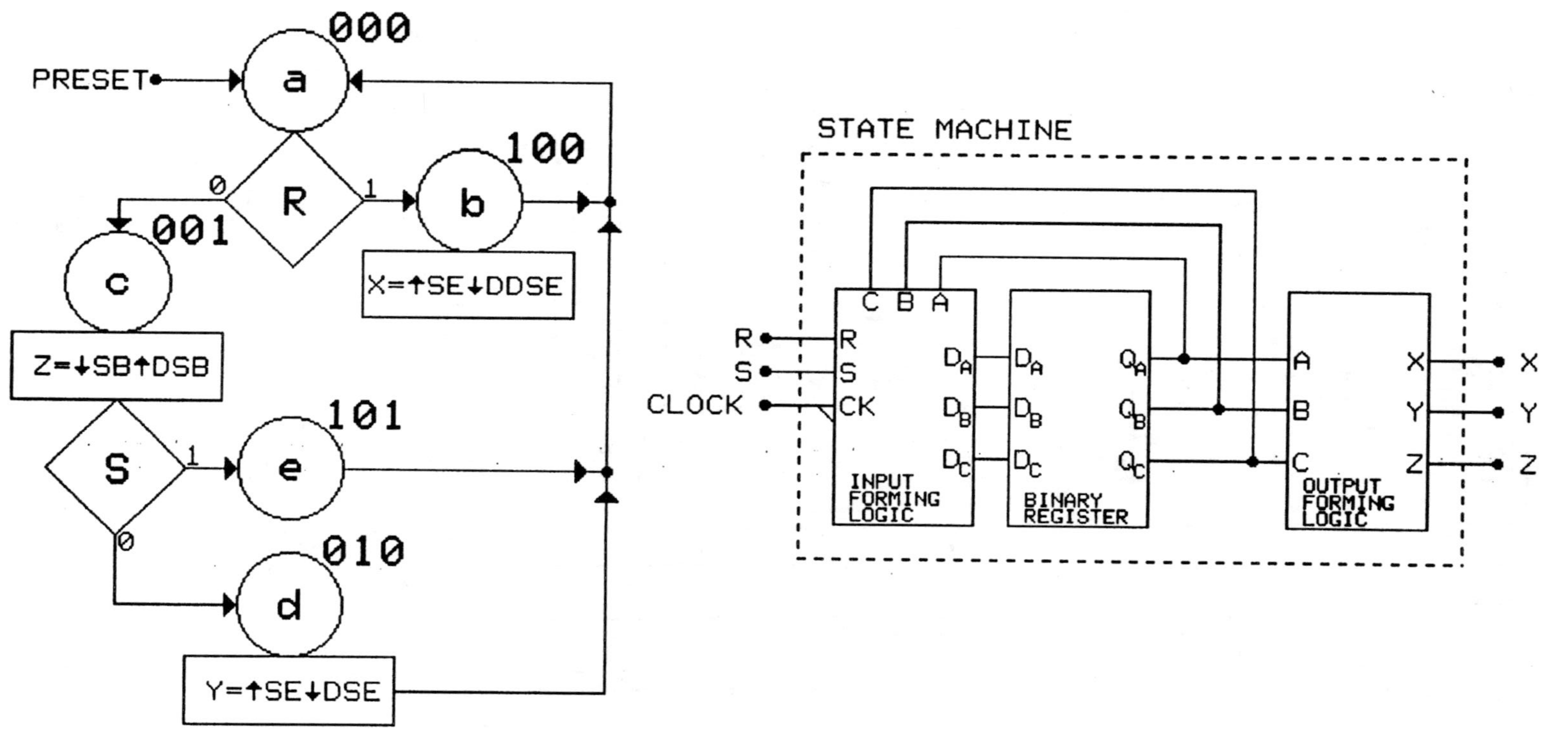

FIGURE 4.17 Practical Exercise 4.1, state diagram and block diagram.

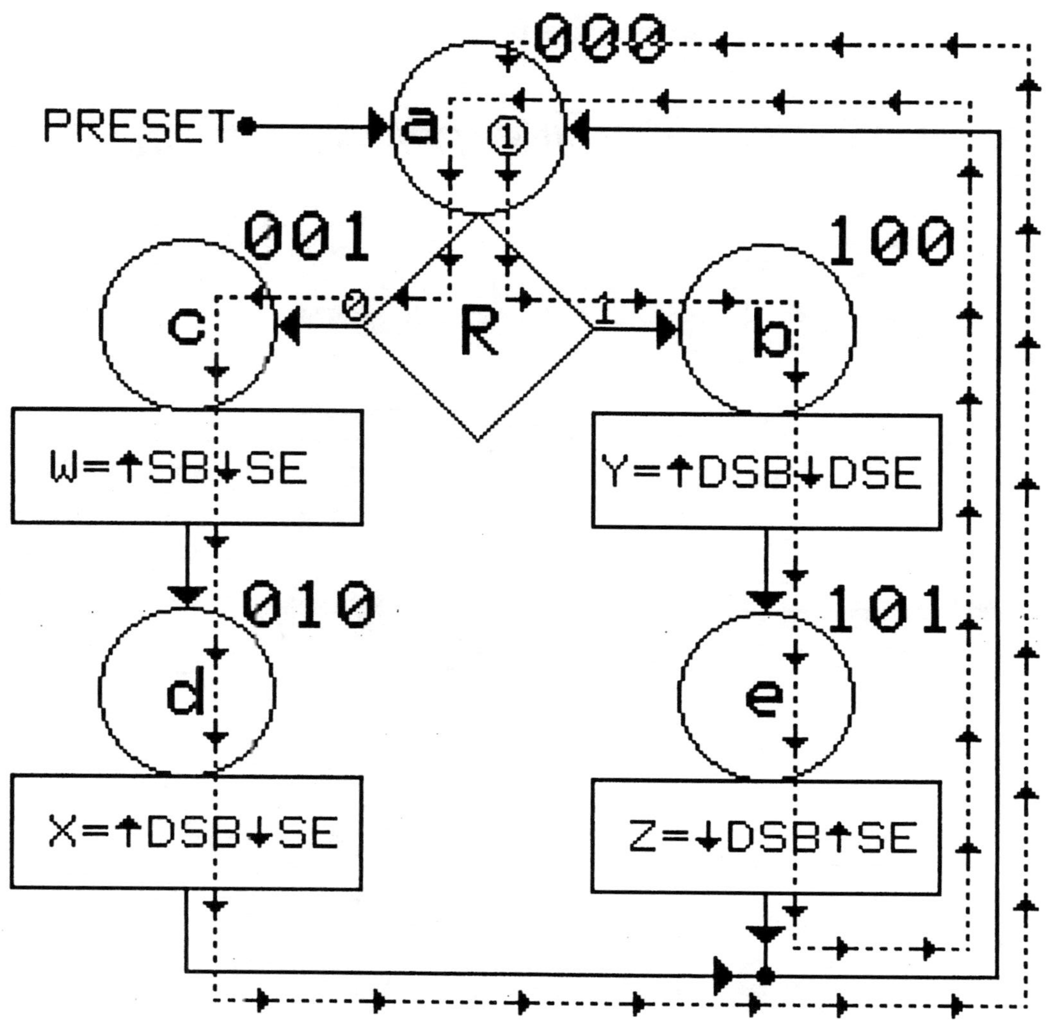

FIGURE 4.18 Practical Example 4.1, parsing algorithm.

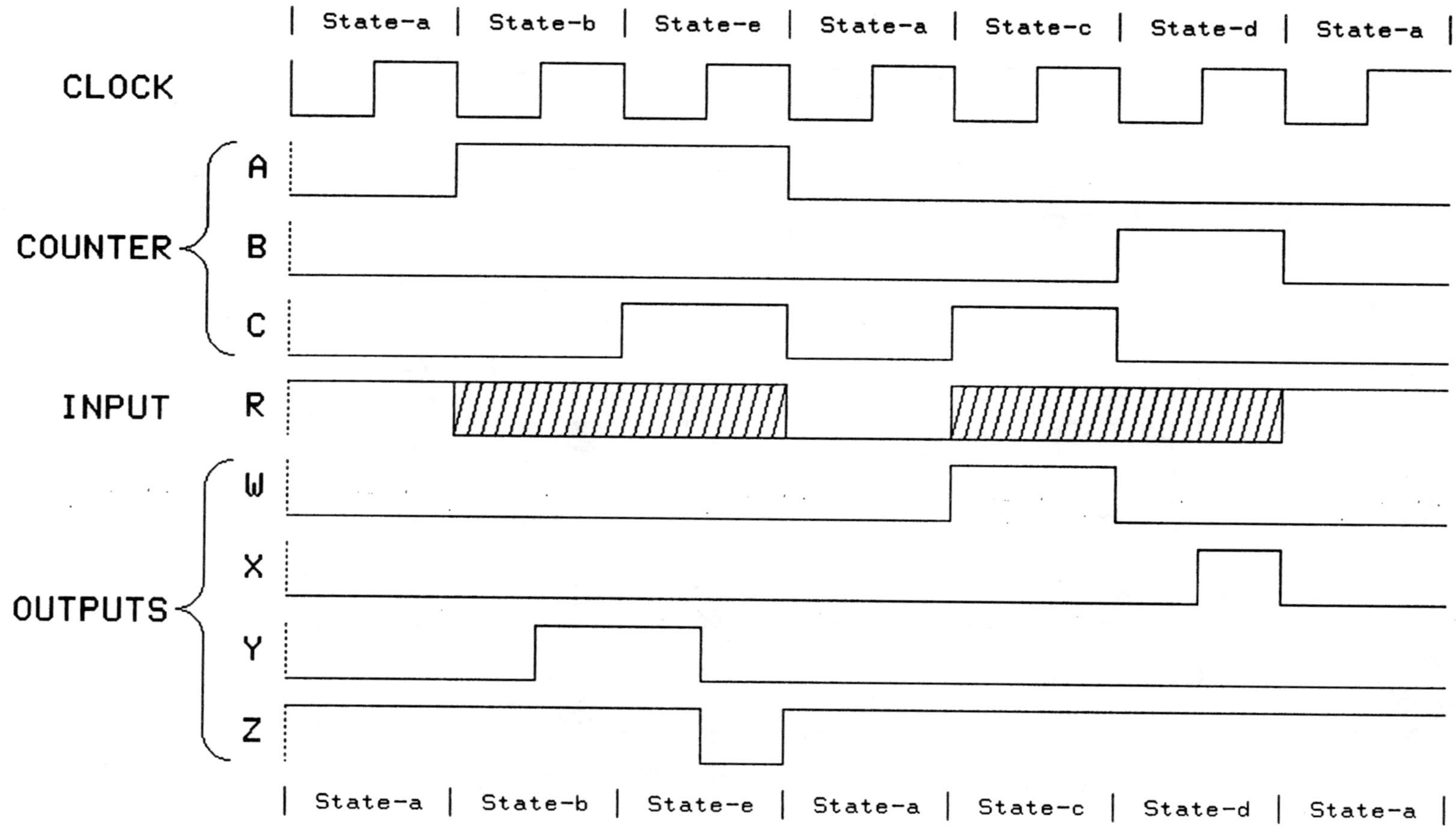

FIGURE 4.19 Practical Example 4.1, timing diagram.

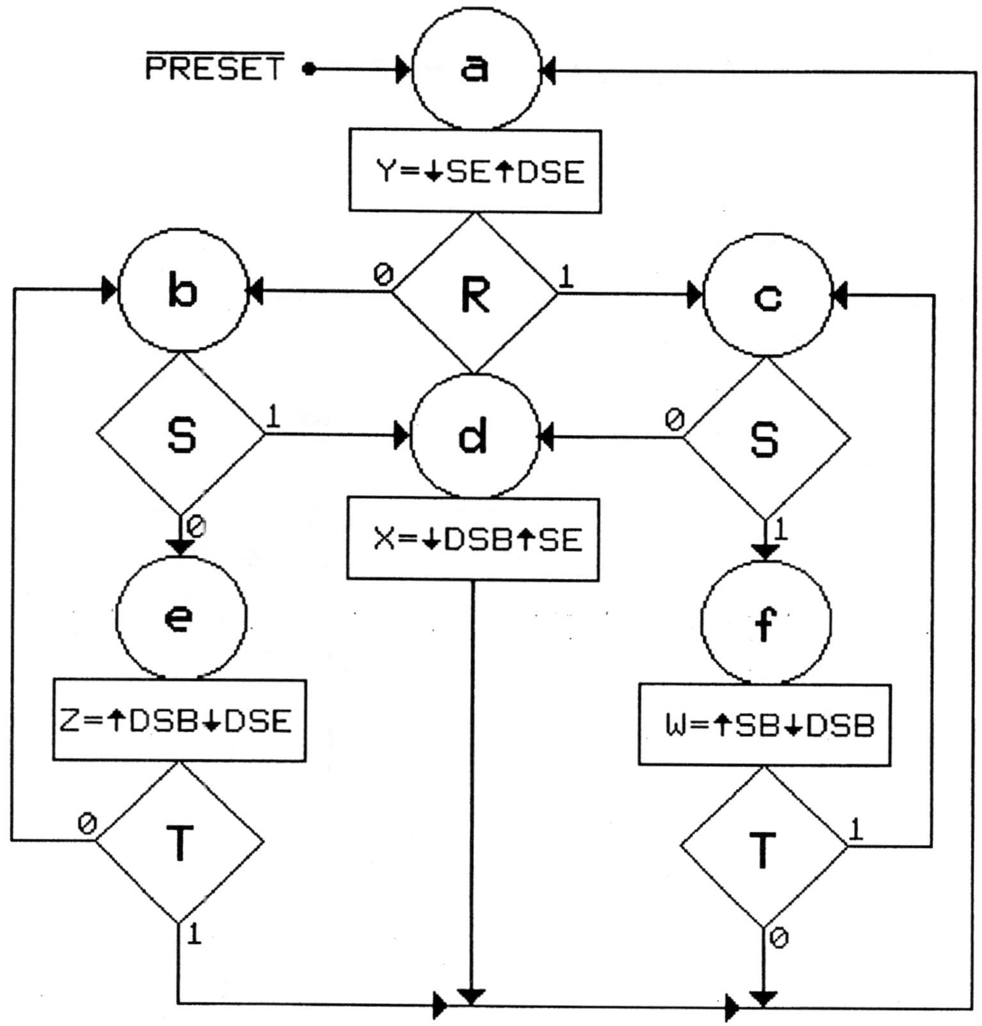

FIGURE 4.20 Practical Exercise 4.3, state diagram.

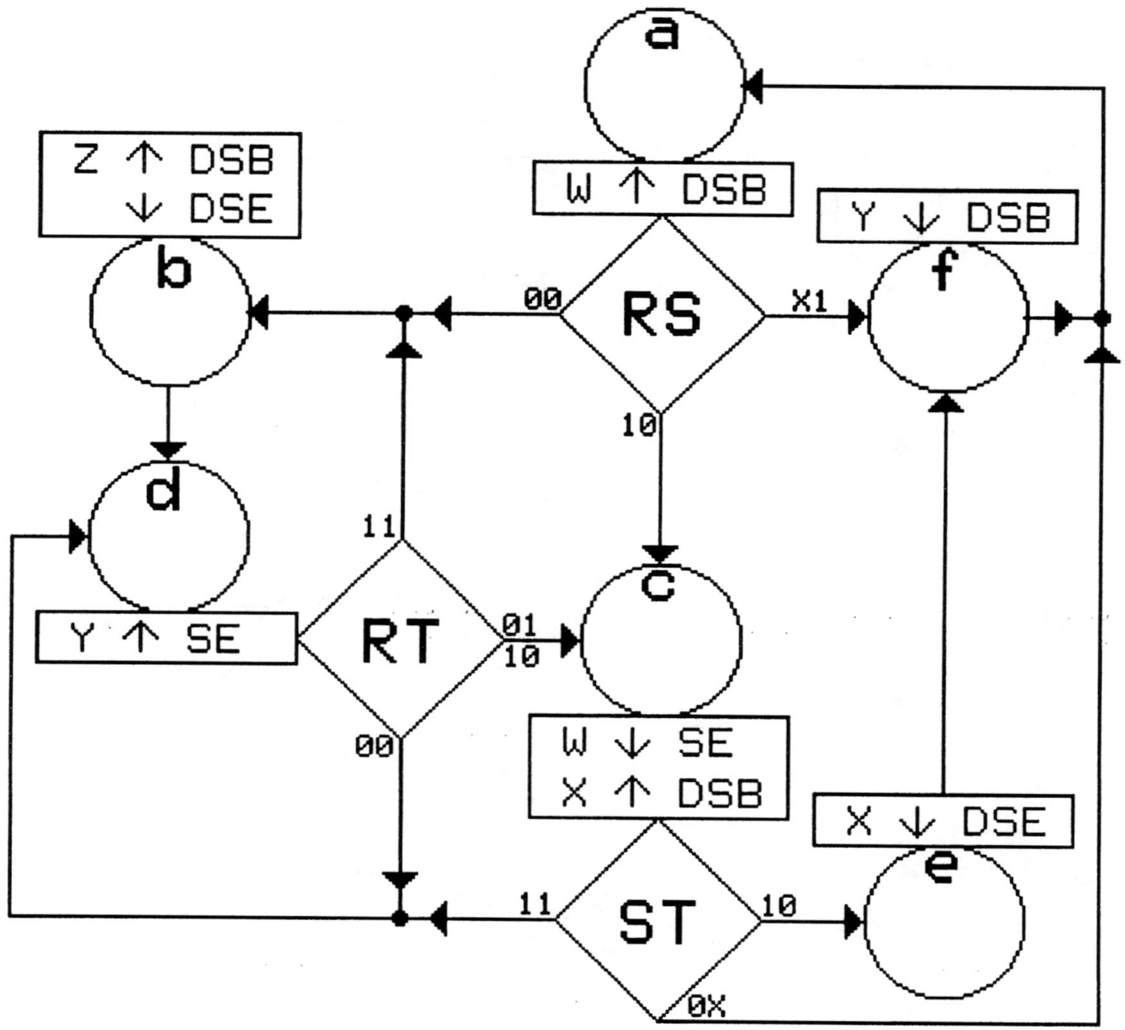

FIGURE 4.21 Practical Example 4.4, state diagram.

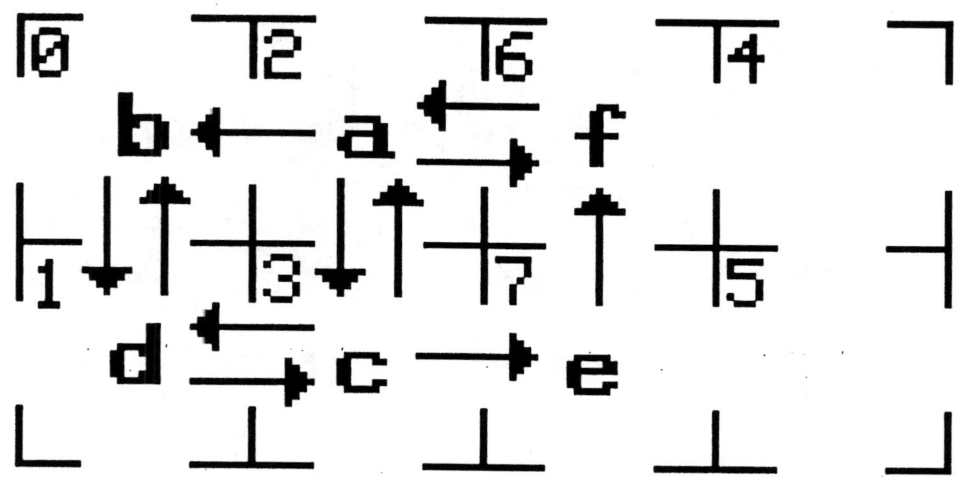

FIGURE 4.22 Practical Example 4.4, next state map.

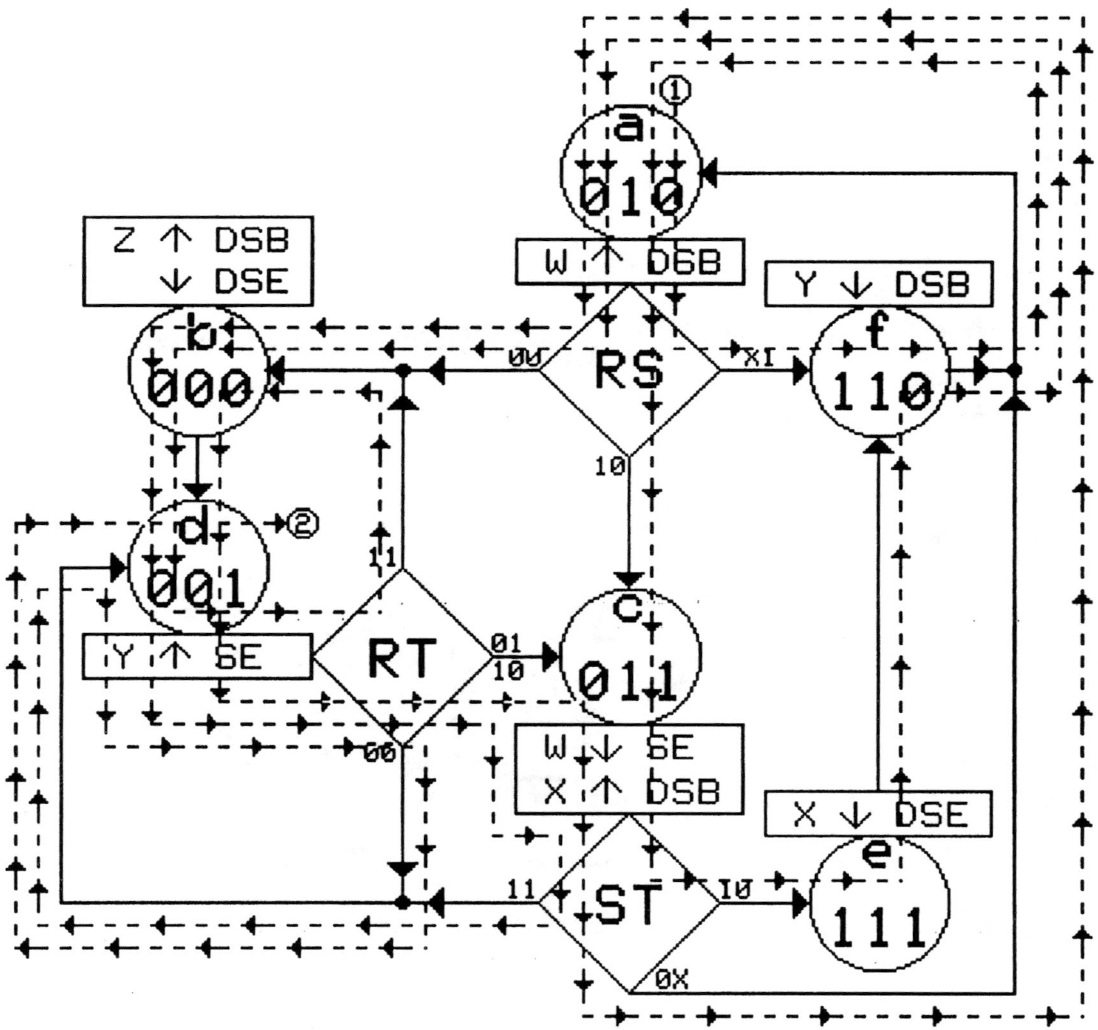

FIGURE 4.23 Practical Example 4.4, parsing algorithm.

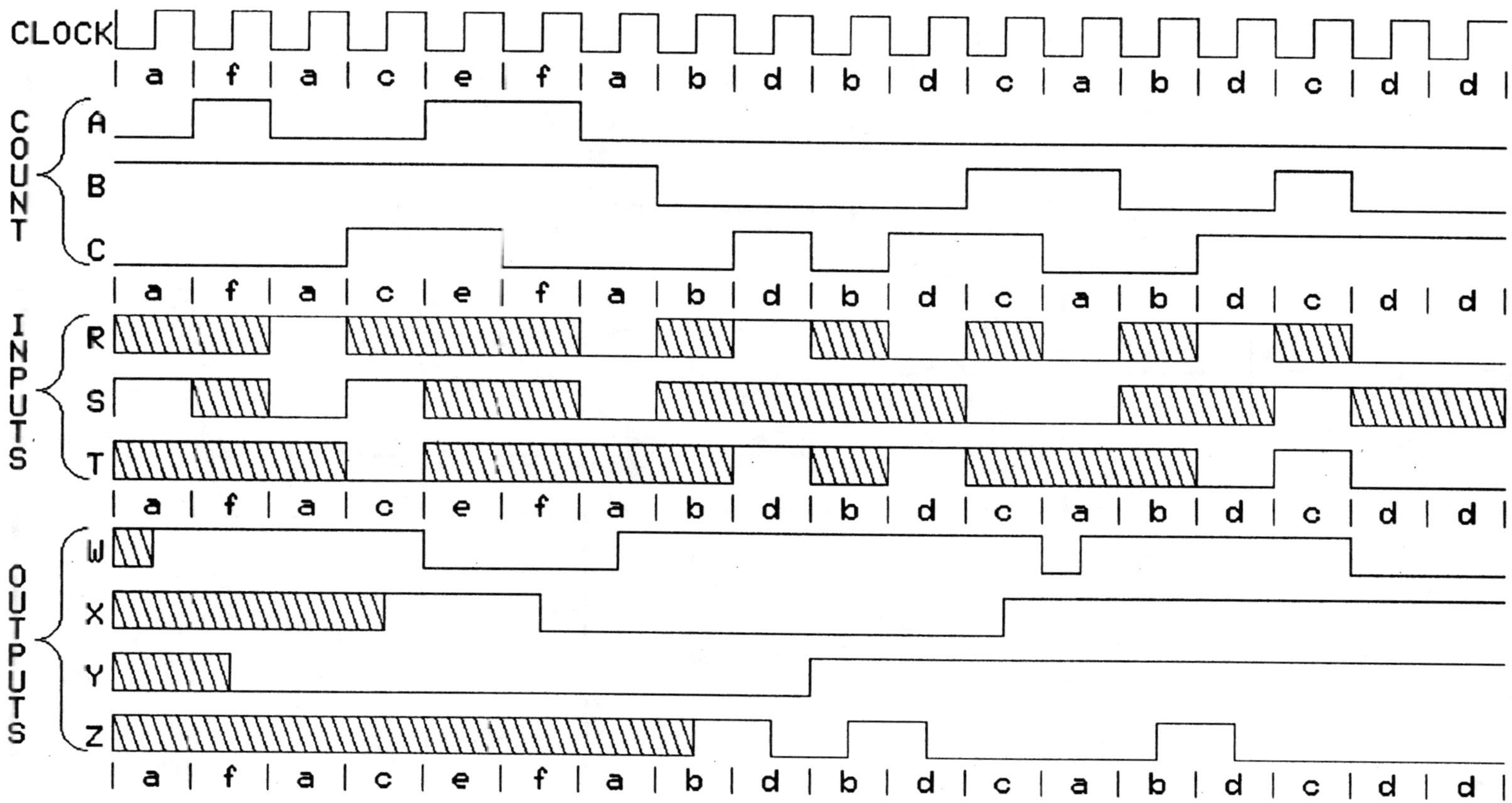

FIGURE 4.24 Practical Example 4.4, timing diagram.

	ABC RST ABC	DA	DB	DC	W	X	Y	Z
b	000 XXX 001	0	0	1				Z↑DSB↓DSE
d	001 0X0 001 001 0X1 011 001 1X0 011 001 1X1 000	0	$\bar{R}T+R\bar{T}$	$\bar{R}+\bar{T}$			Y↑SE	
a	010 00X 000 010 X1X 110 010 10X 011	S	R+S	$R\bar{S}$	W↑DSB			
c	011 X0X 010 011 X10 111 011 X11 001	$S\bar{T}$	$\bar{S}+\bar{T}$	S	W↓SE	X↑DSB		
f	110 XXX 010	0	1	0			Y↓DSB	
e	111 XXX 110	1	1	0		X↓DSE		

FIGURE 4.25 Practical Example 4.4, state table.

	ABC RST ABC	DA	DB	DC
d	001 0X0 0**01** 001 0X1 0**11** 001 1X0 0**11** 001 1X1 0**00**	0	$\bar{R}T+R\bar{T}$	$\bar{R}+\bar{T}$

FIGURE 4.26 State table segment for State-d.

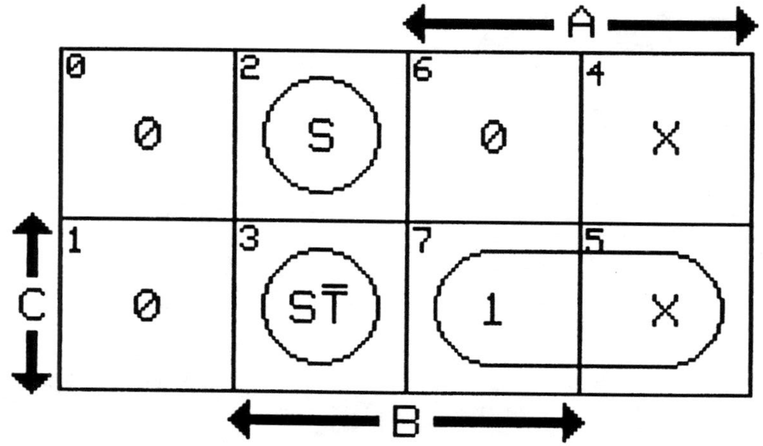

FIGURE 4.27 Practical Example 4.4, karnaugh map for function D_A.

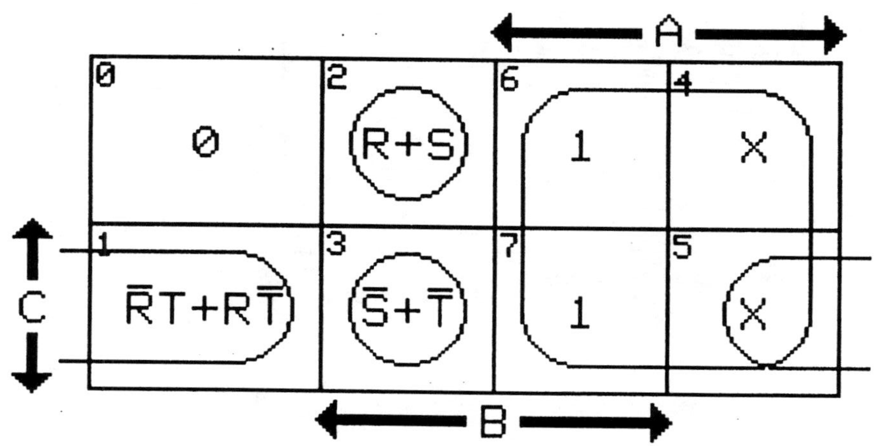

FIGURE 4.28 Practical Example 4.4, karnaugh map for function D_B.

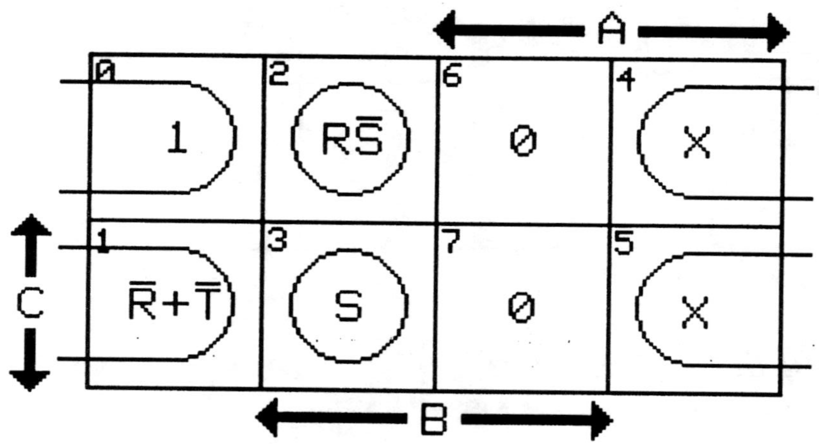

FIGURE 4.29 Practical Example 4.4, karnaugh map for function D_C.

	ABC	W	X	Y	Z
b	000				Z↑DSB↓DSE
d	001 001 001 001			Y↑SE	
a	010 010 010	W↑DSB			
c	011 011 011	W↓SE	X↑DSB		
f	110			Y↓DSB	
e	111		X↓DSE		

FIGURE 4.30 *Practical Example 4.4, OFL portion of state table.*

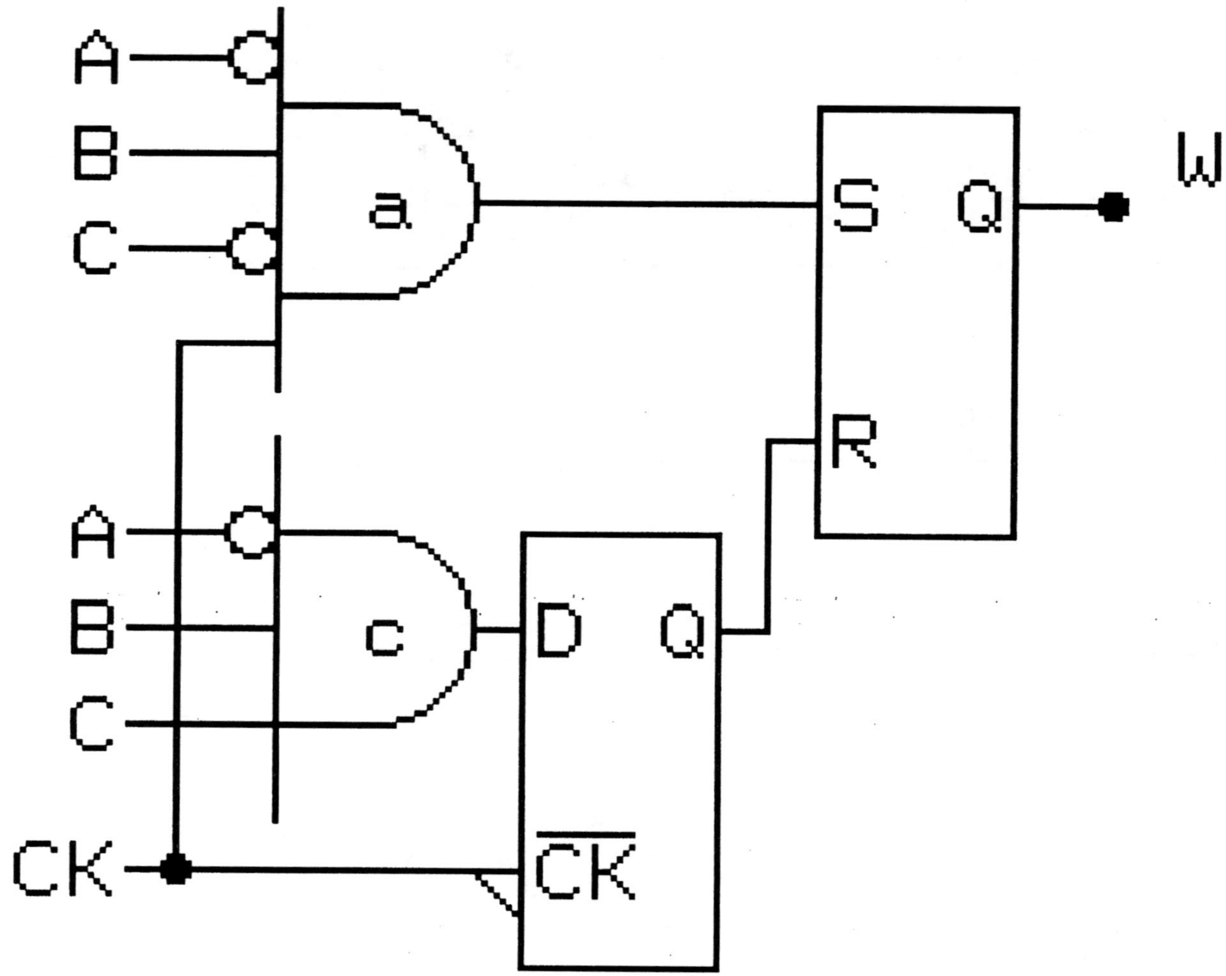

FIGURE 4.31 Practical Example 4.4, W output circuit.

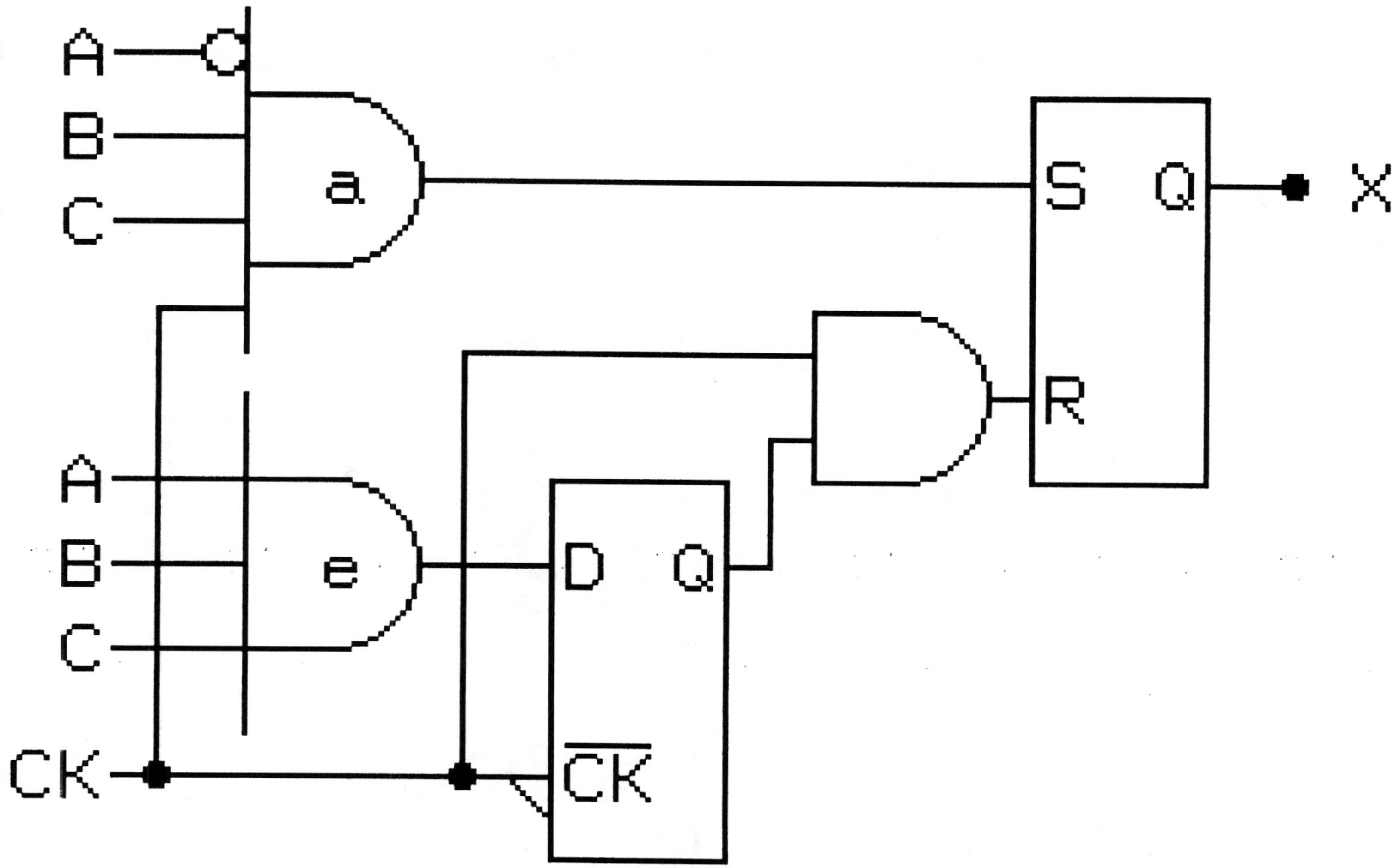

FIGURE 4.32 Practical Example 4.4, X output circuit.

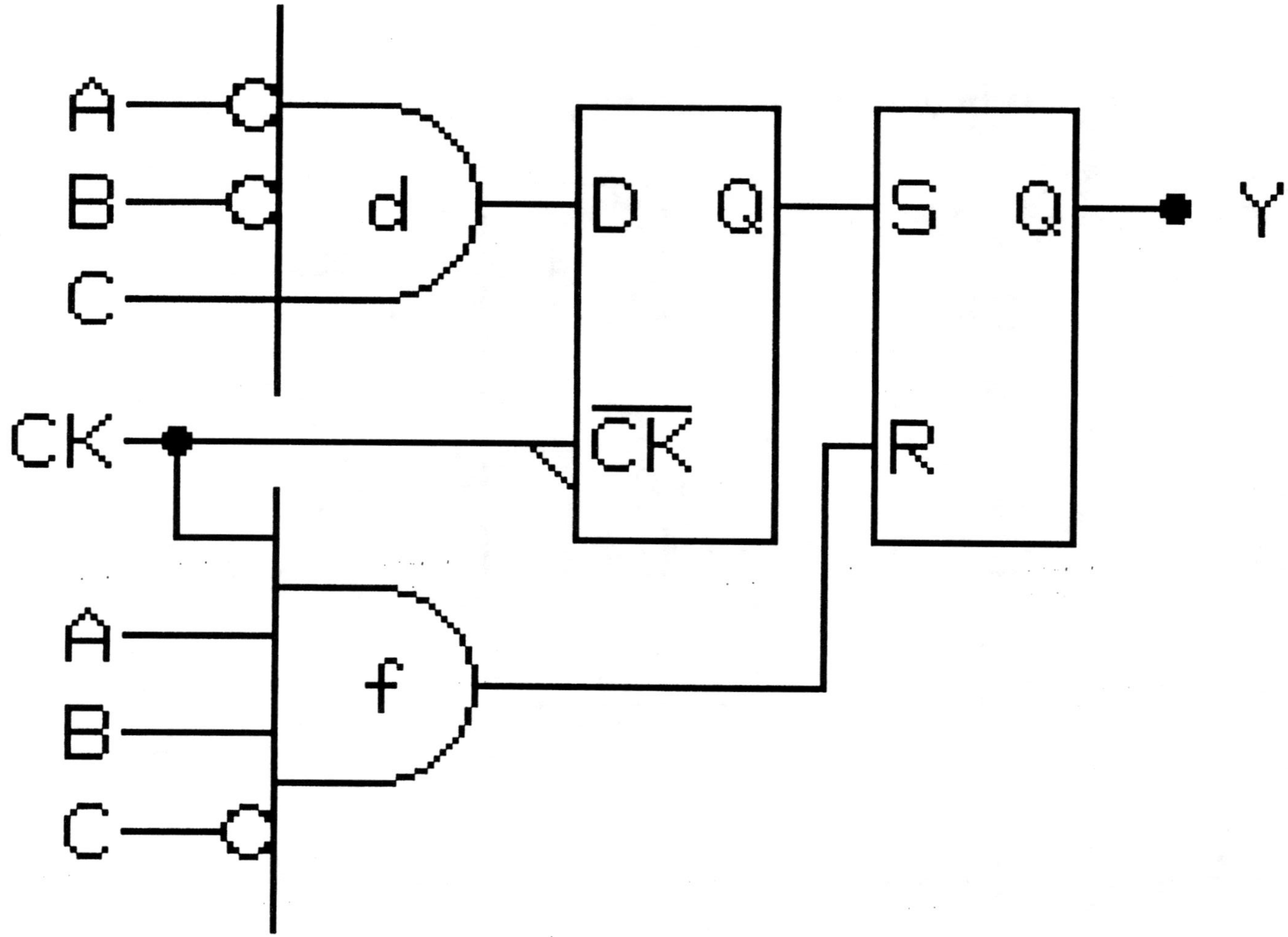

FIGURE 4.33 Practical Example 4.4, Y output circuit.

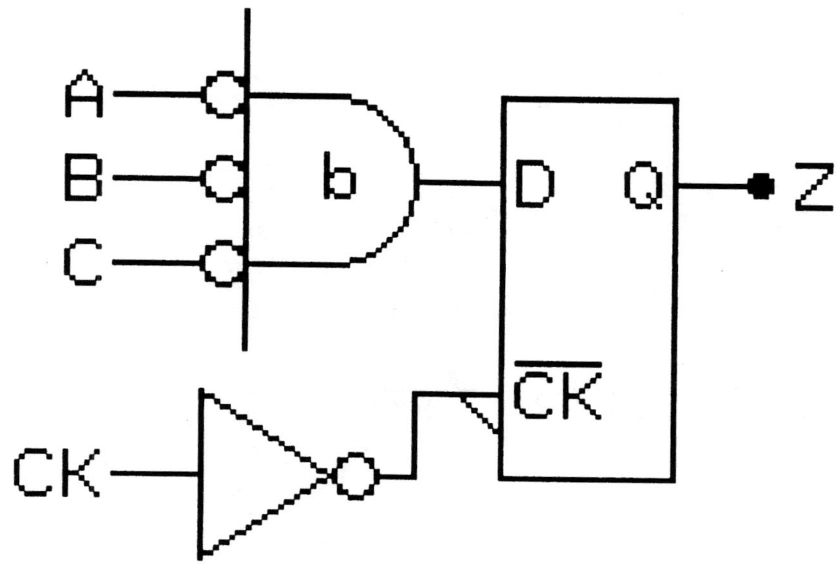

FIGURE 4.34 Practical Example 4.4, Z output circuit.

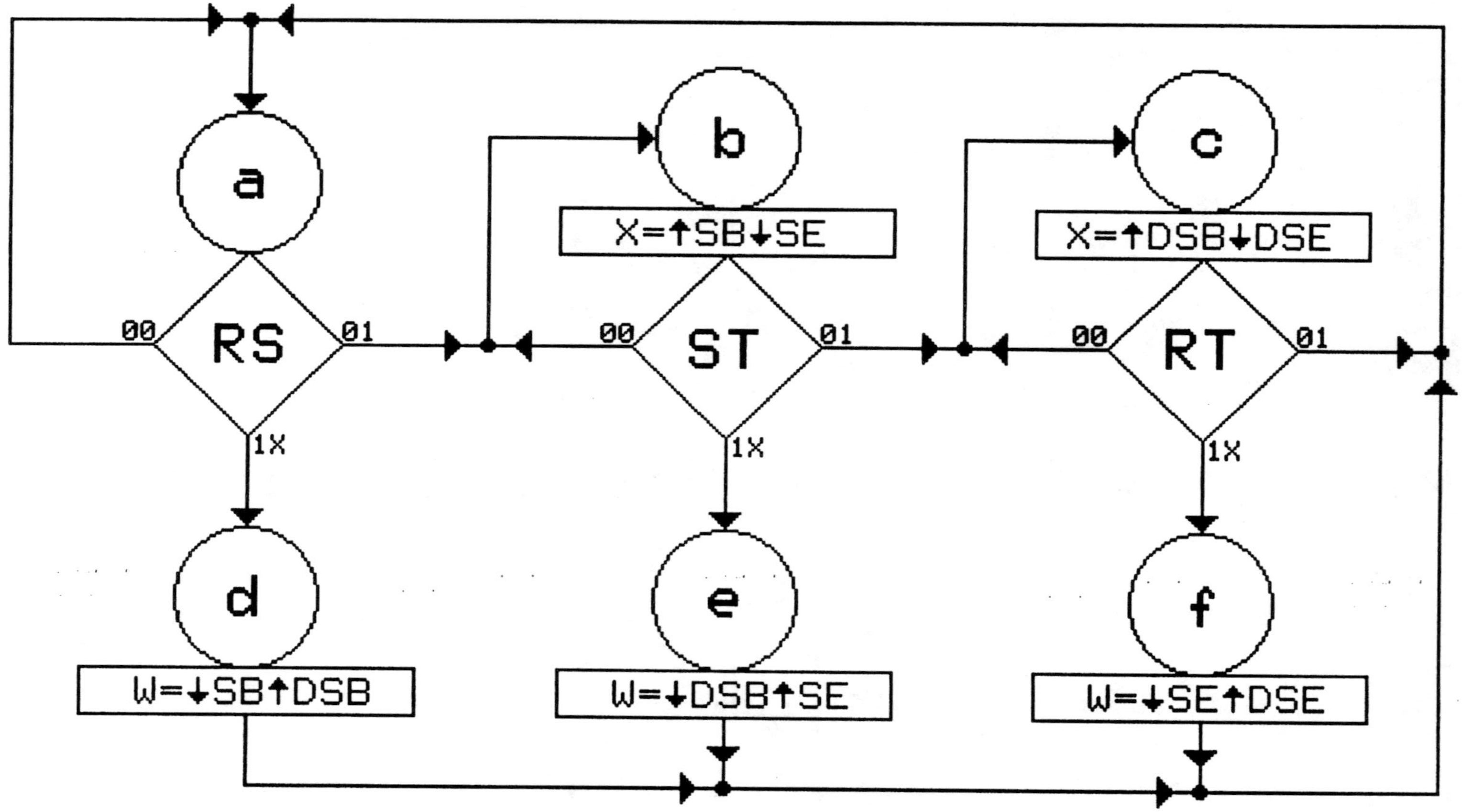

FIGURE 4.35 Practical Example 4.4, state diagram.

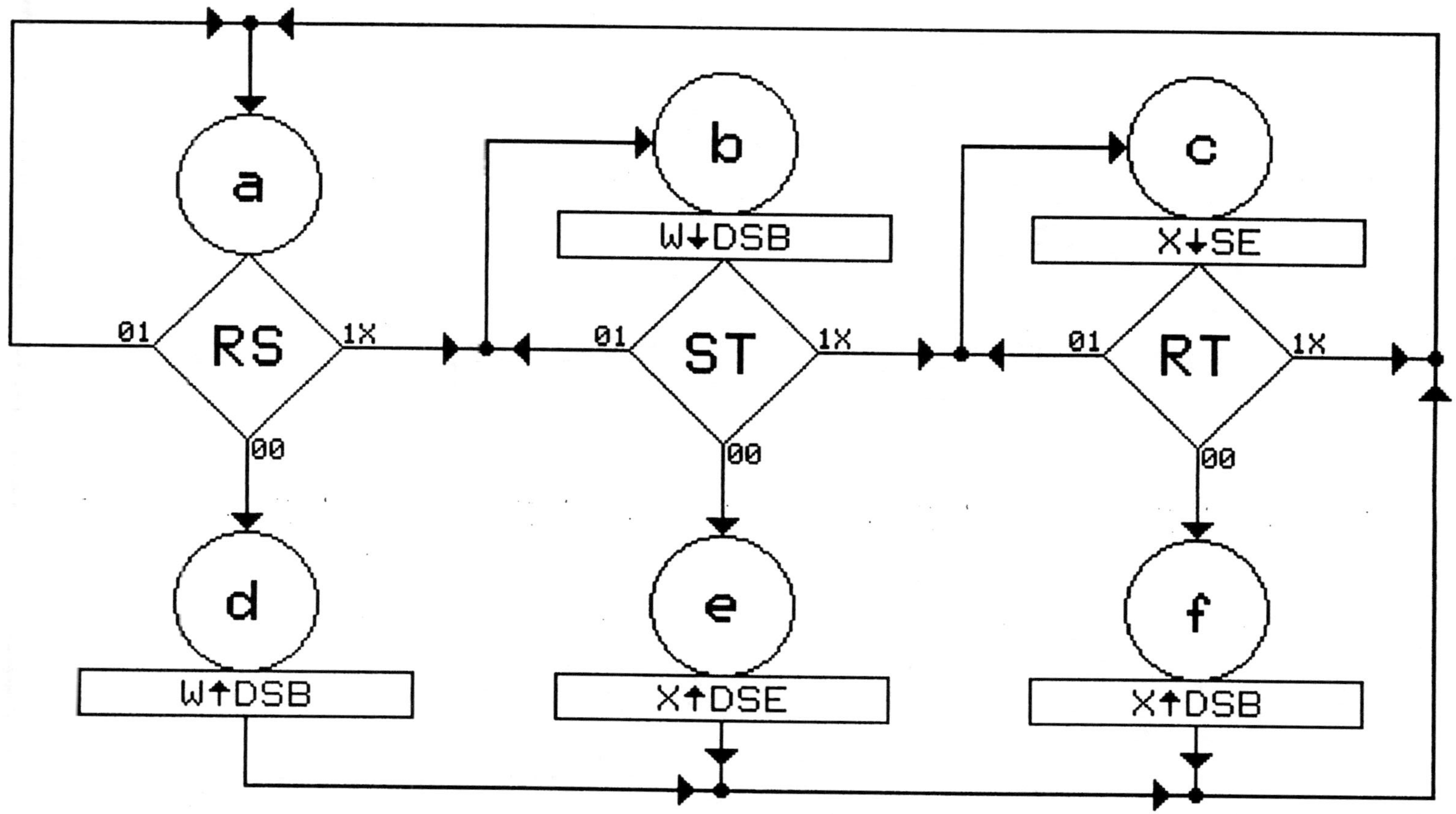

FIGURE 4.36 Practical Example 4.5, state diagram.

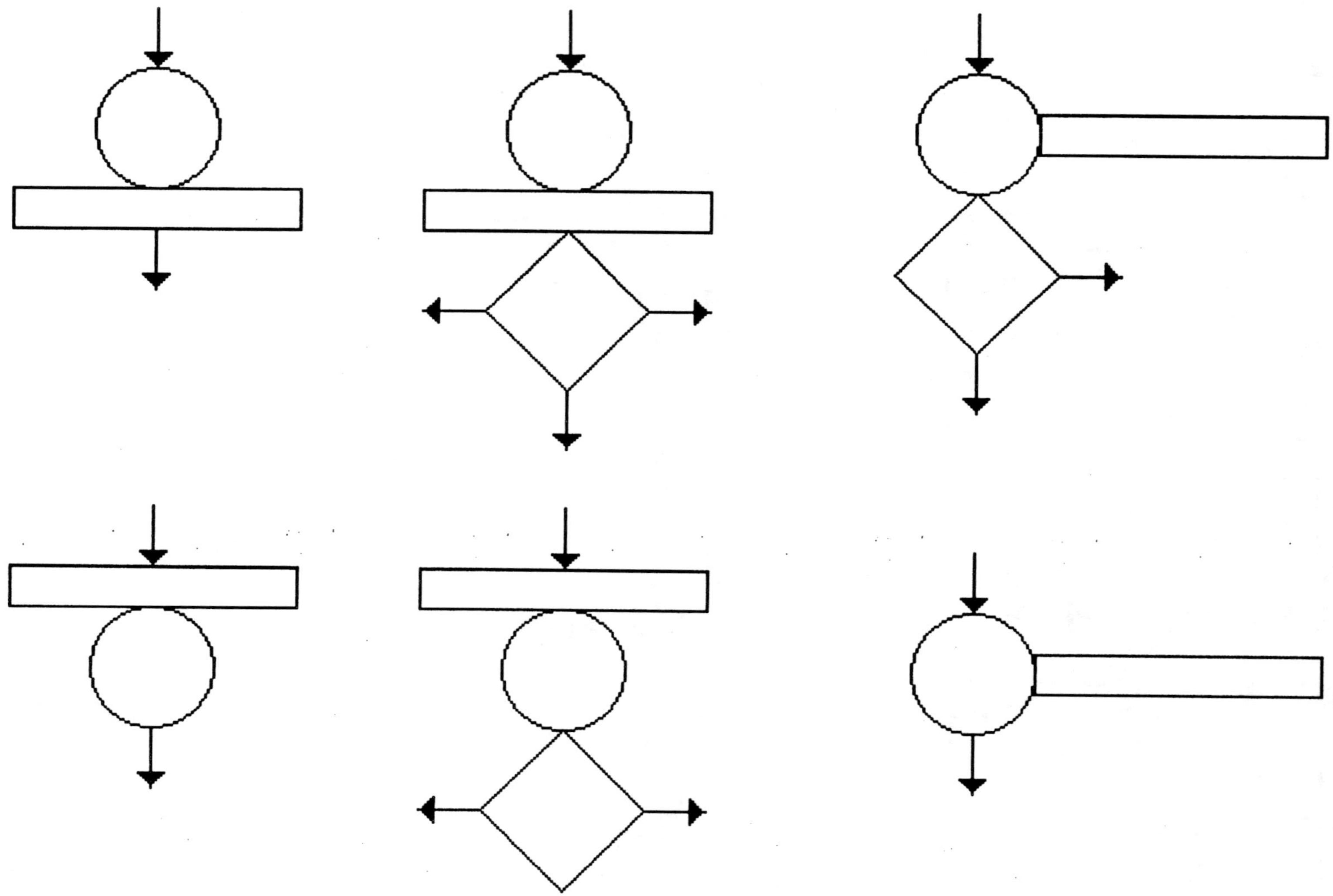

FIGURE 4.37 Unconditional output forming logic.

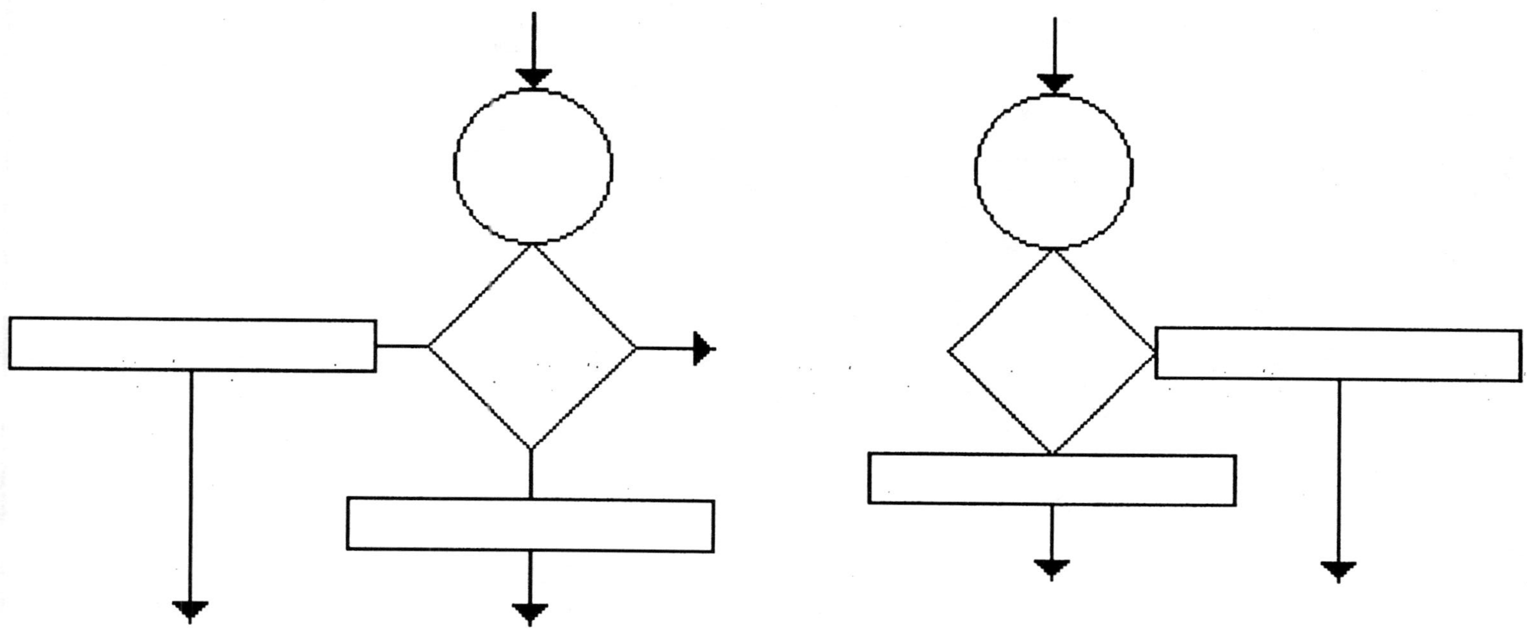

FIGURE 4.38 Conditional output forming logic.

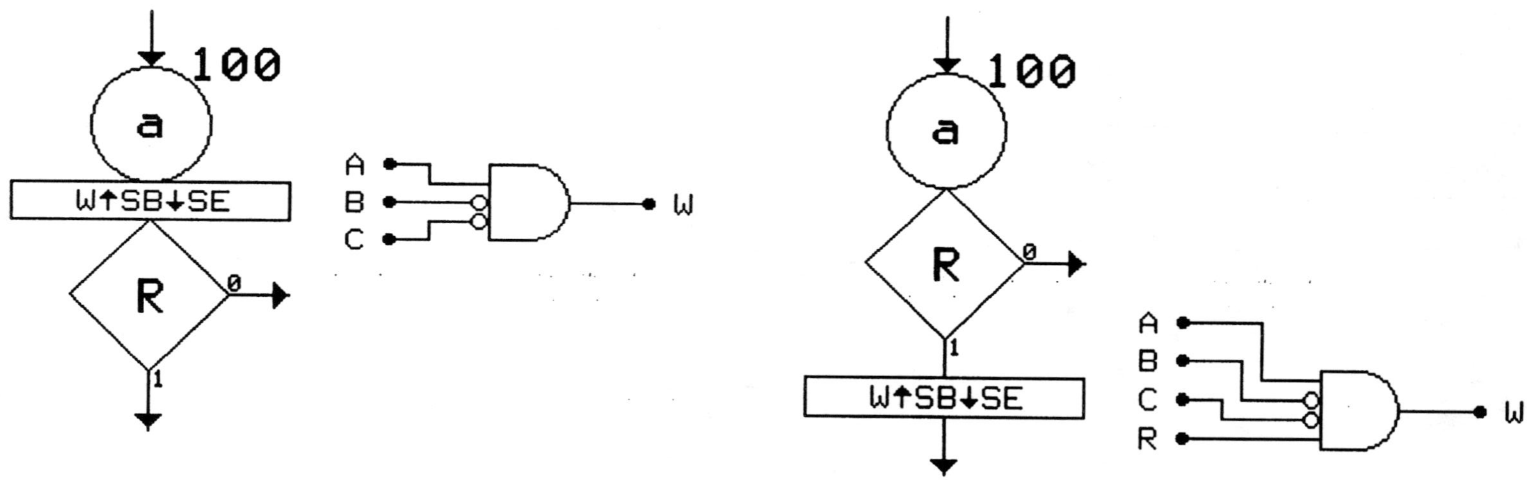

FIGURE 4.39 Unconditional/conditional OFL comparison.

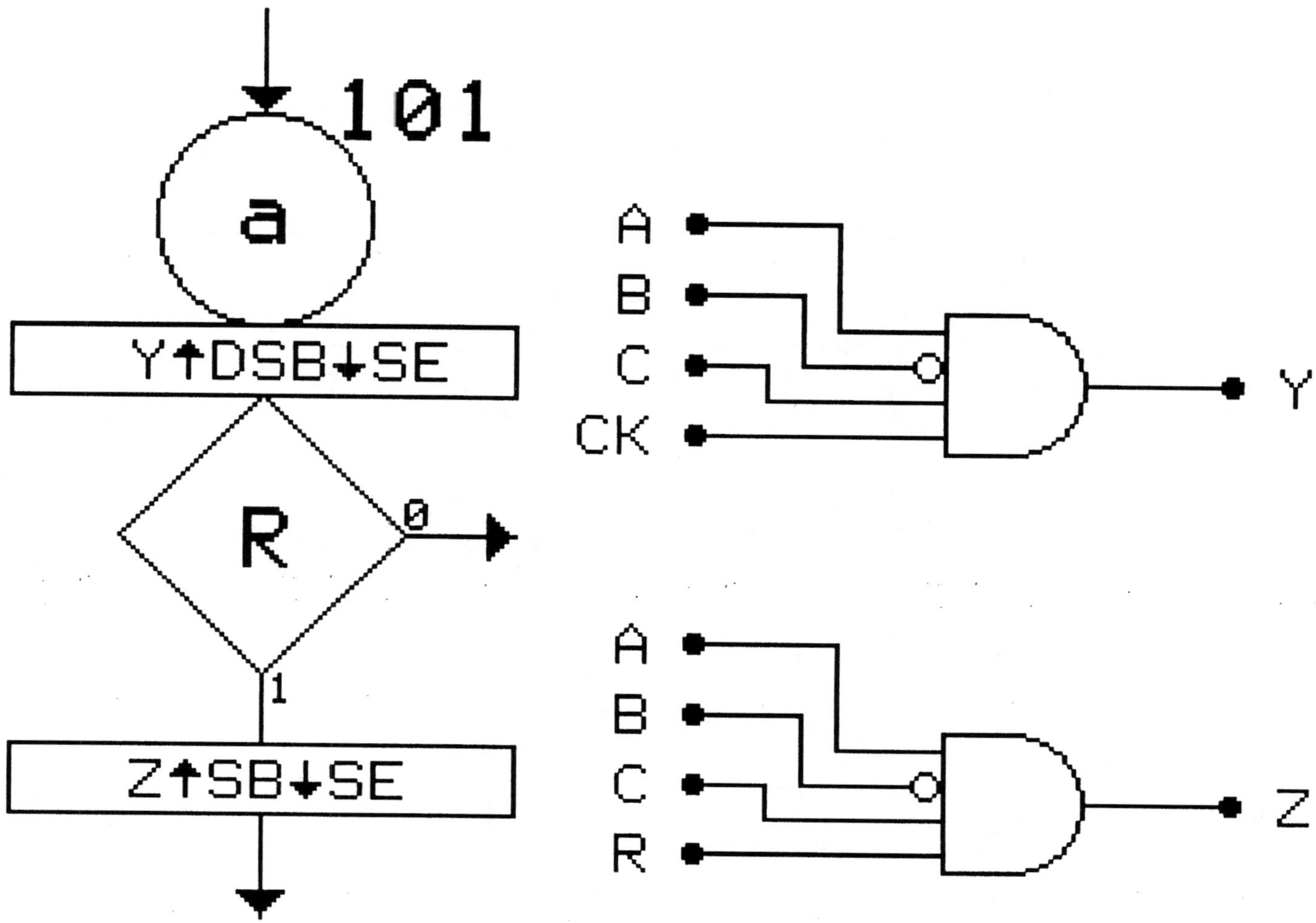

FIGURE 4.40 Mixed OFL.

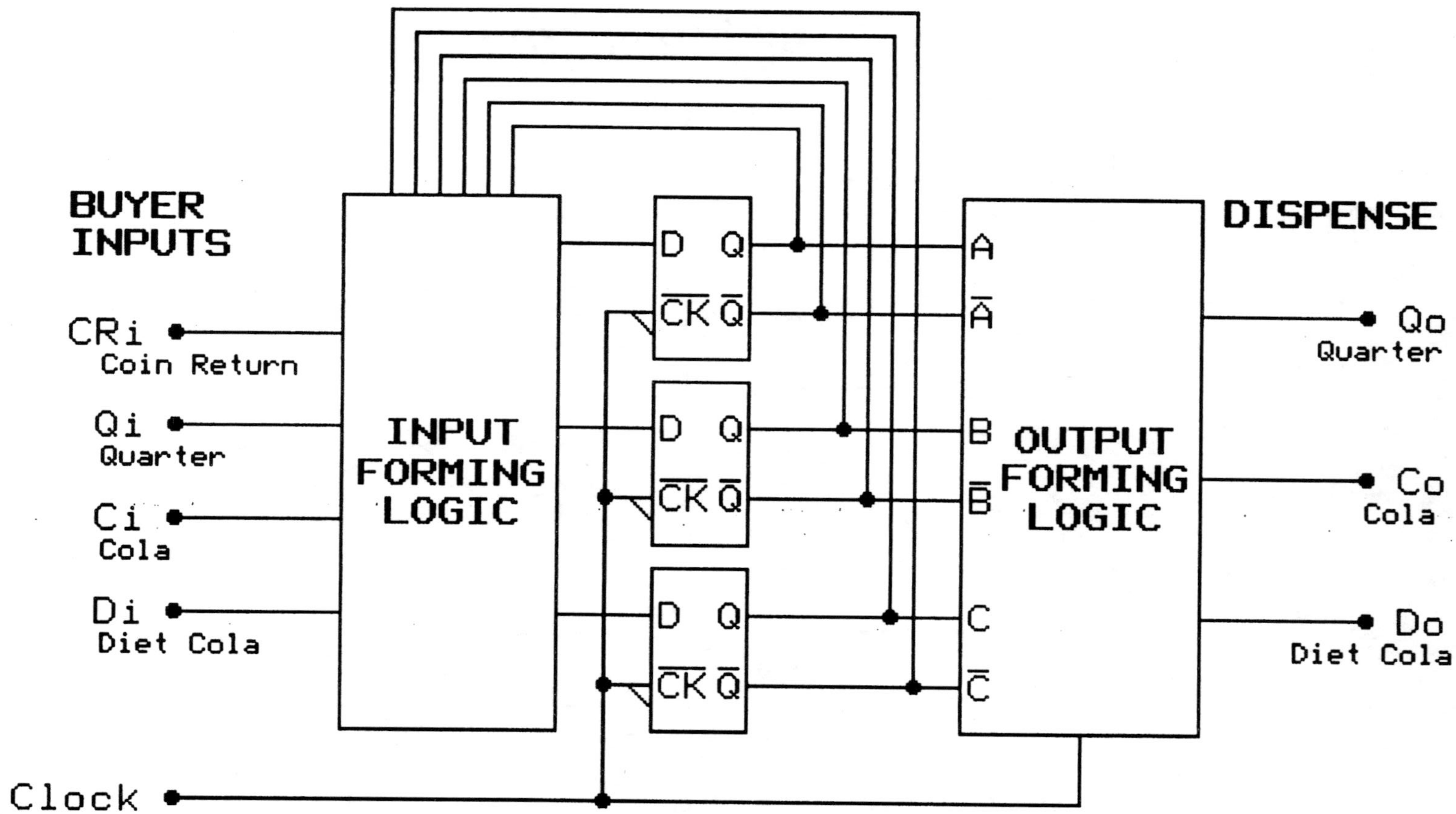

FIGURE 4.41 Practical Example 4.5, block diagram.

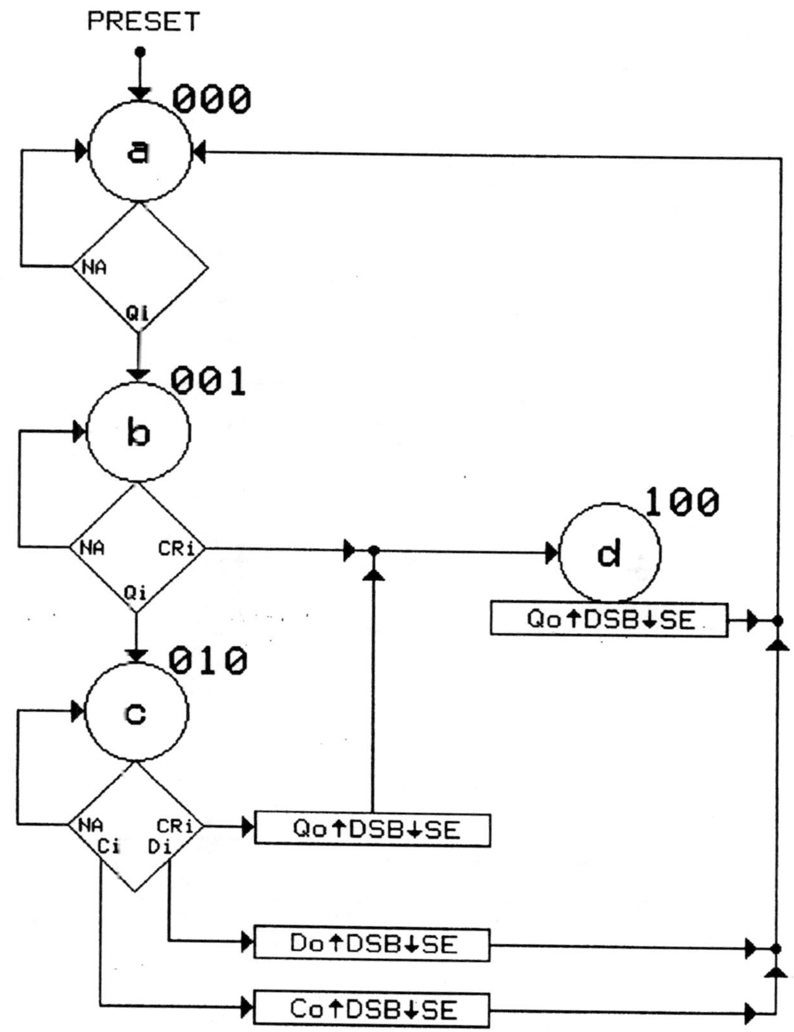

FIGURE 4.42 Practical Example 4.5, state diagram.

	PRES ABC	INPUTS Qi	Cri	Ci	Di	NEXT ABC	D_A	D_B	D_C	Qo	Co	Do
a	ØØØ	Ø	X	X	X	ØØØ	Ø	Ø	Q	Ø	Ø	Ø
		1	X	X	X	ØØ1						
b	ØØ1	Ø	Ø	X	X	ØØ1	Cr	Q	$\overline{Q}\bullet\overline{Cr}$	Ø	Ø	Ø
		1	X	X	X	Ø1Ø						
		X	1	X	X	1ØØ						
c	Ø1Ø	X	Ø	Ø	Ø	Ø1Ø	Cr	$\overline{Cr}\bullet\overline{C}\bullet\overline{D}$	Ø	Ø	Ø	Ø
		X	1	X	X	1ØØ				↑DSB↓SE	Ø	Ø
		X	X	1	X	ØØØ				Ø	↑DSB↓SE	Ø
		X	X	X	1	ØØØ				Ø	Ø	↑DSB↓SE
	Ø11	X	X	X	X	XXX	X	X	X	X	X	X
d	1ØØ	X	X	X	X	ØØØ	Ø	Ø	Ø	↑DSB↓SE	Ø	Ø
	1Ø1	X	X	X	X	XXX	X	X	X	X	X	X
	11Ø	X	X	X	X	XXX	X	X	X	X	X	X
	111	X	X	X	X	XXX	X	X	X	X	X	X

FIGURE 4.43 Practical Example 4.5, state table.

$$D_A = (B \bullet Cr) + (C \bullet Cr)$$

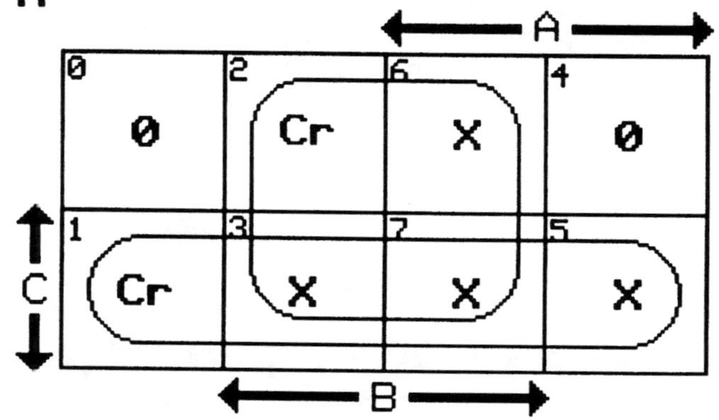

FIGURE 4.44 ***Practical Example 4.5, input function D_A.***

$$D_B = (B \bullet \overline{Cr} \bullet \overline{C} \bullet \overline{D}) + (C \bullet Q)$$

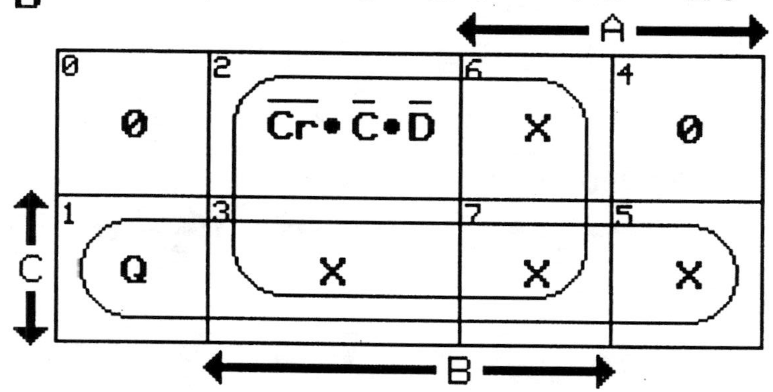

FIGURE 4.45 ***Practical Example 4.5, input function D_B.***

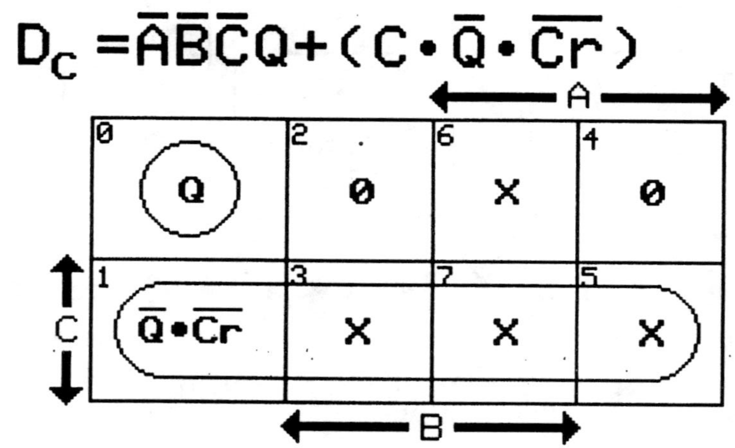

FIGURE 4.46 Practical Example 4.5, input function D_C.

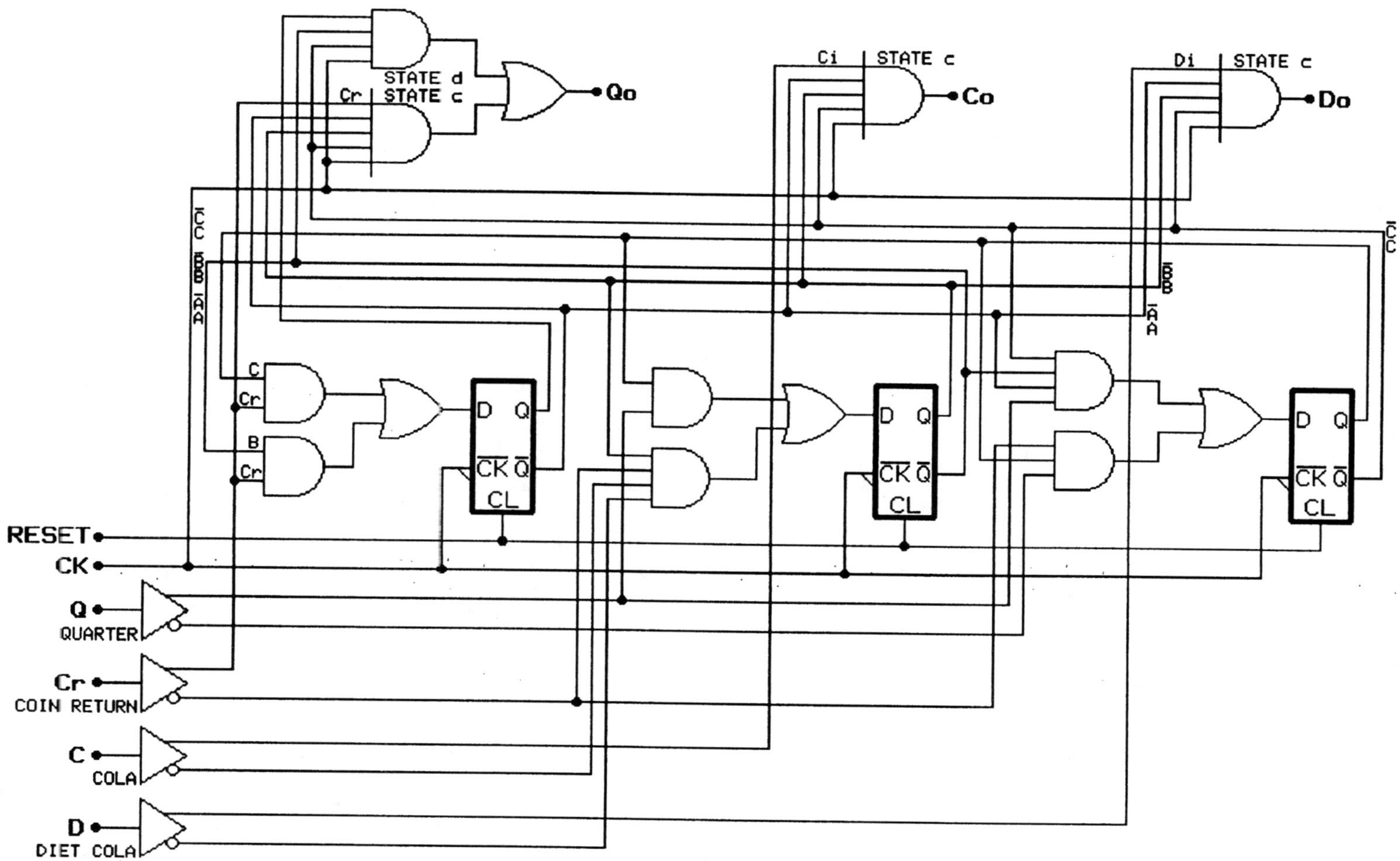

FIGURE 4.47 Practical Example 4.5, schematic diagram.

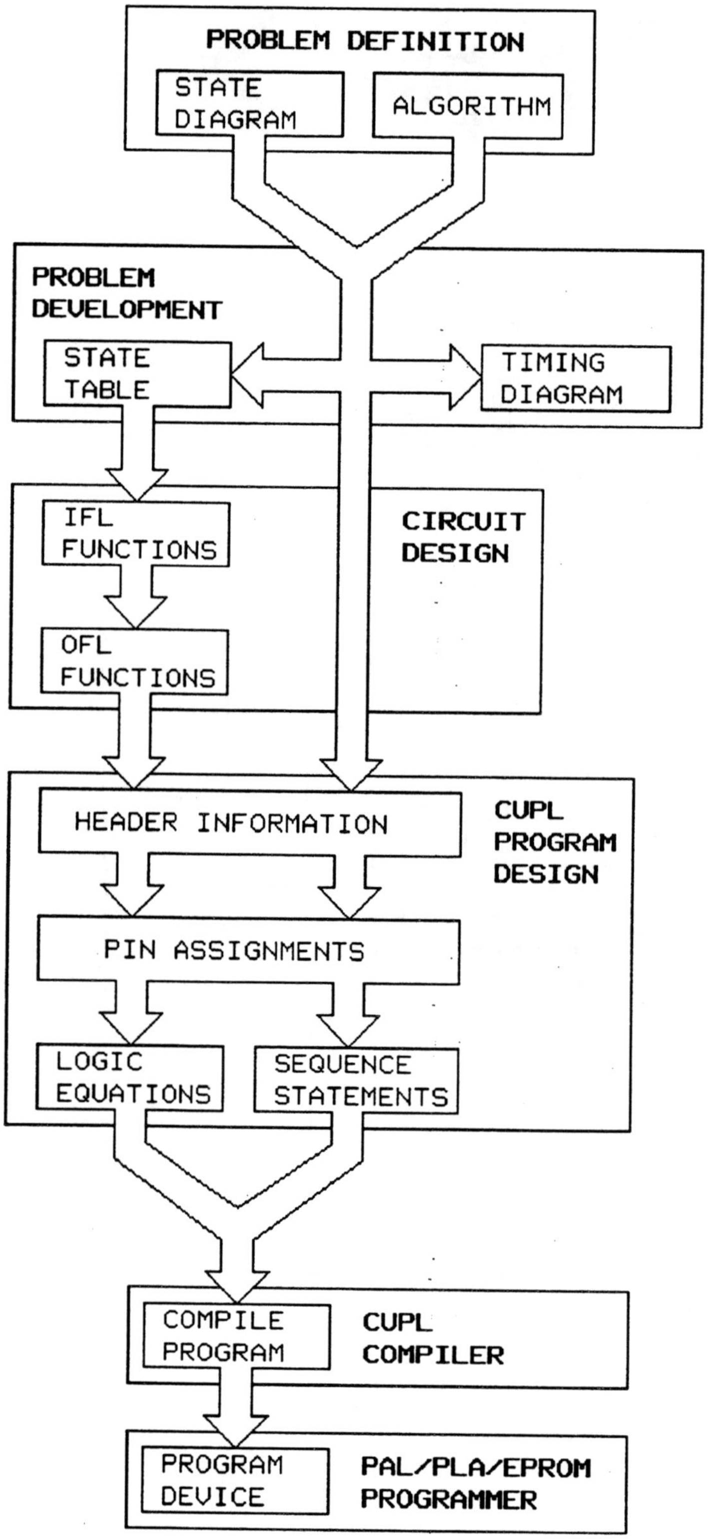

FIGURE 5.1 Programmable logic development sequence.

```
/****************************************************************/
/* Header Information                                           */
/****************************************************************/
Name            Gates;
Partno          CA0001;
Revision        04;
Date            9/12/89;
Designer        G. Woolhiser;
Company         Logical Devices, Inc.;
Location        None;
Assembly        None;
Device          P16L8;

/****************************************************************/
/*      This is a example to demonstrate how CUPL compiles      */
/*      simple gates.                                           */
/*      Target Devices: P16L8, P16LD8, P16P8, EP300, and 82S153 */
/****************************************************************/

/****************************************************************/
/* Pin Assignments                                              */
/****************************************************************/

/* Inputs:  define inputs to build simple gates from */

Pin 1 =  a;
Pin 2 =  b;
```

```
/* Outputs:  define outputs as active HI levels
 * Note: For PAL16L8 and PAL16LD8, DeMorgan's Theorem is applied
 * to invert all outputs due to fixed inverter in the device.  */

Pin 12 = inva;
Pin 13 = invb;
Pin 14 = and;
Pin 15 = nand;
Pin 16 = or;
Pin 17 = nor;
Pin 18 = xor;
Pin 19 = xnor;

/*****************************************************************/
/* Logic Expressions                                             */
/*****************************************************************/

inva = !a;               /* inverters */
invb = !b;
and  = a & b;            /* and gate */
nand = !(a & b);         /* nand gate */
or   = a # b;            /* or gate */
nor  = !(a # b);         /* nor gate */
xor  = a $ b;            /* exclusive or gate */
xnor = !(a $ b);         /* exclusive nor gate */
```

FIGURE 5.2 CUPL program, GATES.PLD.

APPEND	DEVICE	IF	OUT	SEQUENCED
ASSEMBLY	ELSE	JUMP	PARTNO	SEQUENCEJK
ASSY	FIELD	LOC	PIN	SEQUENCERS
COMPANY	FLD	LOCATION	PINNODE	SEQUENCET
CONDITION	FORMAT	MACRO	PRESENT	TABLE
DATE	FUNCTION	MIN	REV	
DEFAULT	FUSE	NAME	REVISION	
DESIGNER	GROUP	NODE	SEQUENCE	

TABLE 5.1 CUPL reserved keywords.

&	#	(	)	-
*	+	[	]	/
:	.	..	/*	*/
;	,	!	'	=
@	$	^		

TABLE 5.2 CUPL reserved symbols.

Operator	Character	Precedence
NOT	!	1
AND	&	2
OR	#	3
XOR	$	4

TABLE 5.3 CUPL logical operators.

Keyword	Description
NAME	The program filename. When using MS-DOS systems, use a valid filename without the extension. That is, limit the filename to eight alphanumeric characters. Do not include a filename extension as the extension for the different files associated with the program vary.
PARTNO	The company's proprietary part number, (not the device number).
REVISION	Start with 01 and increment each time the program is altered.
DATE	Change to the current date each time the program is altered.
DESIGNER	The designer's name
COMPANY	The company name
ASSEMBLY	Identify the assembly on which the PLD will be used.
LOCATION	Identify the PC board reference or board location of the device.
DEVICE	Set the default device type for the compilation. The device type set in the compiler command line overrides this line.
FORMAT	Identify the download output format used by the compiler. Valid values are: h ASCII-Hex output i Signetics HL output j JEDEC output This line overrides any output format option included in the compiler command line.

TABLE 5.4 CUPL header information keywords.

Extension	Description
.AP	Asynchronous preset of a flip-flop
.AR	Asynchronous reset of a flip-flop
.CK	Programmable clock of a flip-flop
.D	D-input of a D-type flip-flop
.DQ	Q-Output of a D-type flip-flop
.INT	Internal feedback path for registered macrocell
.IO	Pin feedback path selection
.J	J input of a JK-type flip-flop
.K	K input of a JK-type flip-flop
.L	D input of a transparent latch
.LE	Programmable latch enable
.LQ	Q Output of a transparent latch
.OE	Programmable output enable
.R	R input of a Set-Reset type flip-flop
.S	S input of a Set-Reset type flip-flop
.SP	Synchronous preset of a flip-flop
.SR	Synchronous reset of a flip-flop
.T	T input of a toggle type flip-flop

TABLE 5.5 CUPL variable extensions.

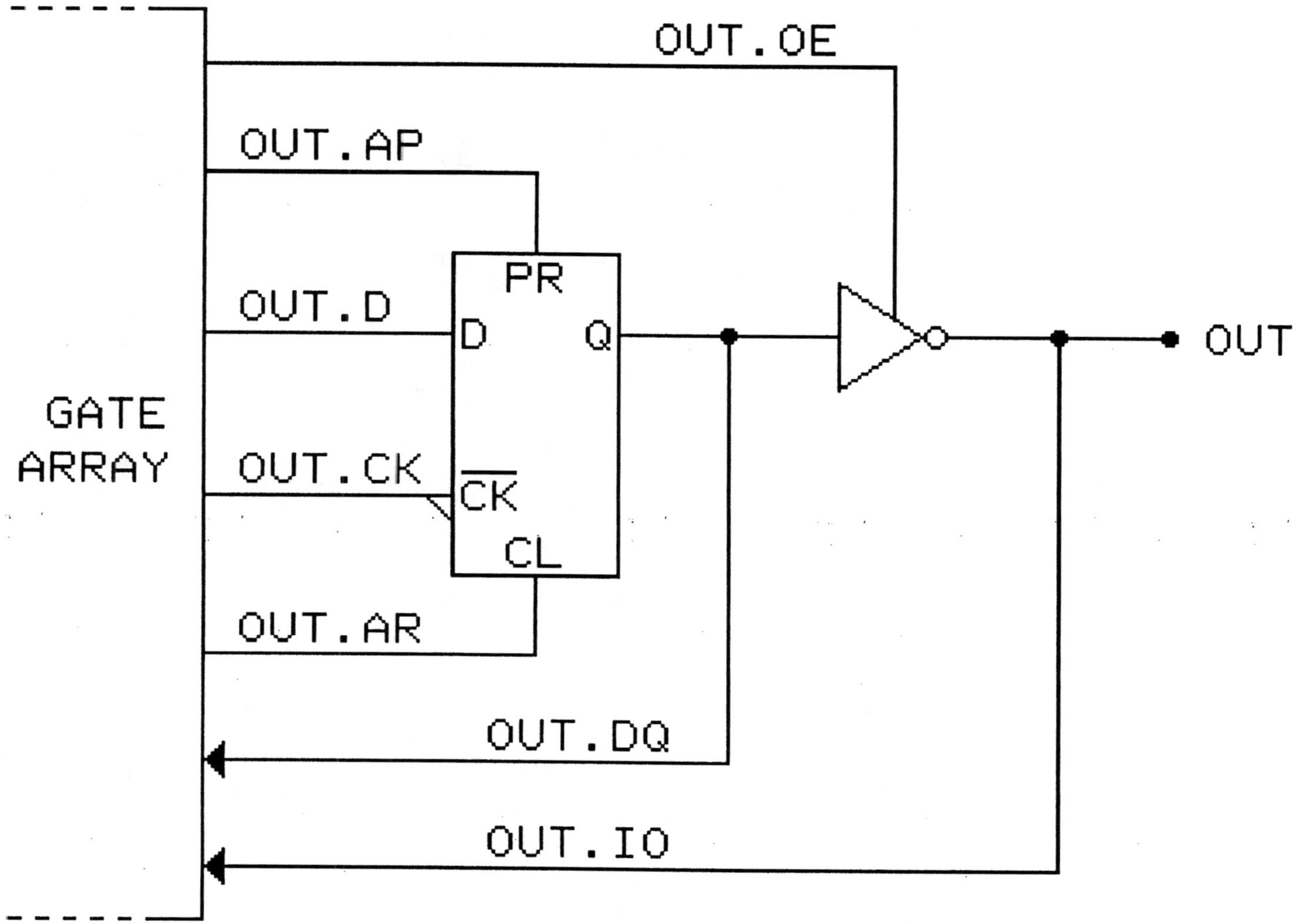

FIGURE 5.3 Typical variable extension application.

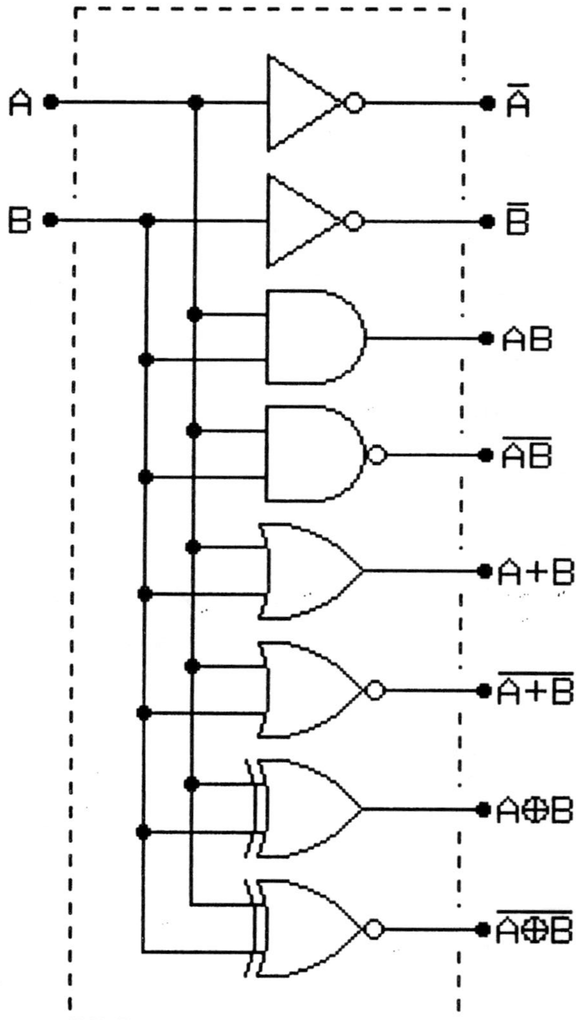

SUM OF PRODUCTS FORM	TYPOGRAPHIC FORM
$\bar{A}$	!A
$\bar{B}$	!B
AB	A&B
$\bar{A}\bar{B}+\bar{A}B+A\bar{B}$	!(A&B)
A+B	A#B
$\bar{A}\bar{B}$	!(A#B)
$\bar{A}B+A\bar{B}$	A$B
$\bar{A}\bar{B}+AB$	!(A$B)

FIGURE 5.4 Combinational logic circuits.

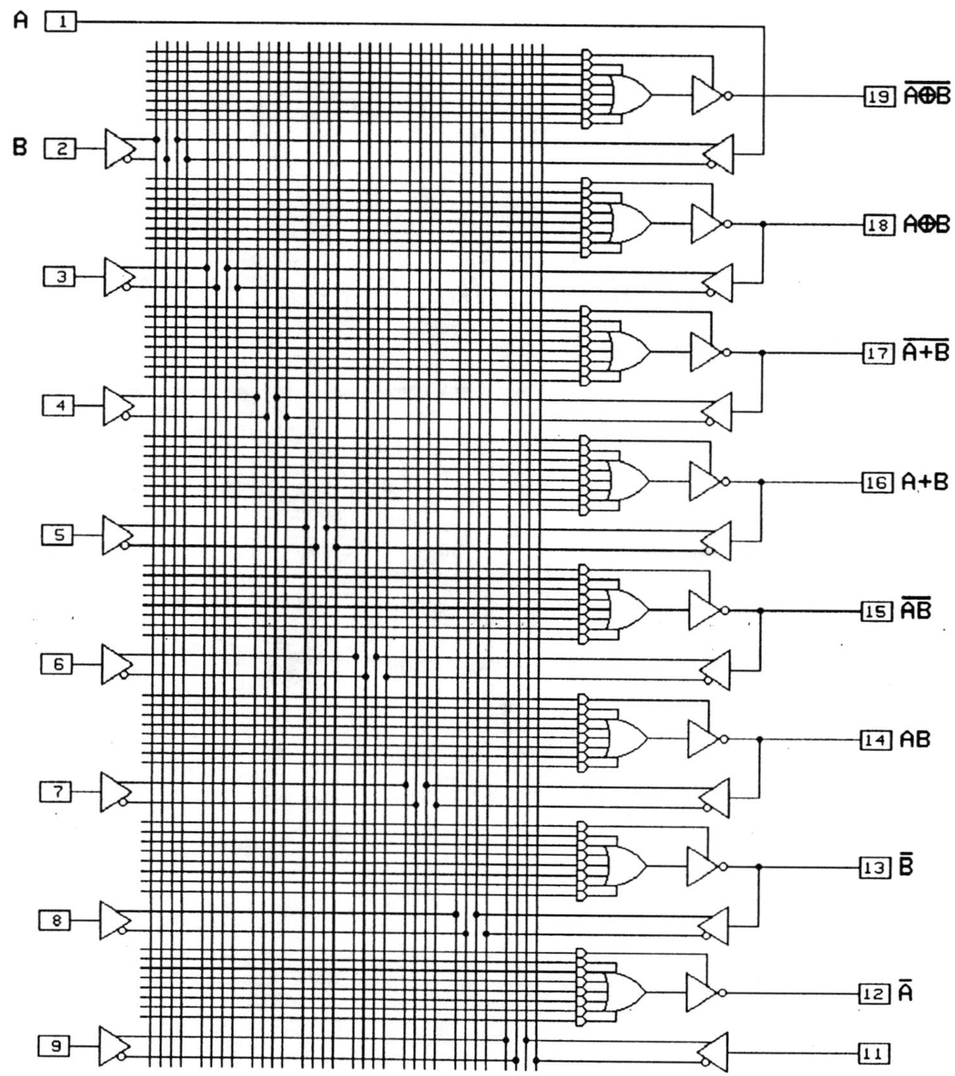

FIGURE 5.5 16L8 PAL circuit diagram with pin assignments.

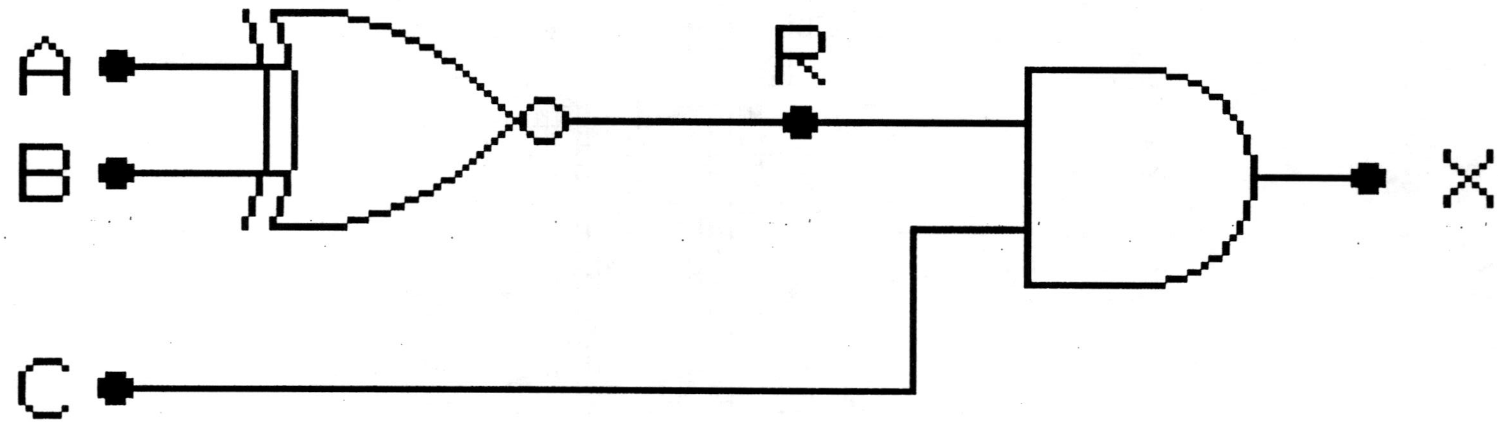

FIGURE 5.6 Circuit with intermediate variable R.

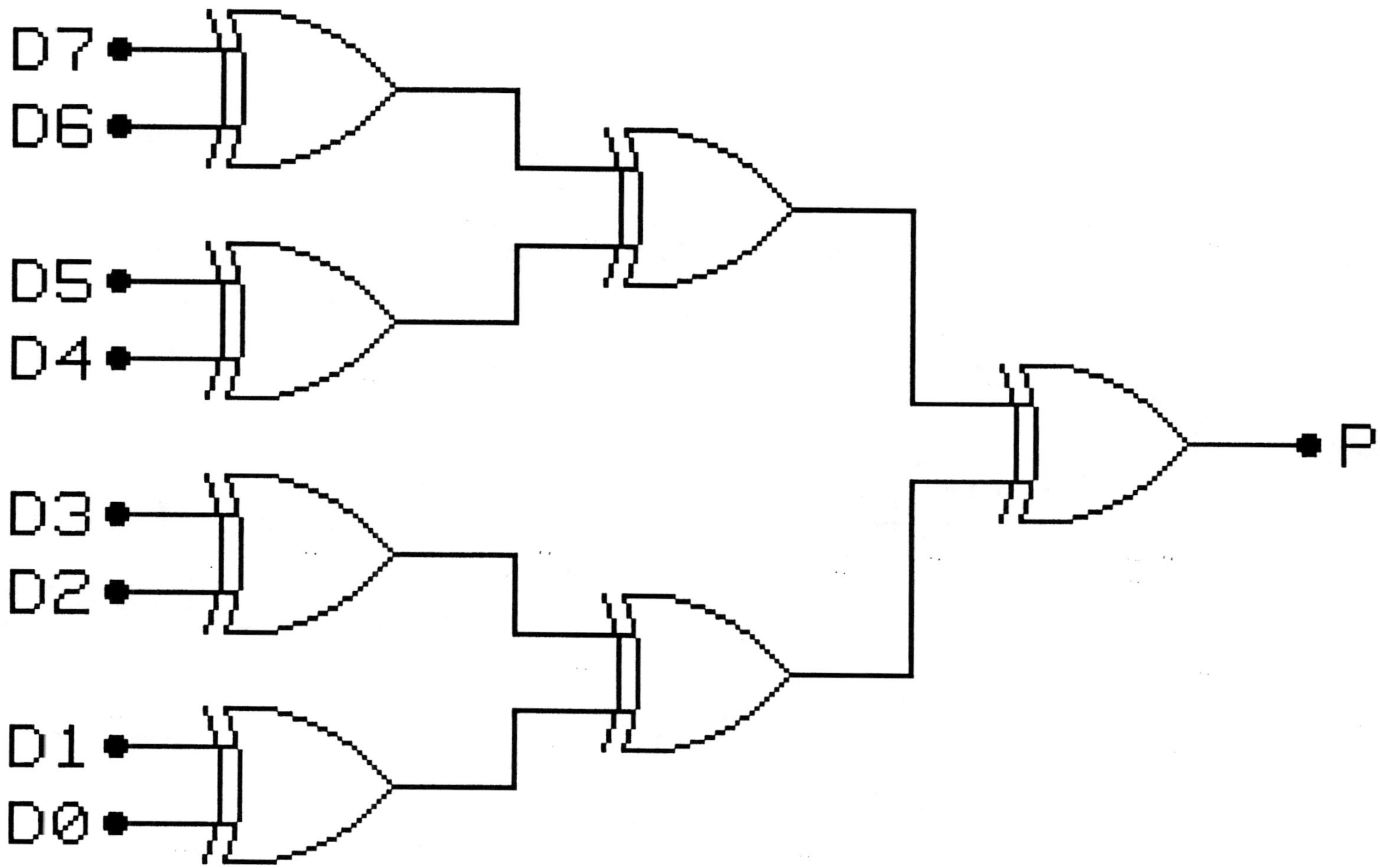

FIGURE 5.7 Practical Example 5.1, 8-bit parity generator.

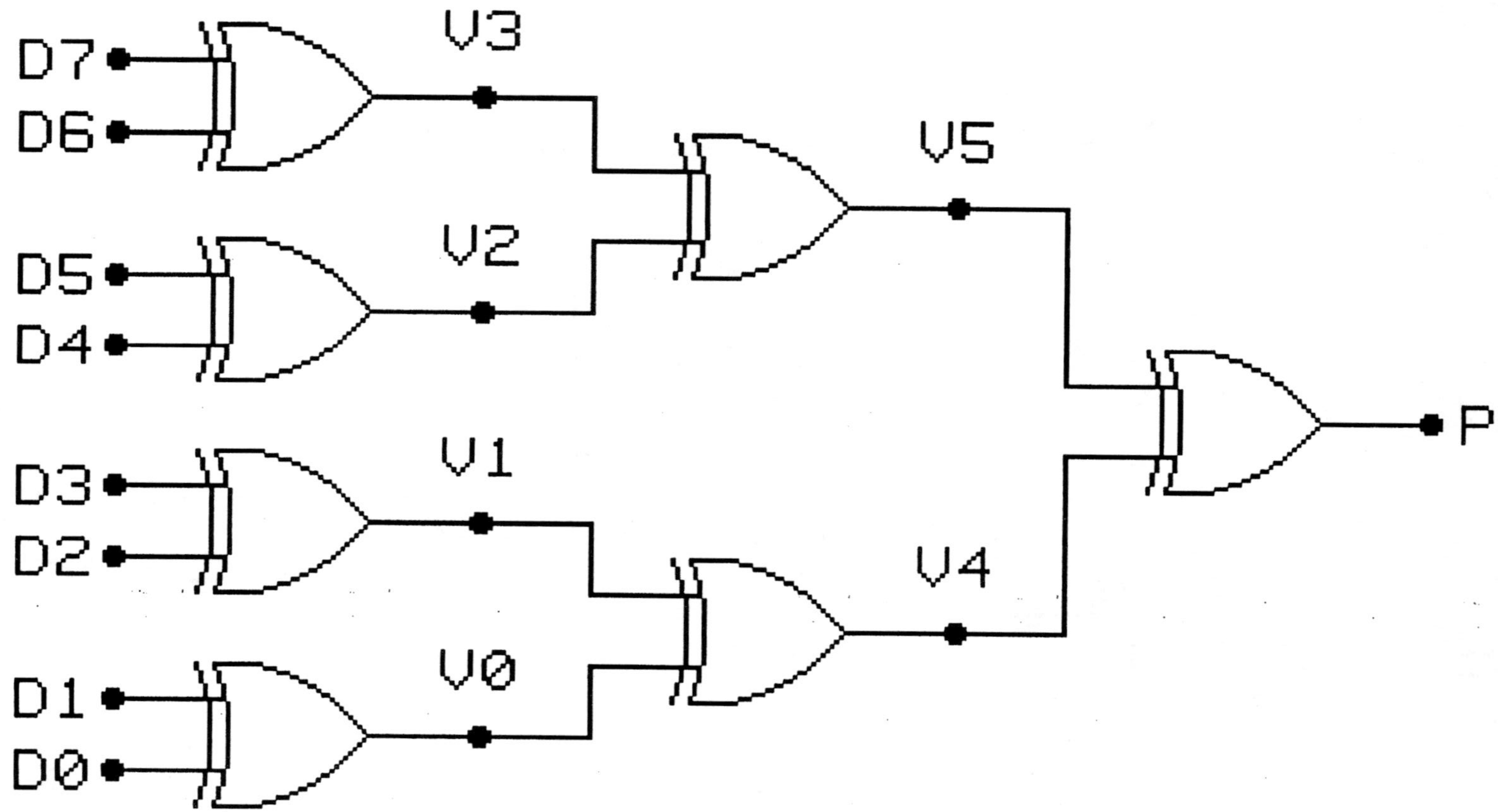

FIGURE 5.8 Practical Example 5.1, intermediate variable assignments.

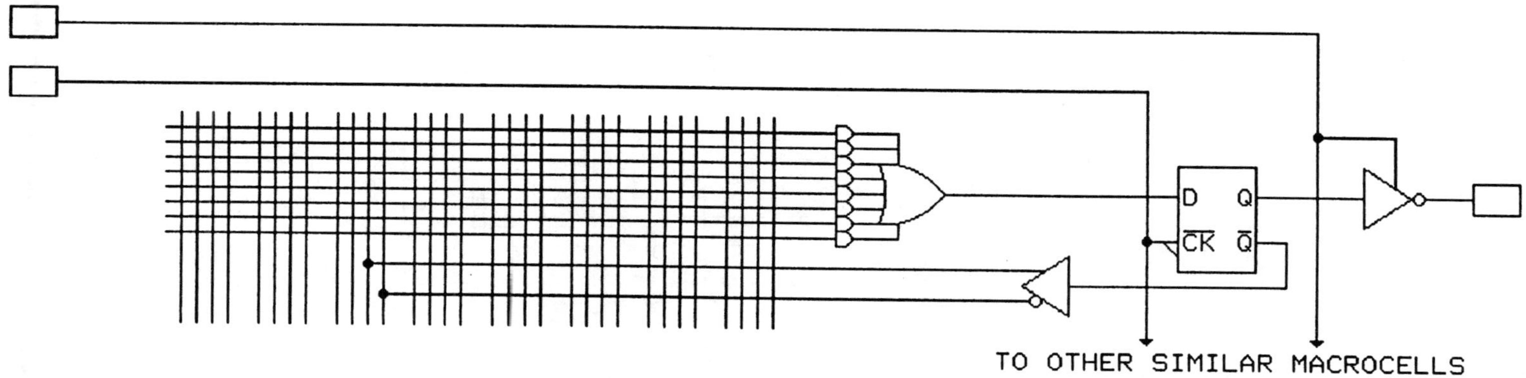

FIGURE 5.9 Typical R-type macrocell.

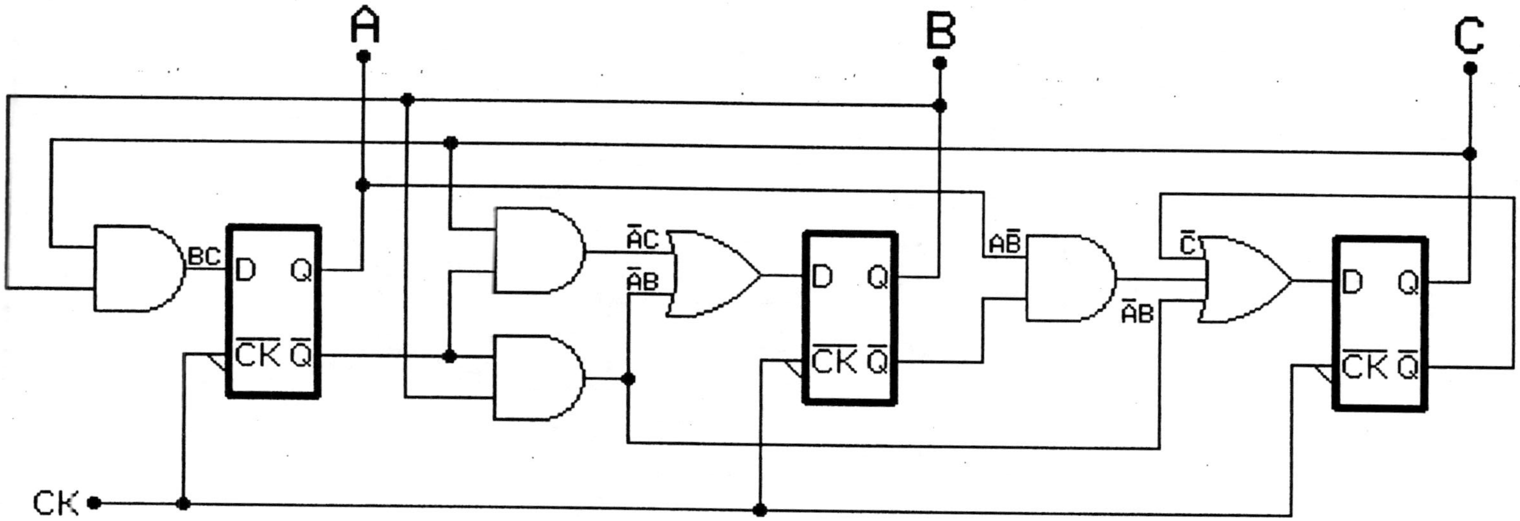

FIGURE 5.10 Example sequential circuit.

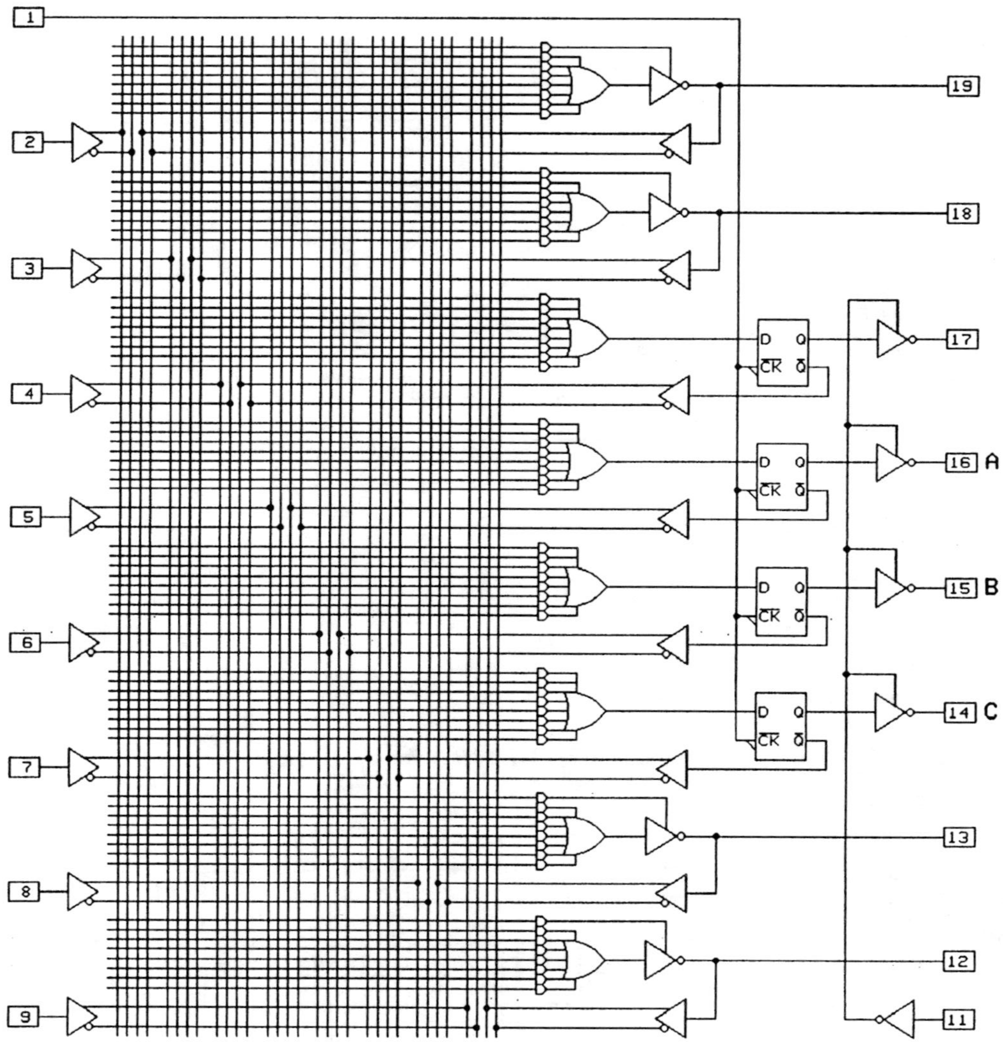

FIGURE 5.11 16R4 PAL with pin assignments for a synchronous three-bit counter.

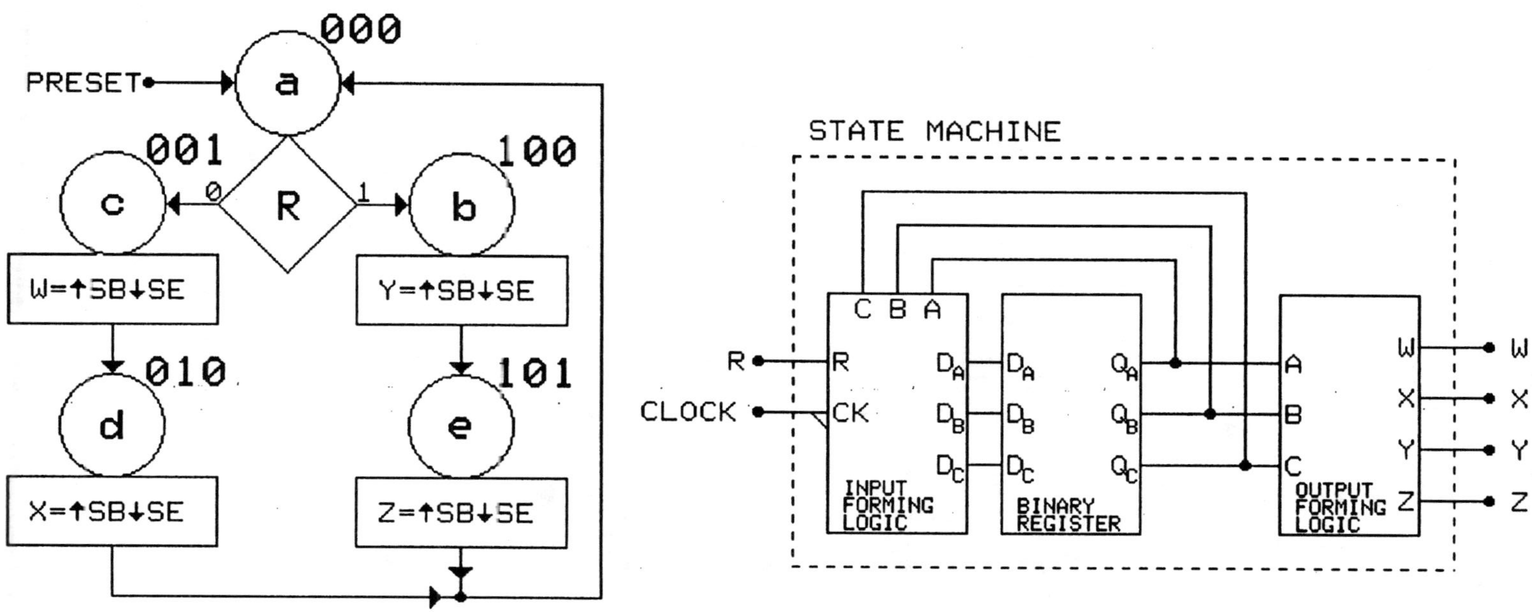

FIGURE 5.12 Practical Example 5.2, state diagram and block diagram.

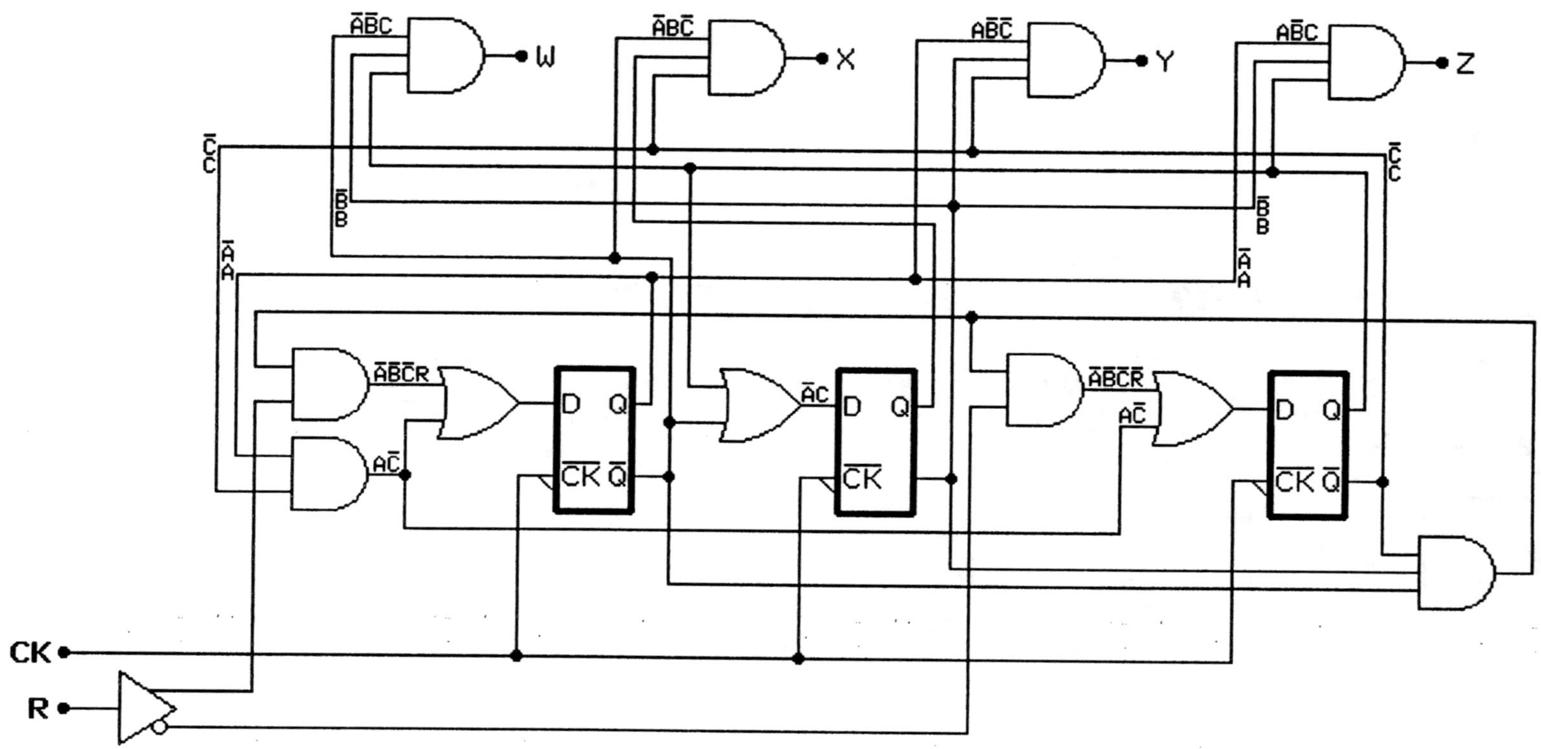

FIGURE 5.13 Practical Example 5.2, schematic diagram.

```
Name            EXP0502;
Revision        01;
Date            02/14/95;
Designer        J.W. Carter;
Company         University of North Carolina, Charlotte;
Location        None;
Assembly        None;
Device          P16R4;

/*******************************************************************/
/*      Practical Example No. 5.2  Complete State Machine         */
/*******************************************************************/

Pin  1 = R;    /* Input R */
Pin 16 = A;    /*  Counter Bits A, B and C */
Pin 15 = B;
Pin 14 = C;
Pin 19 = W;    /* Output Forming Logic Bits W, X, Y and Z */
Pin 18 = X;
Pin 13 = Y;
Pin 12 = Z;

/*******************************************************************/
/* State Machine Program                                           */
/*******************************************************************/

SEQUENCE [A,B,C] {
     PRESENT `b`000
            IF !R   NEXT `b`001;
            DEFAULT NEXT `b`100;
     PRESENT `b`001 NEXT `b`010  OUT W;
     PRESENT `b`010 NEXT `b`000  OUT X;
     PRESENT `b`100 NEXT `b`101  OUT Y;
     PRESENT `b`101 NEXT `b`000  OUT Z; }
```

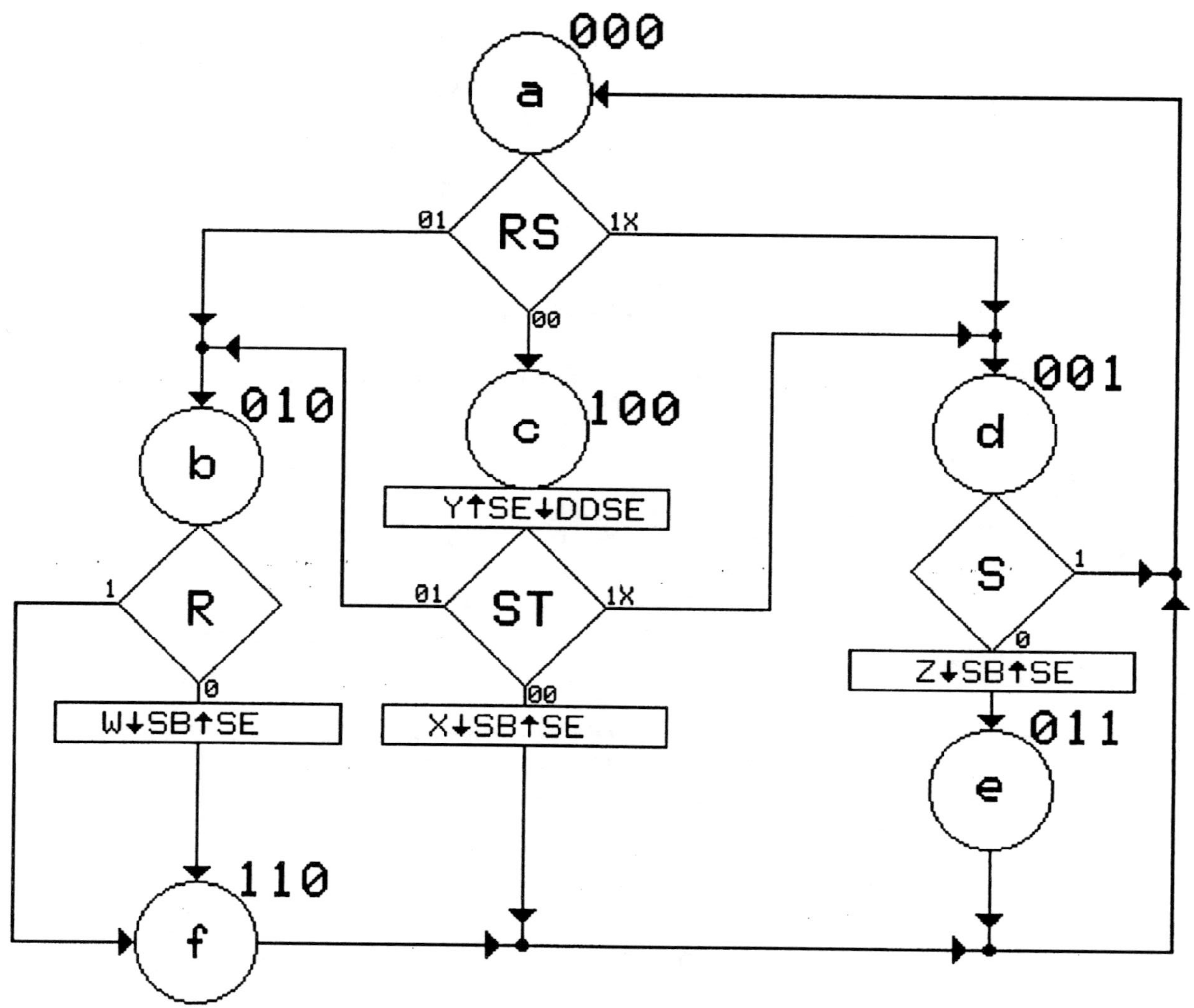

FIGURE 5.14 Practical Exercise 5.3, state diagram

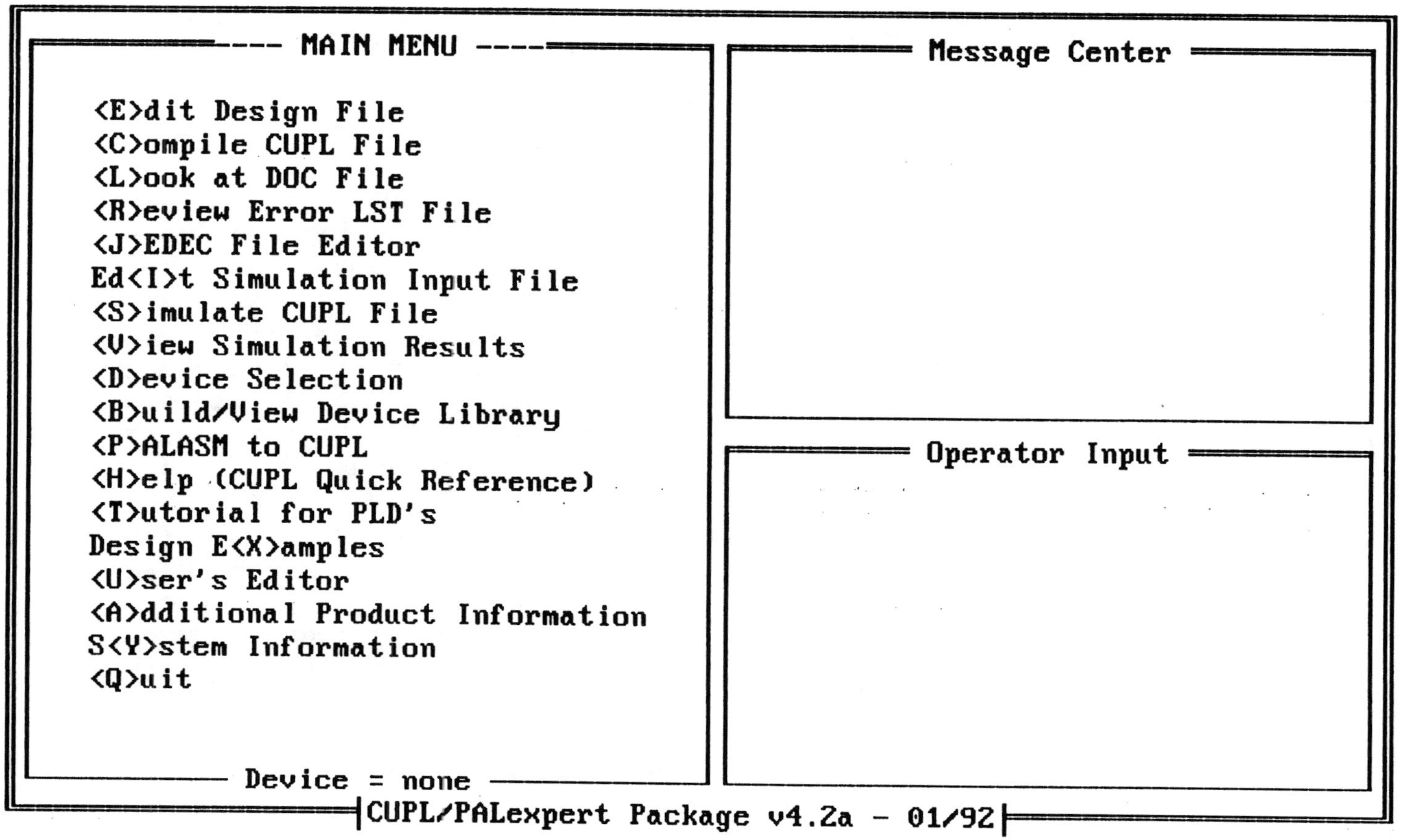

FIGURE 5.15 CUPL/PALexpert® menu shell display.

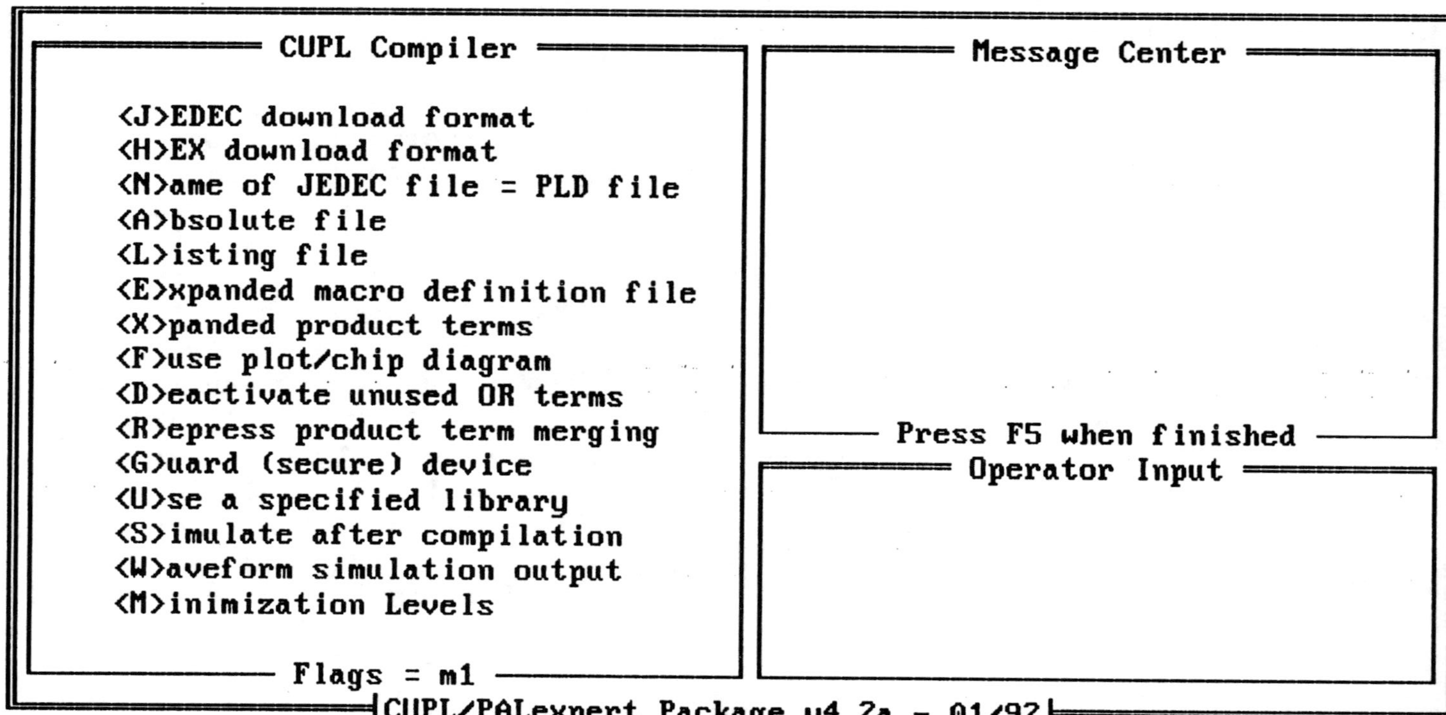

FIGURE 5.16 CUPL/PALexpert® compiler menu display.

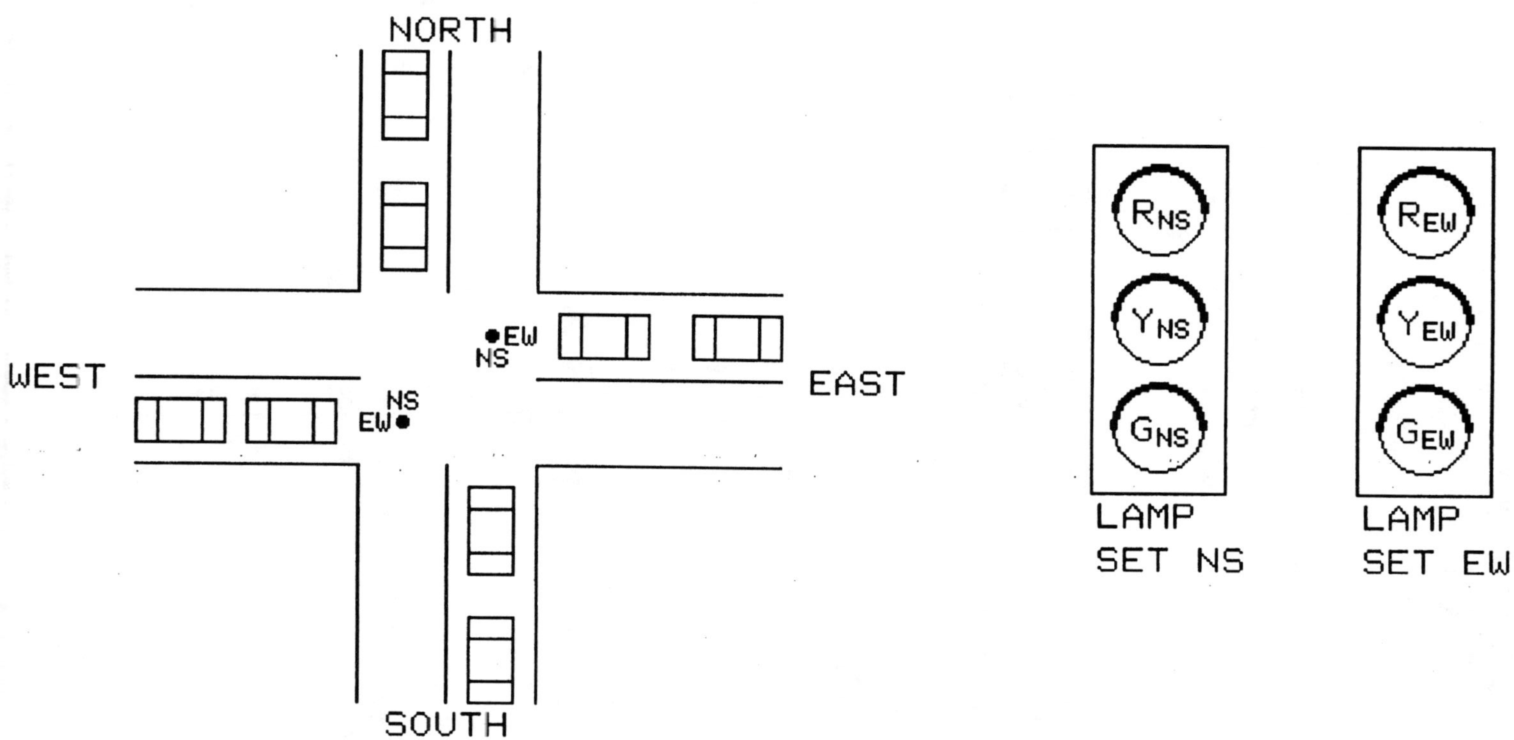

FIGURE 5.17 Practical Exercise 5.4, simple traffic intersection.

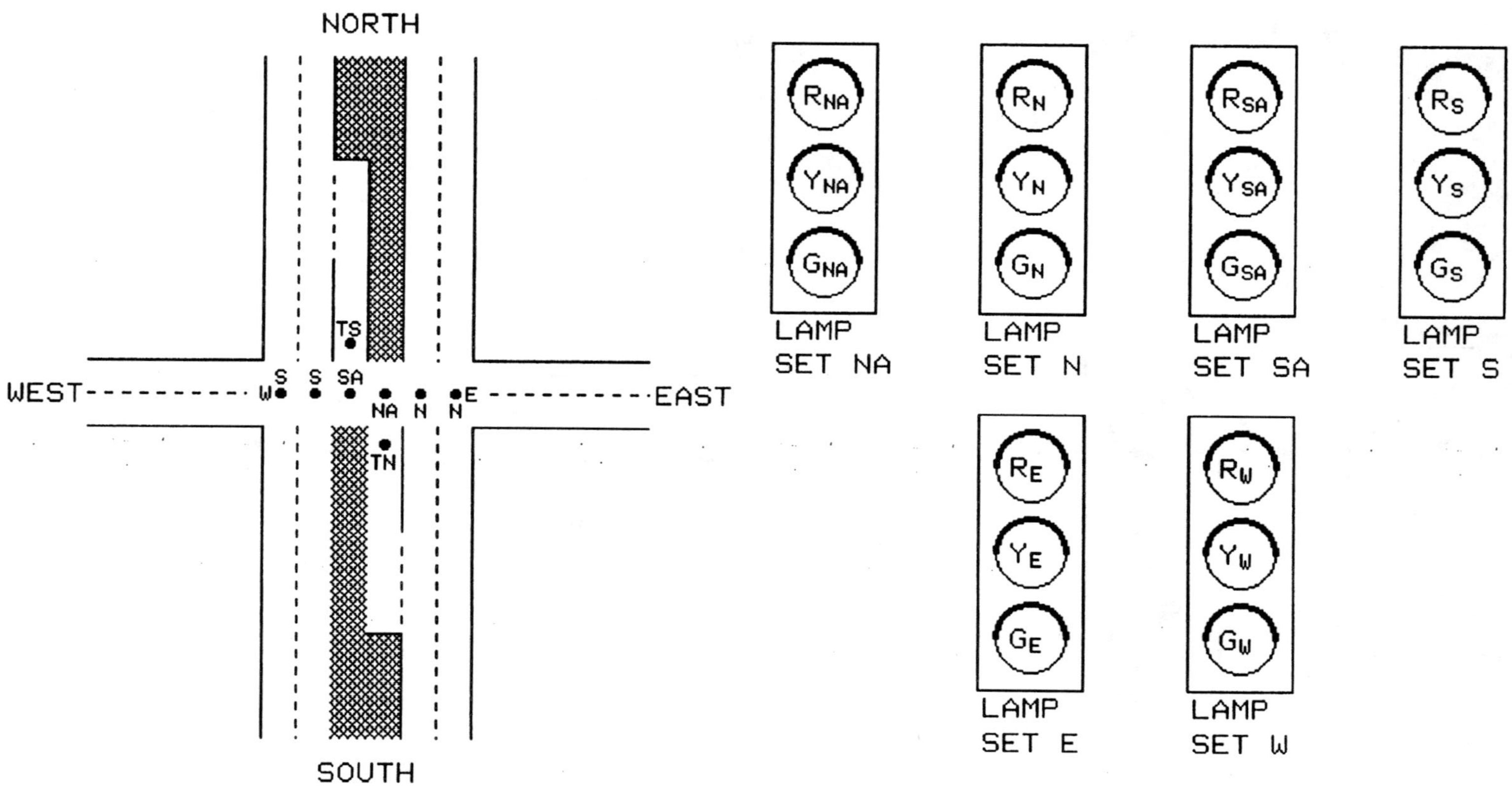

FIGURE 5.18 Practical Exercise 5.5, moderate traffic intersection.

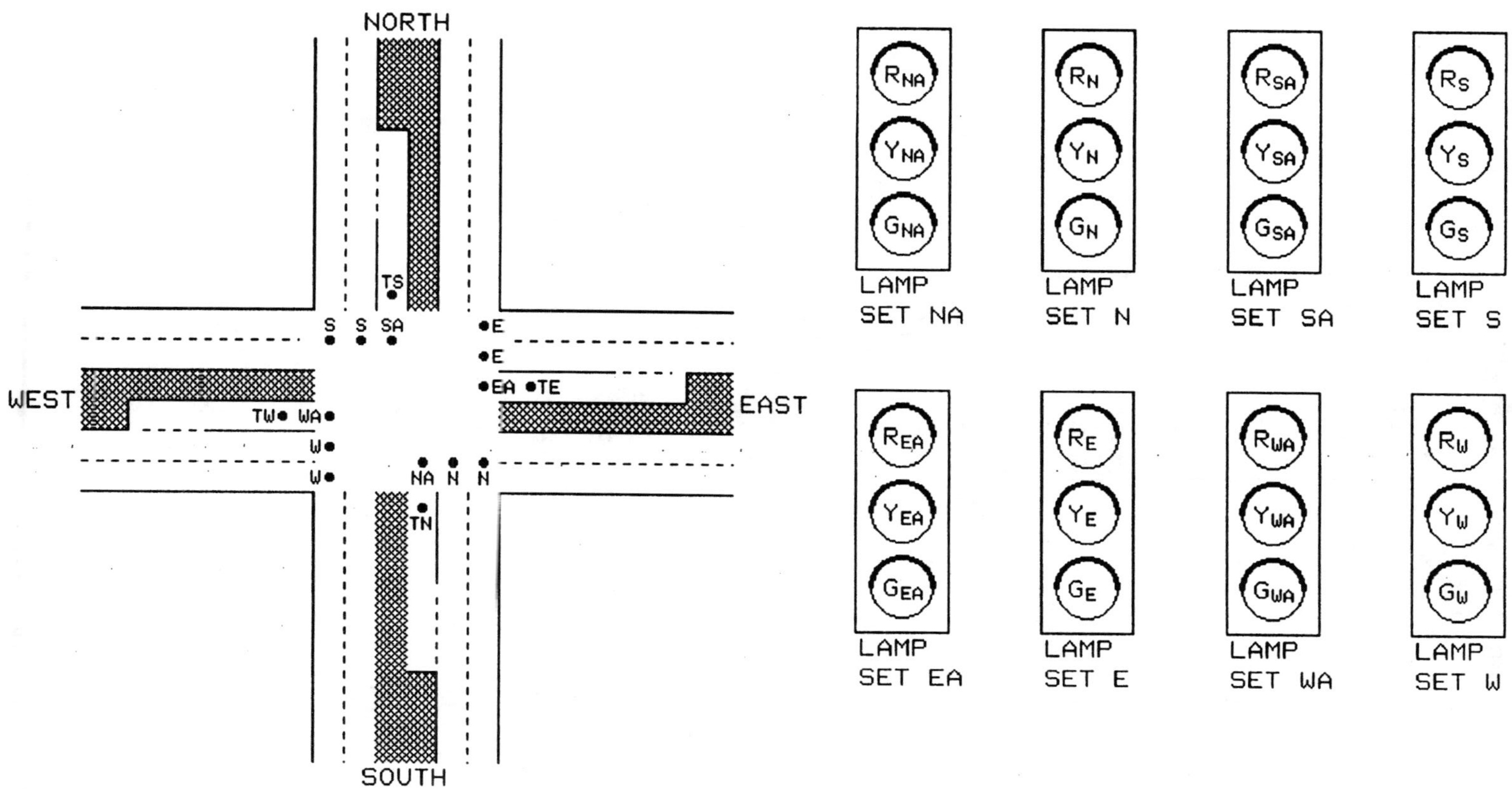

FIGURE 5.19 Practical Exercise 5.6, complete traffic intersection.

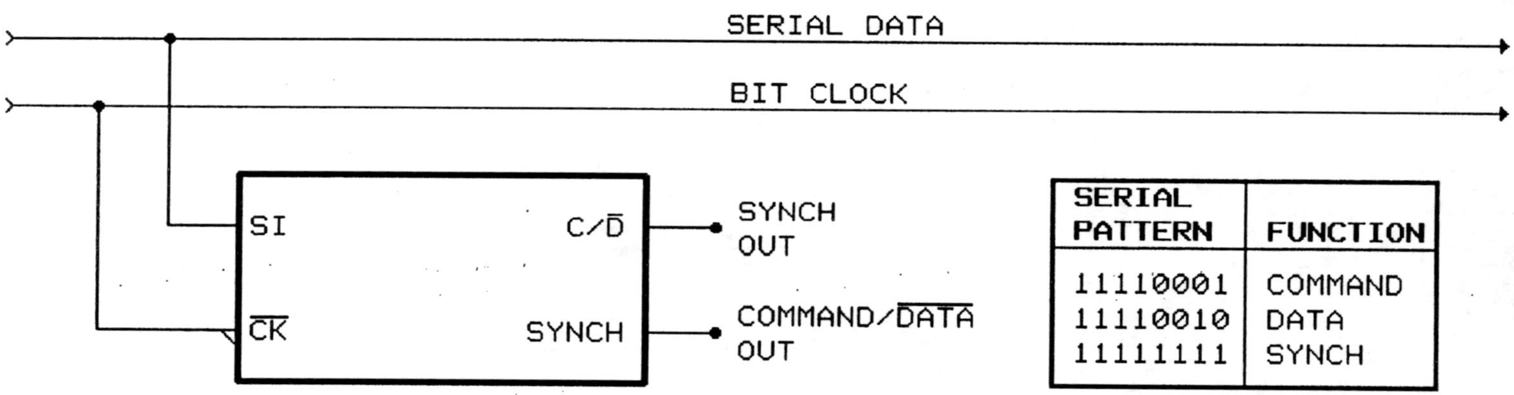

SERIAL PATTERN	FUNCTION
11110001	COMMAND
11110010	DATA
11111111	SYNCH

FIGURE 6.1 Serial pattern recognition circuit block diagram

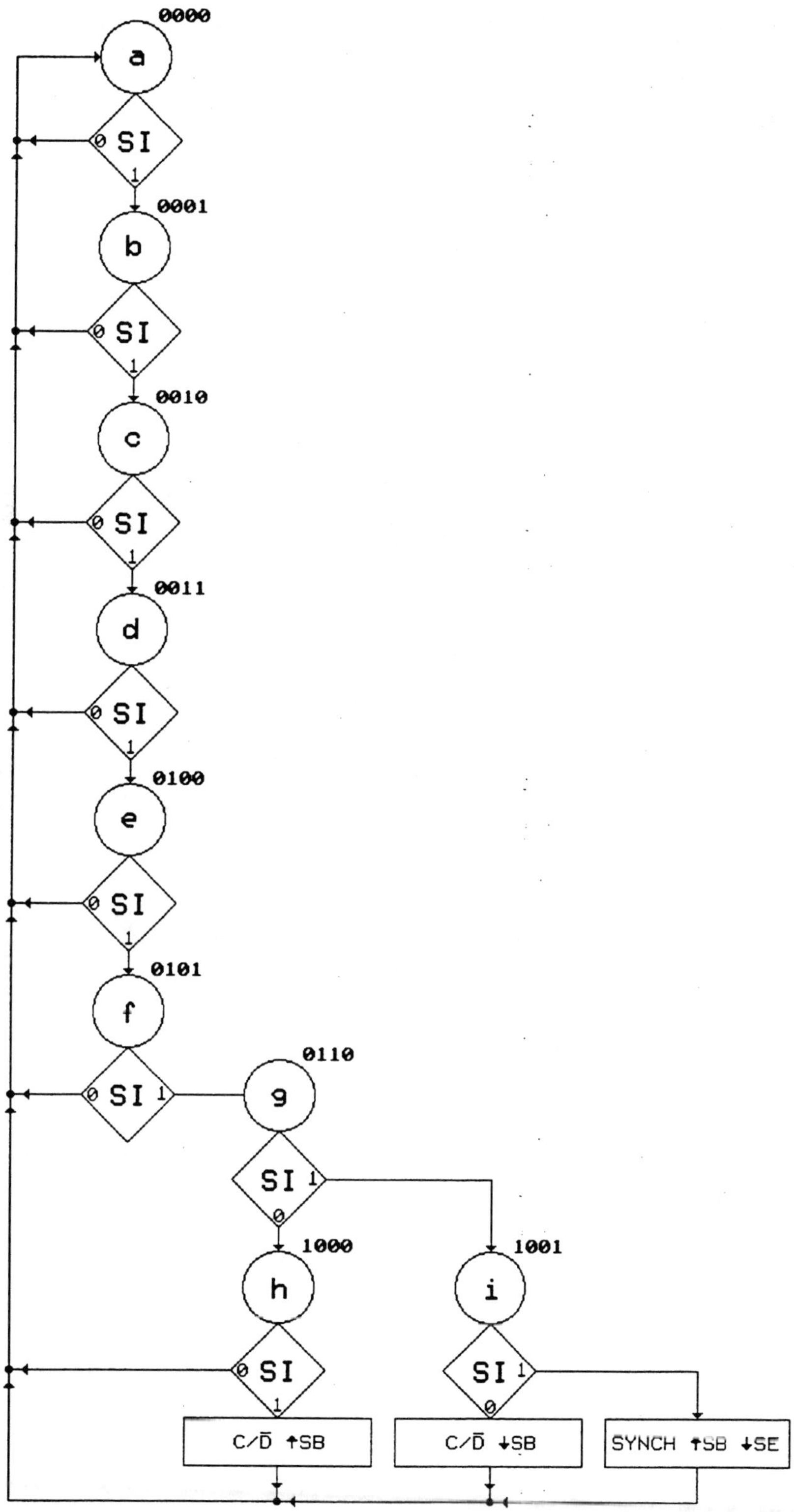

FIGURE 6.2 State diagram, serial pattern recognition

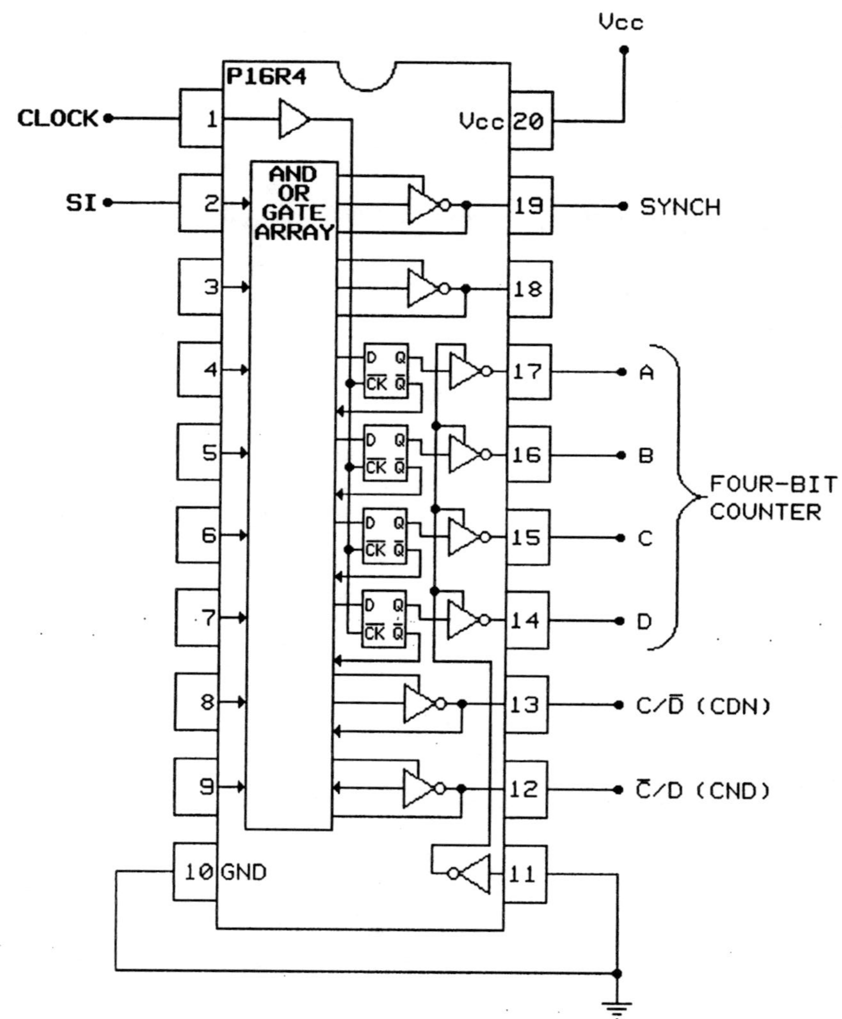

FIGURE 6.3 16R4 PAL implementation of serial pattern recognition circuit.

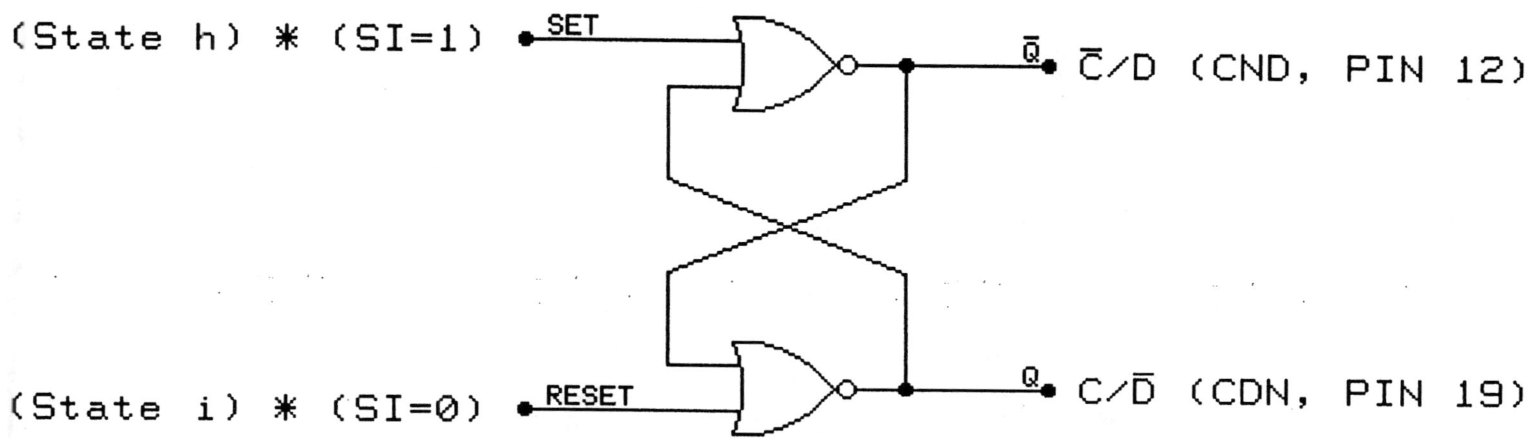

FIGURE 6.4 Two 16R4 combinational outputs configured as a Set-Reset flip-flop.

	ABCD	SI	ABCD	DA	DB	DC	DD	CDN	SYNCH
a	0000	0	0000	0	0	0	SI		
		1	0001						
b	0001	0	0000	0	0	SI	0		
		1	0010						
c	0010	0	0000	0	0	SI	SI		
		1	0011						
d	0011	0	0000	0	SI	0	0		
		1	0100						
e	0100	0	0000	0	SI	0	SI		
		1	0101						
f	0101	0	0000	0	SI	SI	0		
		1	0110						
g	0110	0	1000	1	0	0	SI		
		1	1001						
	0111	X	0000	0	0	0	0		
h	1000	0	0000	0	0	0	0		
		1	0000					↑SB	
i	1001	0	0000	0	0	0	0	↓SB	
		1	0000						↑SB ↓SE
	1010	X	0000	0	0	0	0	Note: Unused states go to State a to avoid hanging the counter.	
	1011	X	0000	0	0	0	0		
	1100	X	0000	0	0	0	0		
	1101	X	0000	0	0	0	0		
	1110	X	0000	0	0	0	0		
	1111	X	0000	0	0	0	0		

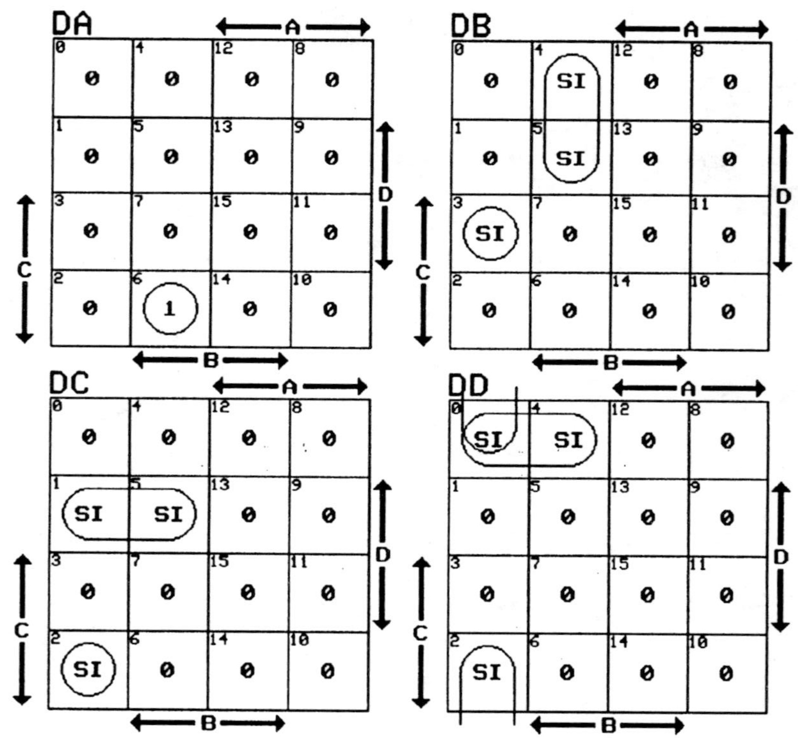

$DA = \bar{A}BC\bar{D}$

$DB = \bar{A}B\bar{C}\bullet SI + \bar{A}\bar{B}CD\bullet SI$

$DC = \bar{A}\bar{C}D\bullet SI + \bar{A}\bar{B}C\bar{D}\bullet SI$

$DD = \bar{A}\bar{C}\bar{D}\bullet SI + \bar{A}\bar{B}\bar{D}\bullet SI$

FIGURE 6.5 State table solution of serial pattern recognition circuit.

```
Name        EXP0601A;
Date        01/01/96;
Revision    01;
Designer    J. Carter;
Device      P16R4;
Format      j;
/*********************************************************************/
/* SERIAL PATTERN RECOGNITION CIRCUIT                                */
/* Example 6.1A    Using counter input logic functions.              */
/*********************************************************************/

Pin 2       = SI       ;     /* Serial Input, SI                     */
Pin 19      = SYNCH    ;     /* SYNCH output                         */
Pin 17      = A        ;     /* Four-Bit Counter                     */
Pin 16      = B        ;     /* "                                    */
Pin 15      = C        ;     /* "                                    */
Pin 14      = D        ;     /* "                                    */
Pin 13      = CDN      ;     /* Command/Data Output                  */
Pin 12      = CND      ;     /* Complement of Pin 13                 */

/*  Input Forming Logic */

  A.D = (!A &  B &  C & !D);
  B.D = (!A &  B & !C & SI) + (!A & !B &  C &  D & SI);
  C.D = (!A & !C &  D & SI) + (!A & !B &  C & !D & SI);
  D.D = (!A 7 !C & !D & SI) + (!A & !B & !D & SI);

/*  Output Forming Logic */

  CDN = !((!SI &  A & !B & !C &  D) # CND);
  CND = !(( SI &  A & !B & !C & !D) # CDN);
  SYNCH =    SI &  A & !B & !C & D;
```

```
Name        EXP0601B;
Date        01/01/96;
Revision    01;
Designer    J. Carter;
Device      P16R4;
Format      j;

/******************************************************************/
/* SERIAL PATTERN RECOGNITION CIRCUIT                             */
/* Example 6.1B    Using counter sequence statement.              */
/******************************************************************/

Pin 2        = SI        ;      /* Serial Input, SI               */

Pin 19       = SYNCH     ;      /* SYNCH output                   */
Pin 17       = A         ;      /* Four-Bit Counter               */
Pin 16       = B         ;      /* "                              */
Pin 15       = C         ;      /* "                              */
Pin 14       = D         ;      /* "                              */
Pin 13       = CDN       ;      /* Command/Data Output            */
Pin 12       = CND       ;      /* Complement of Pin 13           */

sequence[A,B,C,D] {
   present 'b'0000
             if SI next 'b'0001;  default next 'b'0000;
   present 'b'0001
             if SI next 'b'0010;  default next 'b'0000;
   present 'b'0010
             if SI next 'b'0011;  default next 'b'0000;
```

```
present 'b'0011
          if SI next 'b'0100;  default next 'b'0000;
present 'b'0100
          if SI next 'b'0101;  default next 'b'0000;
present 'b'0101
          if SI next 'b'0110;  default next 'b'0000;
present 'b'0110
          if SI next 'b'1001;  default next 'b'1000;
present 'b'0111 next 'b'0000;
present 'b'1000 next 'b'0000;
present 'b'1001
          if SI next 'b'0000 OUT SYNCH;  default next 'b'0000 ;
present 'b'1010 next 'b'0000;
present 'b'1011 next 'b'0000;
present 'b'1100 next 'b'0000;
present 'b'1101 next 'b'0000;
present 'b'1110 next 'b'0000;
present 'b'1111 next 'b'0000;
}

CDN = !((!SI &  A & !B & !C &  D) # CND);
CND = !(( SI &  A & !B & !C & !D) # CDN);
```

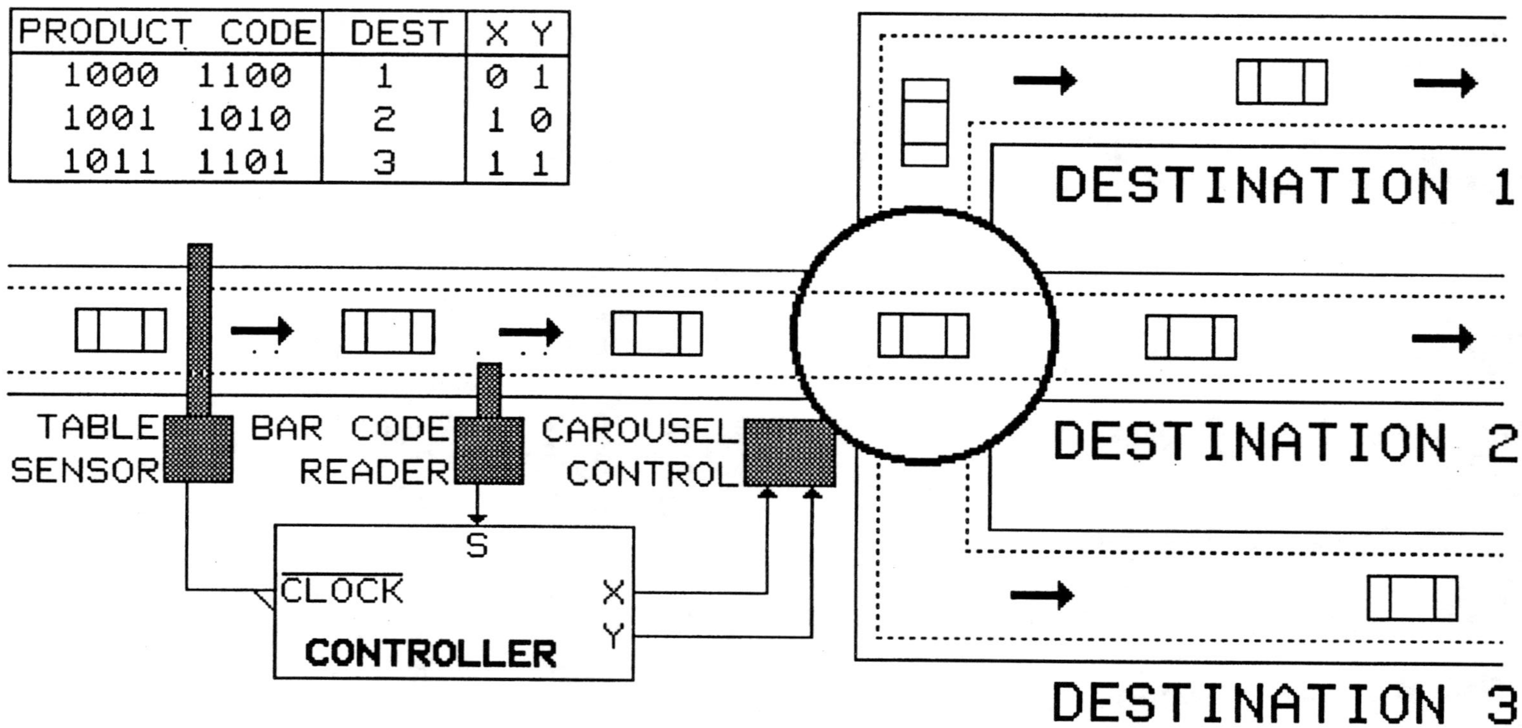

FIGURE 6.6 Assembly line.

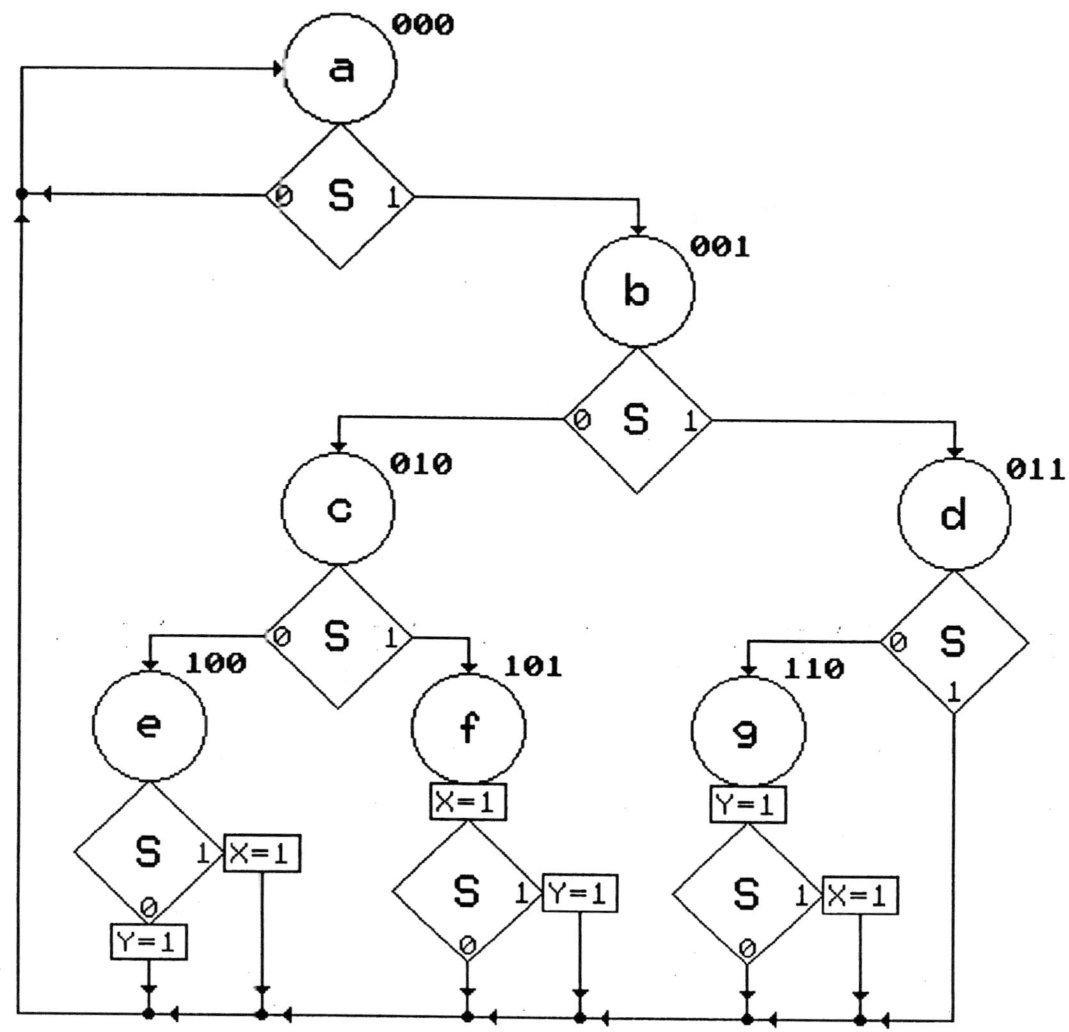

FIGURE 6.7 State diagram, assembly line carousel controller

	ABC	S	ABC	DA	DB	DC	X	Y
a	000	0 1	000 001	0	0	S	0	0
b	001	0 1	010 011	0	1	S	0	0
c	010	0 1	100 101	1	0	S	0	0
d	011	0 1	110 000	$\bar{S}$	$\bar{S}$	0	0	0
e	100	0 1	000 000	0	0	0	0 1 =S	1 0 =$\bar{S}$
f	101	0 1	000 000	0	0	0	1 1	0 1 =S
g	110	0 1	000 000	0	0	0	0 1 =S	1 1
	111	X	000	0	0	0	0	0

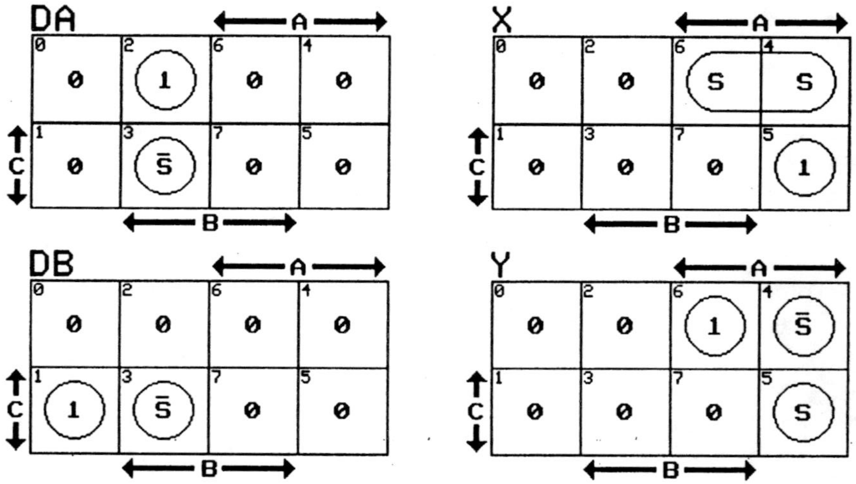

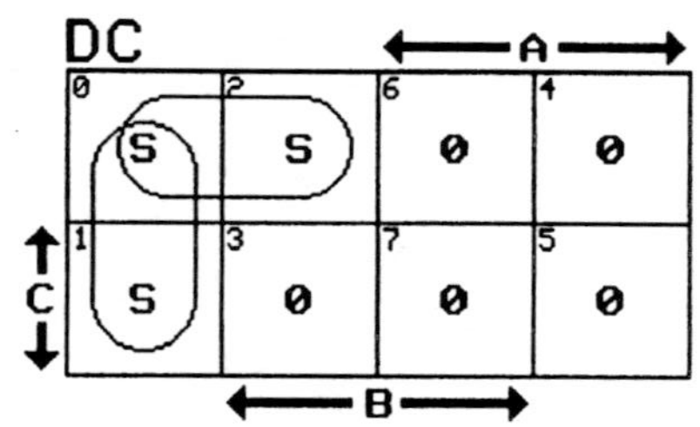

$$DA = \bar{A}B\bar{C} + \bar{A}BC\bar{S}$$

$$DB = \bar{A}BC\bar{S} + \bar{A}\bar{B}C$$

$$DC = \bar{A}\bar{C}S + \bar{A}\bar{B}S$$

$$X = A\bar{C}S + A\bar{B}C$$

$$Y = AB\bar{C} + A\bar{B}\bar{C}\bar{S} + A\bar{B}CS$$

FIGURE 6.8 Truth table solution, assembly line carousel controller.

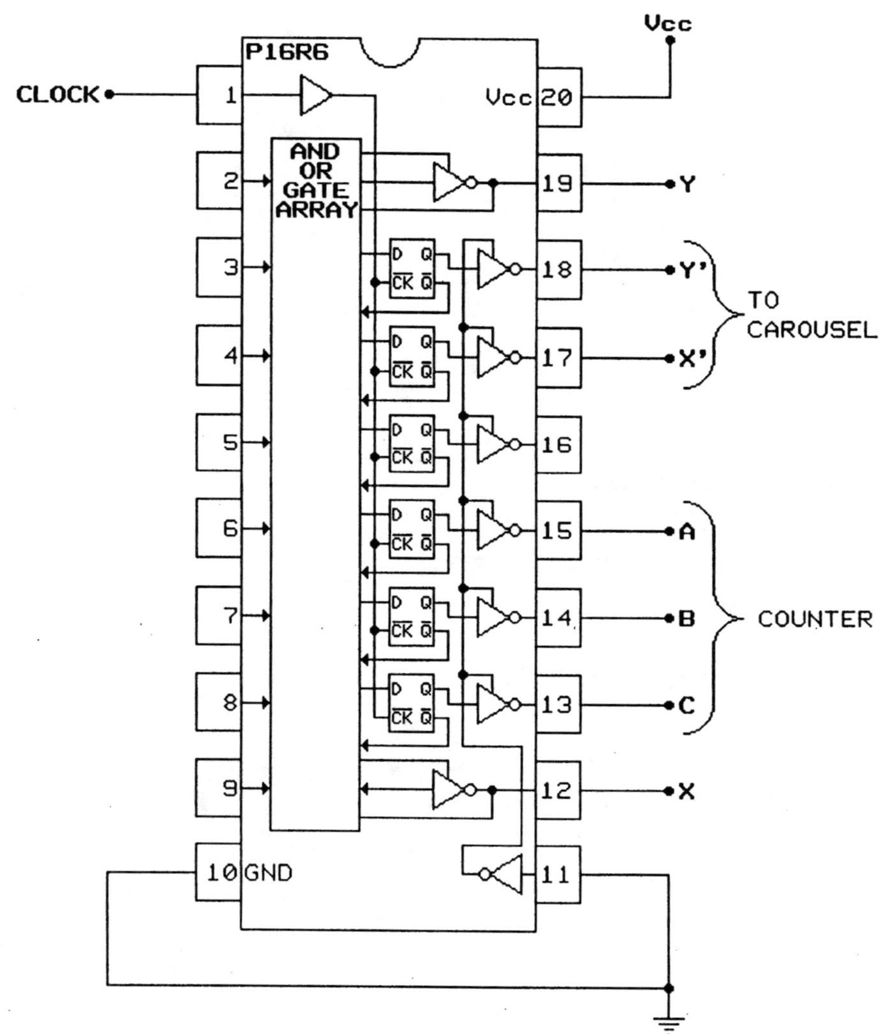

FIGURE 6.9 P16R6 PAL configured for the assembly line carousel controller.

```
Name        EXP0602A;
Date        01/01/96;
Revision    01;
Designer    J. Carter;
Device      P16R6;
Format      j;

/********************************************************************/
/* ASSEMBLY LINE CAROUSEL CONTROLLER                                */
/* Example 6.2A    Using counter input logic functions.             */
/********************************************************************/

Pin 2        = S        ;        /* Serial Input from Bar Code      */

Pin 18       = YP       ;        /* X Output to Carousel            */
Pin 17       = XP       ;        /* Y Output to Carousel            */
Pin 19       = Y        ;        /* Intermediate Y Calculation      */
Pin 12       = X        ;        /* Intermediate X Calculation      */
Pin 15       = A        ;        /* Three-Bit Counter               */
Pin 14       = B        ;        /* "                               */
Pin 13       = C        ;        /* "                               */

/*  Input Forming Logic */

  A.D = (!A &  B & !C) + (!A &  B &  C & !S);
  B.D = (!A & !B &  C) + (!A &  B &  C & !S);
  C.D = (!A & !C &  S) + (!A & !B &  S);
```

```
/*  Output Forming Logic */

   X = (A & !C &  S) + (!A & !B &  C);
   Y = (A &  B & !C) + ( A & !B & !C & !S) + (A & !B & C & S);
   XP.D = X;
   YP.D = Y;
```

```
sequence[A,B,C] {
   present 'b'000 if S next 'b'001;  default next 'b'000;
   present 'b'001 if S next 'b'011;  default next 'b'010;
   present 'b'010 if S next 'b'101;  default next 'b'100;
   present 'b'011 if S next 'b'000;  default next 'b'110;
   present 'b'100 if S next 'b'101 OUT X; default next 'b'000 OUT Y;
   present 'b'101 next `b'000 OUT X; if S OUT Y;
   present 'b'110 next `b'000 OUT Y; if S OUT X;
   present 'b'111 next 'b'000;
   }
XP.D = X;
YP.D = Y;
```

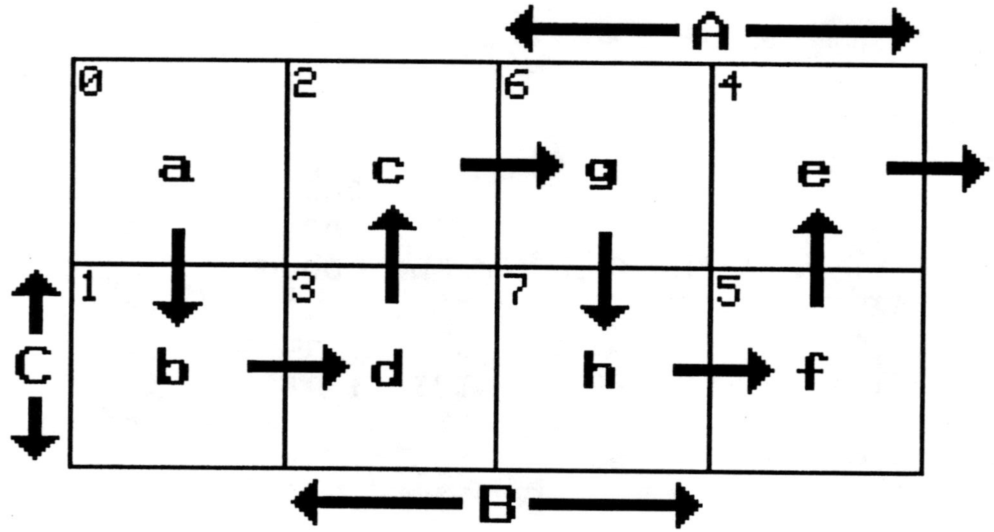

FIGURE 6.10 Next state map for glitch-free counter.

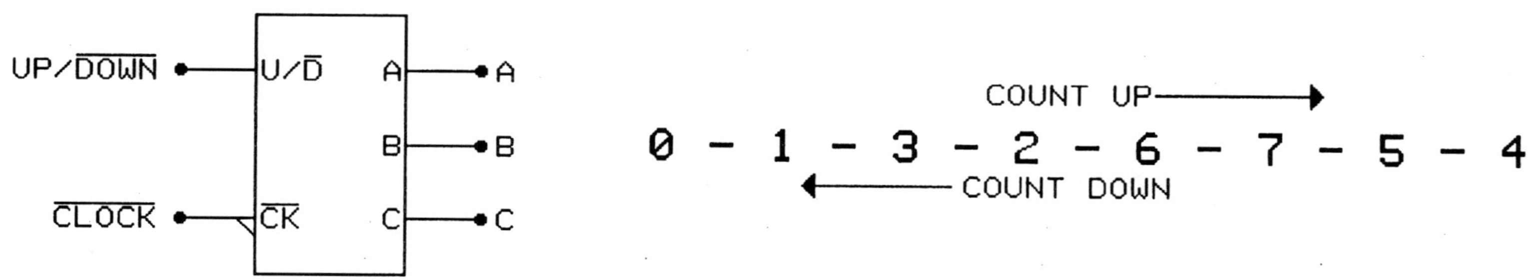

FIGURE 6.11 Functional diagram of a glitch-free up-down binary counter.

	ABC	U	ABC	DA	DB	DC
a	000	0	100	$\bar{U}$	0	U
		1	001			
b	001	0	000	0	U	U
		1	011			
c	010	0	011	U	1	$\bar{U}$
		1	110			
d	011	0	001	0	U	$\bar{U}$
		1	010			
e	100	0	101	$\bar{U}$	0	$\bar{U}$
		1	000			
f	101	0	111	1	$\bar{U}$	$\bar{U}$
		1	100			
g	110	0	010	U	1	U
		1	111			
h	111	0	110	1	$\bar{U}$	U
		1	101			

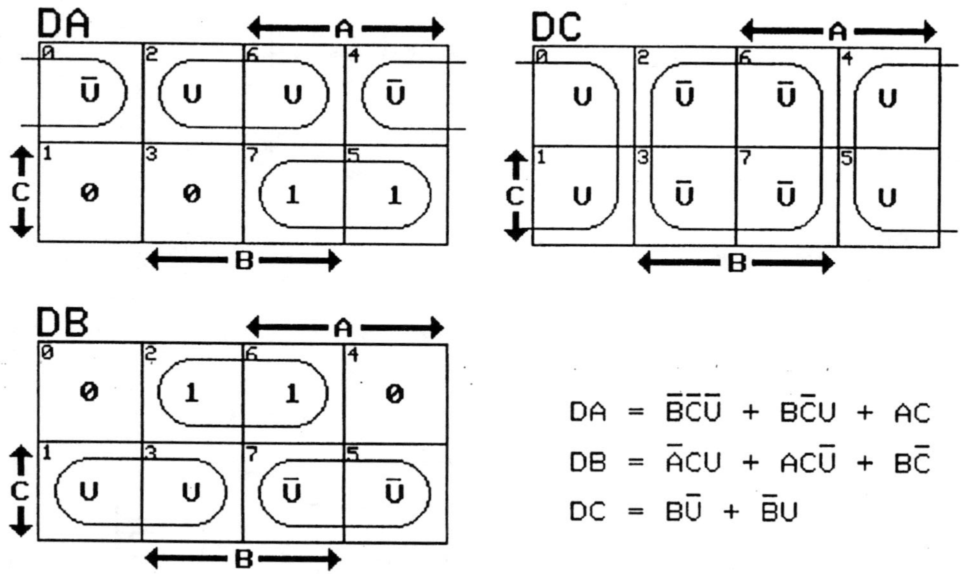

$$DA = \bar{B}\bar{C}\bar{U} + B\bar{C}U + AC$$

$$DB = \bar{A}CU + AC\bar{U} + B\bar{C}$$

$$DC = B\bar{U} + \bar{B}U$$

FIGURE 6.12 State table solution for glitch-free up-down counter.

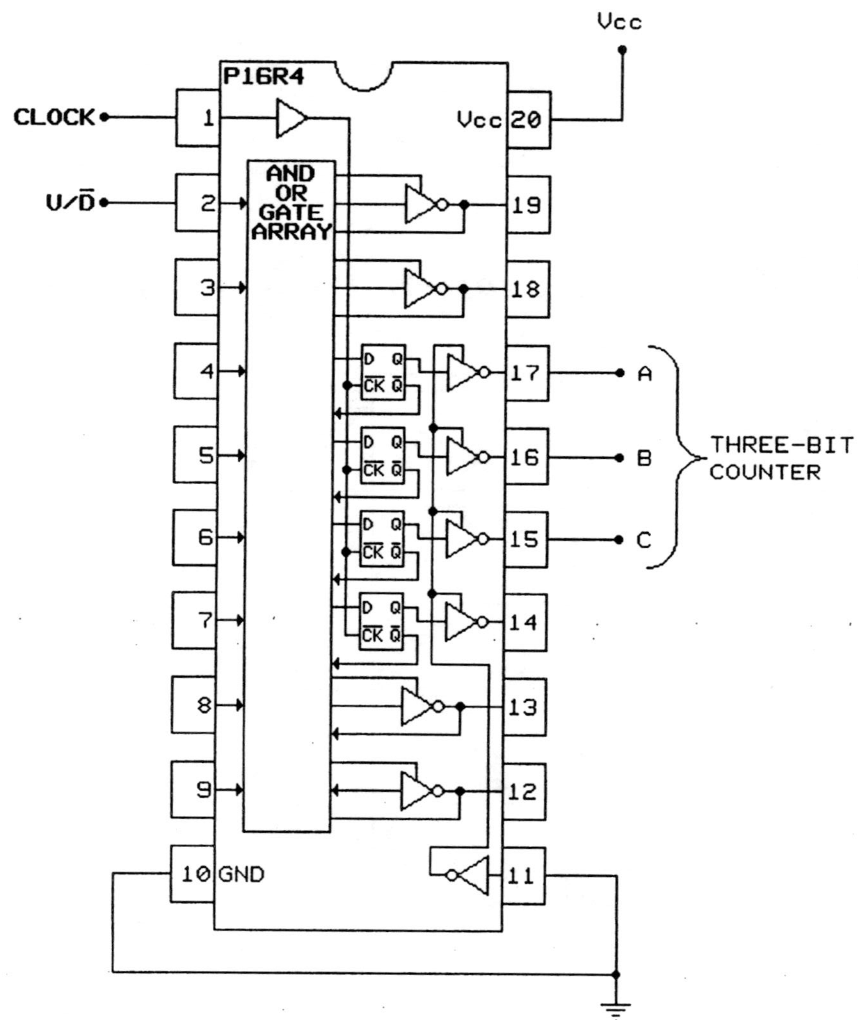

FIGURE 6.13 16R4 PAL configured as a three-bit up-down counter.

```
Name        EXP0603A;
Date        01/01/96;
Revision    01;
Designer    J. Carter;
Device      P16R4;
Format      j;
/*****************************************************************/
/* GLITCH-FREE THREE-BIT UP/DOWN BINARY COUNTER                  */
/* Example 6.3A    Using counter input logic functions.          */
/*****************************************************************/
Pin 2        = UDN      ;       /* UP/DOWN'   (1=UP, 0=DOWN)     */
Pin 16       = A        ;       /* Three-Bit Counter             */
Pin 15       = B        ;       /* "                             */
Pin 14       = C        ;       /* "                             */

/*  Input Forming Logic */
  A.D = (!B & !C & !U) + (B & !C &  U) + (A &  C);
  B.D = (!A &  C &  U) + (A &  C & !U) + (B & !C);
  C.D = ( B & !U) + (!B & U);

sequence[A,B,C] {
   present 'b'000 if UDN next 'b'001;  default next 'b'100;
   present 'b'001 if UDN next 'b'011;  default next 'b'000;
   present 'b'010 if UDN next 'b'110;  default next 'b'011;
   present 'b'011 if UDN next 'b'010;  default next 'b'001;
   present 'b'100 if UDN next 'b'101;  default next 'b'101;
   present 'b'101 if UDN next 'b'100;  default next 'b'111;
   present 'b'110 if UDN next 'b'111;  default next 'b'010;
   present 'b'111 if UDN next 'b'101;  default next 'b'110;}
```

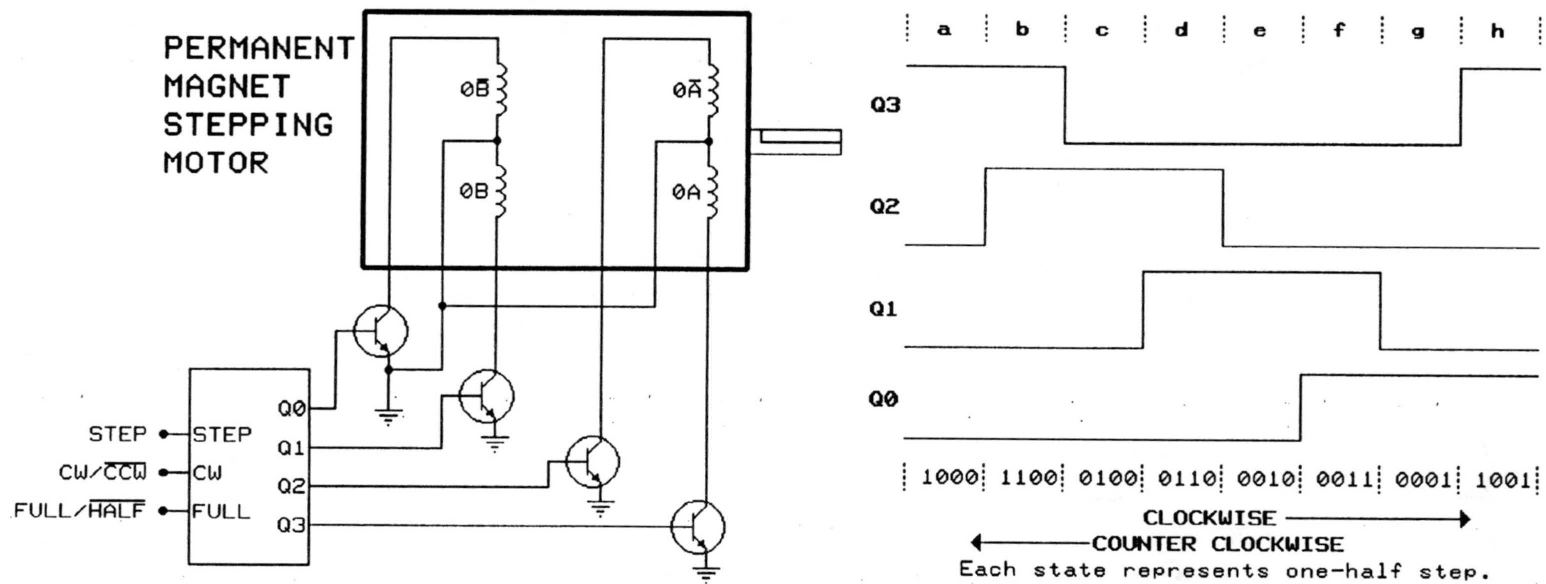

FIGURE 6.14 Stepper motor controller schematic and timing diagram.

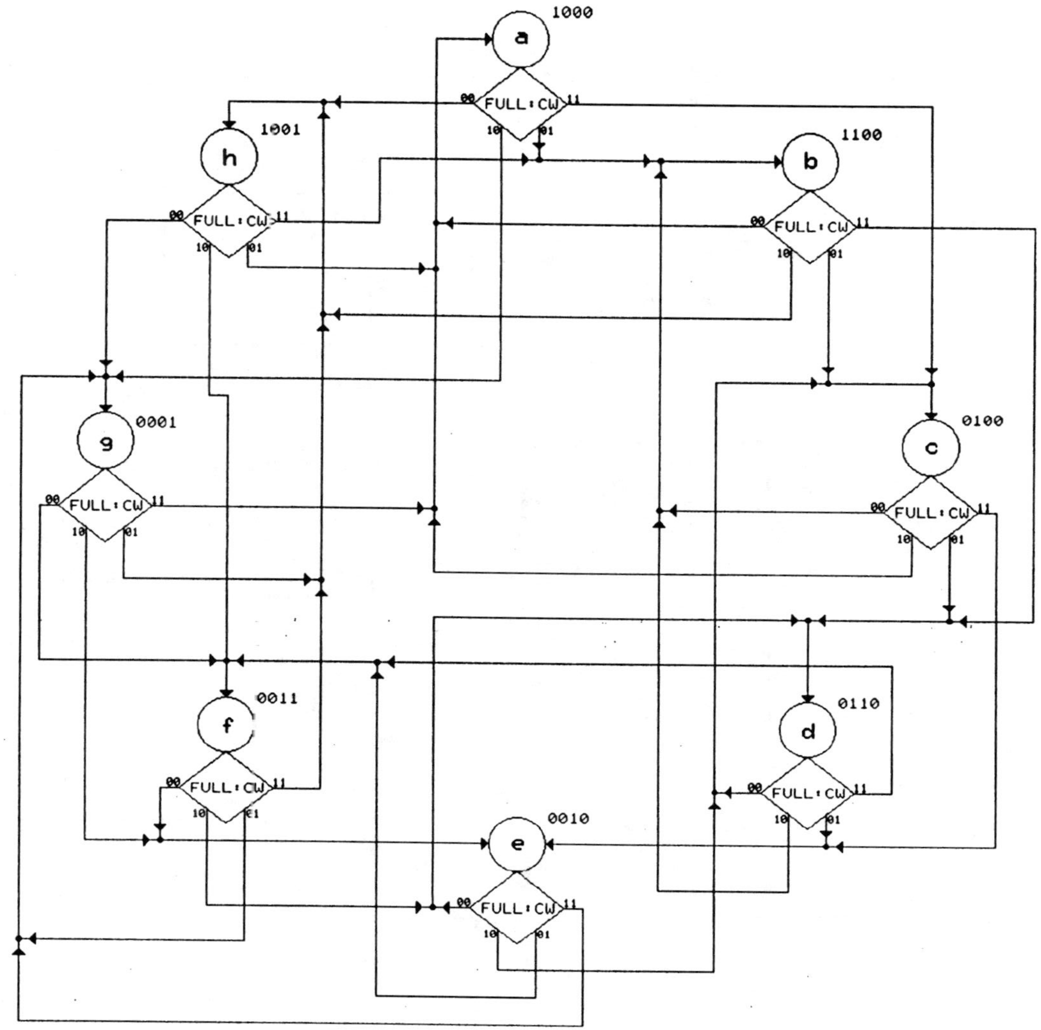

FIGURE 6.15 Stepper motor state diagram.

	ABCE	F	G	ABCE	DA	DB	DC	DE
	0000	0	0	0000	0	0	0	1
g	0001	0	0	0011	G	0	$\bar{G}$	$\bar{F}$
		0	1	1001				
		1	0	0010				
		1	1	1000				
e	0010	0	0	0110	0	$\bar{G}$	$\bar{F}$	G
		0	1	0011				
		1	0	0100				
		1	1	0001				
f	0011	0	0	0010	FG	$F\bar{G}$	$\bar{G}$	G
		0	1	0001				
		1	0	0110				
		1	1	1001				
c	0100	0	0	1100	$\bar{G}$	$\bar{F}$	G	0
		0	1	0110				
		1	0	1000				
		1	1	0010				
	0101	0	0	0000	0	0	0	0
d	0110	0	0	0100	$F\bar{G}$	$\bar{G}$	G	FG
		0	1	0010				
		1	0	1100				
		1	1	0011				
	0111	0	0	0000	0	0	0	0
a	1000	0	0	1001	$\bar{F}$	G	0	$\bar{G}$
		0	1	1100				
		1	0	0001				
		1	1	0100				
h	1001	0	0	0001	G	FG	$F\bar{G}$	$\bar{G}$
		0	1	1000				
		1	0	0011				
		1	1	1100				
	1010	0	0	0000	0	0	0	0
	1011	0	0	0000	0	0	0	0
b	1100	0	0	1000	$\bar{G}$	G	FG	$F\bar{G}$
		0	1	0100				
		1	0	1001				
		1	1	0110				
	1101	0	0	0000	0	0	0	0
	1110	0	0	0000	0	0	0	0
	1111	0	0	0000	0	0	0	0

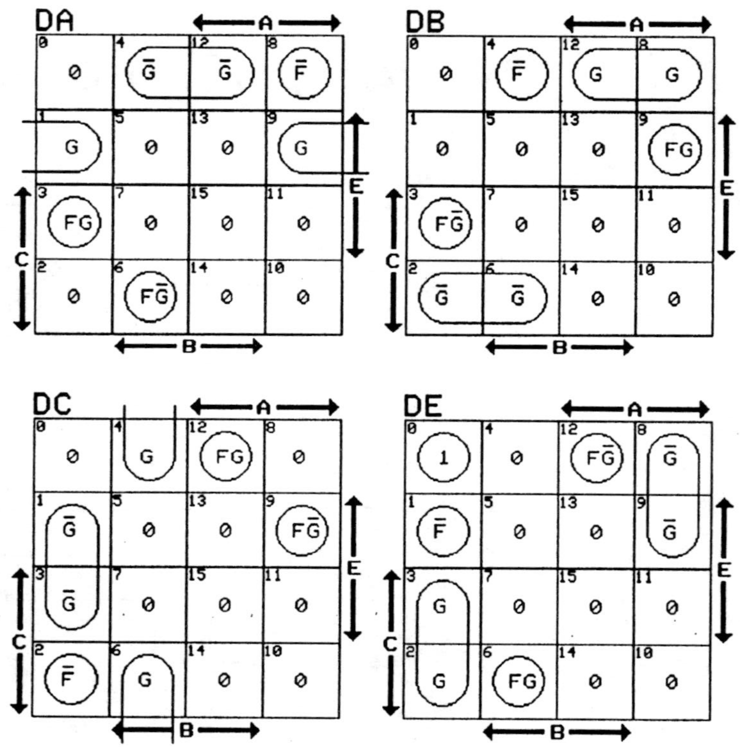

F = FULL
G = CW

$$DA = B\bar{C}\bar{E}\bar{G} + \bar{B}\bar{C}EG + A\bar{B}\bar{C}\bar{E}\bar{F} + \bar{A}\bar{B}CEFG + \bar{A}BC\bar{E}\bar{F}\bar{G}$$

$$DB = A\bar{C}\bar{E}G + \bar{A}C\bar{E}\bar{G} + \bar{A}B\bar{C}\bar{E}\bar{F} + A\bar{B}\bar{C}EFG + \bar{A}\bar{B}CEF\bar{G}$$

$$DC = \bar{A}\bar{B}E\bar{G} + \bar{A}B\bar{E}G + AB\bar{C}\bar{E}FG + \bar{A}\bar{B}C\bar{E}\bar{F} + A\bar{B}\bar{C}EF\bar{G}$$

$$DE = A\bar{B}\bar{C}\bar{G} + \bar{A}\bar{B}CG + AB\bar{C}\bar{E}F\bar{G} + \bar{A}\bar{B}\bar{C}E\bar{F} + \bar{A}BC\bar{E}FG + \bar{A}\bar{B}\bar{C}\bar{E}$$

FIGURE 6.16 State table solution for stepper motor controller.

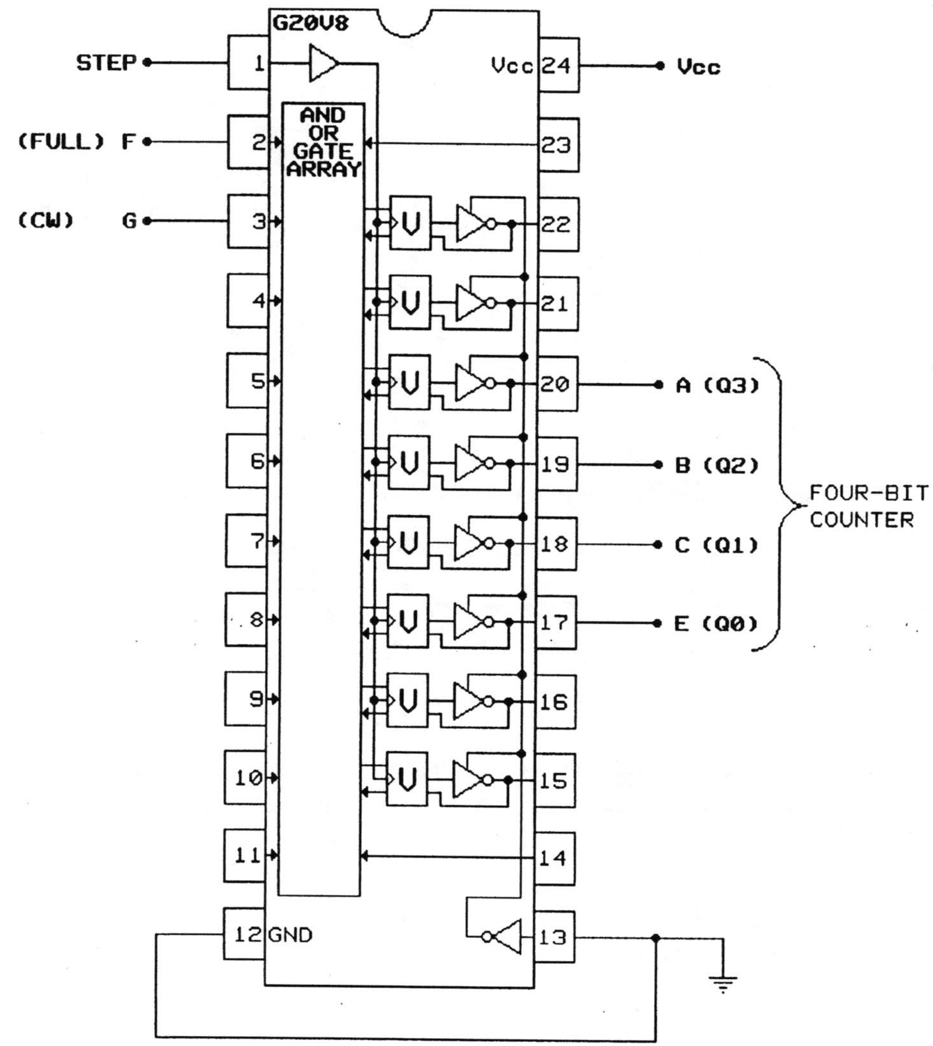

FIGURE 6.17 20V8 GAL configuration for stepper motor controller.

```
Name        EXP0604A;
Date        01/01/96;
Revision    01;
Designer    J. Carter;
Device      G20V8;
Format      J;

/*******************************************************************/
/* PM TYPE STEPPER MOTOR CONTROLLER WITH HALF STEP CONTROL        */
/* Example 6.4A    Using counter input logic functions.           */
/*******************************************************************/

/* Inputs */
Pin 2 = F;                    /* 1=Full Step  0=Half Step          */
PIN 3 = G;                    /* 1=Clockwise  0=Counterclockwise   */

/* Outputs */
Pin 20 = A;                   /* Motor Coil Control Lines Q3       */
Pin 19 = B;                   /*                          Q2       */
Pin 18 = C;                   /*                          Q1       */
Pin 17 = E;                   /*                          Q0       */

/* Input Forming Logic */
A.D = (      B & !C & !E &      !G)
    # (     !B & !C &  E &       G)
    # ( A & !B & !C & !E & !F)
    # (!A & !B &  C &  E &  F &  G)
    # (!A &  B &  C & !E & !F & !G);
```

```
B.D = ( A &      !C & !E &        G)
    # (!A &       C & !E &       !G)
    # (!A &  B & !C & !E & !F)
    # ( A & !B & !C &  E &  F &  G)
    # (!A & !B &  C &  E &  F & !G);

C.D = (!A & !B &       E &       !G)
    # (!A &  B &      !E &        G)
    # ( A &  B & !C & !E &  F &  G)
    # (!A & !B &  C & !E & !F)
    # ( A & !B & !C &  E &  F & !G);

E.D = ( A & !B & !C &            !G)
    # (!A & !B &  C &             G)
    # ( A &  B & !C & !E &  F & !G)
    # (!A & !B & !C &  E & !F)
    # (!A &  B &  C & !E &  F &  G)
    # (!A & !B & !C & !E);
```

```
Sequence [A,B,C,E] {
        Present 'b'0000     next 'b'0001;
        Present 'b'0001                            /*  STATE g */
                if !F & !G next 'b'0011;
                if !F &  G next 'b'1001;
                if  F & !G next 'b'0010;
                if  F &  G next 'b'1000;
        Present 'b'0010                            /*  STATE e */
                if !F & !G next 'b'0110;
                if !F &  G next 'b'0011;
                if  F & !G next 'b'0100;
                if  F &  G next 'b'0001;
        Present 'b'0011                            /*  STATE f */
                if !F & !G next 'b'0010;
                if !F &  G next 'b'0001;
                if  F & !G next 'b'0110;
                if  F &  G next 'b'1001;
        Present 'b'0100                            /*  STATE c */
                if !F & !G next 'b'1100;
                if !F &  G next 'b'0110;
                if  F & !G next 'b'1000;
                if  F &  G next 'b'0010;
        Present 'b'0101     next 'b'0000;
        Present 'b'0110                            /*  STATE d */
                if !F & !G next 'b'0100;
                if !F &  G next 'b'0010;
                if  F & !G next 'b'1100;
                if  F &  G next 'b'0011;
        Present 'b'0111     next 'b'0000;
```

```
Present 'b'1000                                   /*  STATE a */
        if !F & !G next 'b'1001;
        if !F &  G next 'b'1100;
        if  F & !G next 'b'0001;
        if  F &  G next 'b'0100;
Present 'b'1001                                   /*  STATE h */
        if !F & !G next 'b'0001;
        if !F &  G next 'b'1000;
        if  F & !G next 'b'0011;
        if  F &  G next 'b'1100;
Present 'b'1010    next 'b'0000;
Present 'b'1011    next 'b'0000;
Present 'b'1100                                   /*  STATE b */
        if !F & !G next 'b'1000;
        if !F &  G next 'b'0100;
        if  F & !G next 'b'1001;
        if  F &  G next 'b'0110;
Present 'b'1101    next 'b'0000;
Present 'b'1110    next 'b'0000;
Present 'b'1111    next 'b'0000; }
```

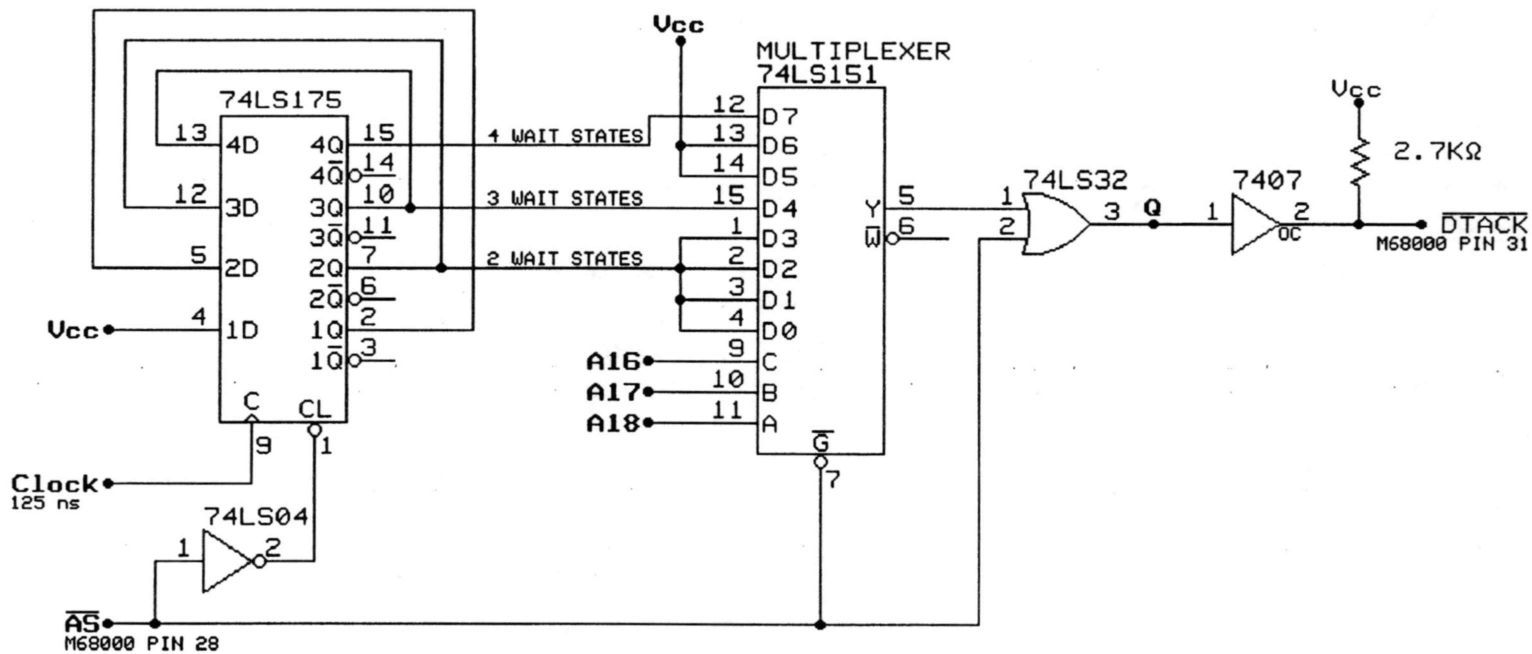

FIGURE 6.18 Wait state generation circuitry.

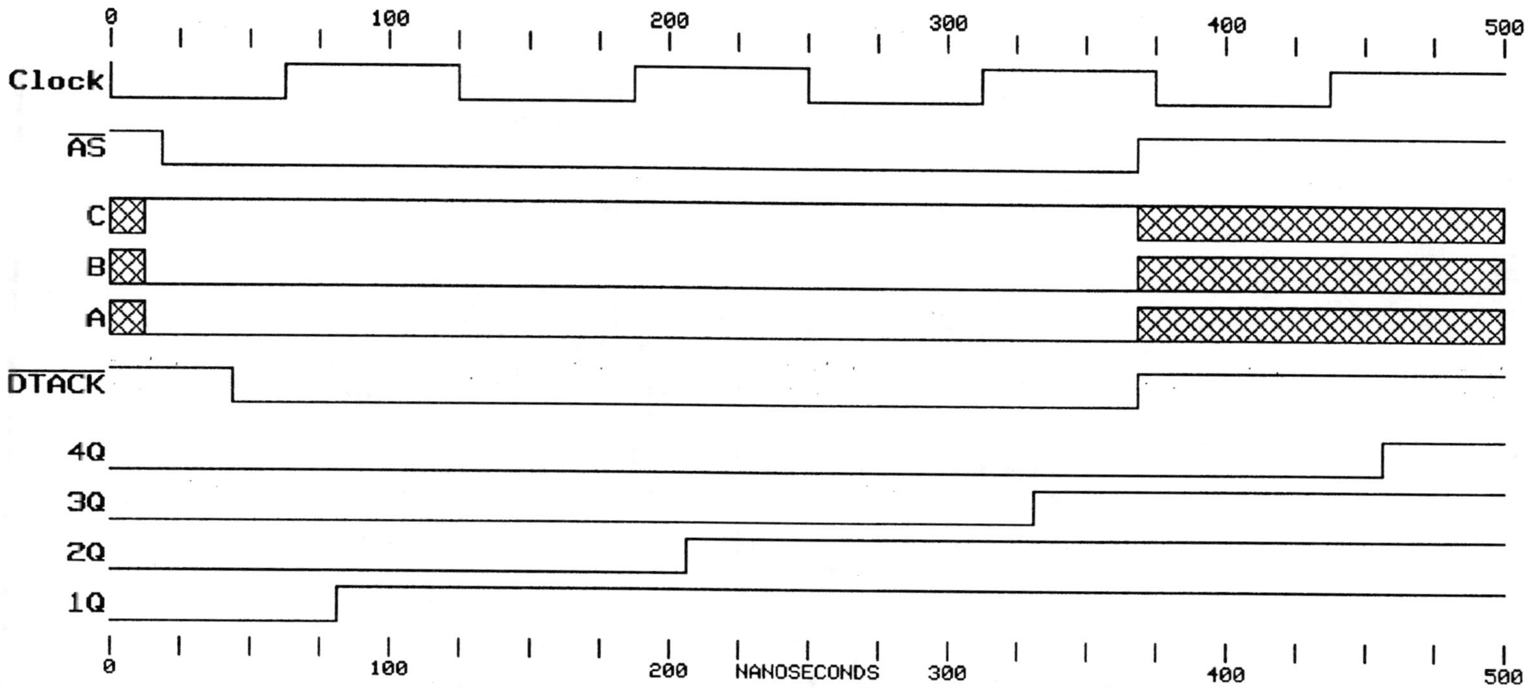

FIGURE 6.19 Wait state generator timing.

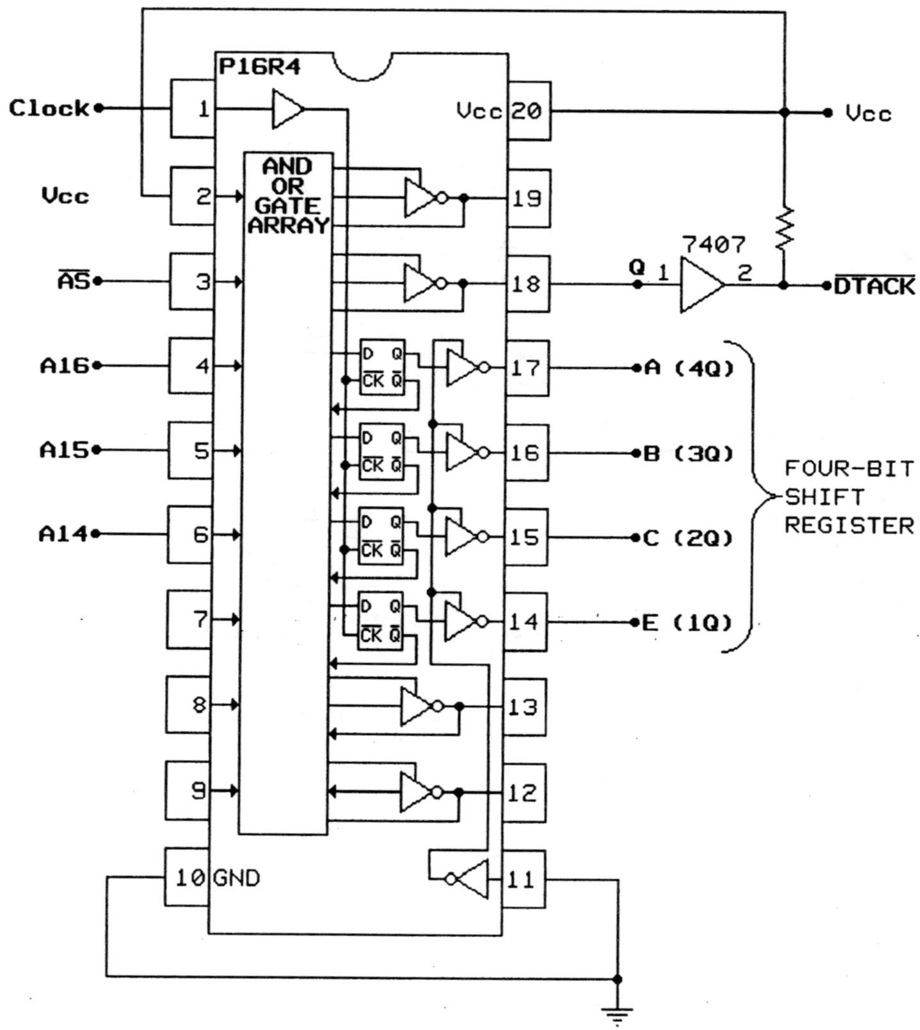

FIGURE 6.20 16R4 PAL implementation of the wait state generator.

```
Name      EXP0605;
Date      01/01/96;
Revision  01;
Designer  J. Carter;
Device    P16R4;
Format    J;

/*****************************************************************/
/* Example 6.5  DTACK Generator Circuit                          */
/*****************************************************************/

/* Inputs */
Pin 2 = Vcc;            /* Input to D-Type Shift Register     */
Pin 3 = AS;             /* Address Strobe                     */
Pin 4 = A16;            /* Address Bus                        */
Pin 5 = A15;
Pin 6 = A14;

/* Outputs */
Pin 18 = Q;             /* Generator Output to Buffer         */
Pin 17 = A;             /* Shift Register Bit 1Q              */
Pin 16 = B;             /*                    2Q              */
Pin 15 = C;             /*                    3Q              */
Pin 14 = E;             /*                    4Q              */

/*  Shift Register */
A.D = Vcc;
B.D = A;
C.D = B;
E.D = C;

/* Multiplexer */
Q = ( (!A16 & B)                   /* D3, D2, D1, D0          */
    # ( A16 & !A15 & !A14 & C)     /* D4                      */
    # ( A16 & !A15 &  A14)         /* D5                      */
    # ( A16 &  A15 & !A14)         /* D6                      */
    # ( A16 &  A15 &  A14 & E) )   /* D7                      */
    # AS;
```

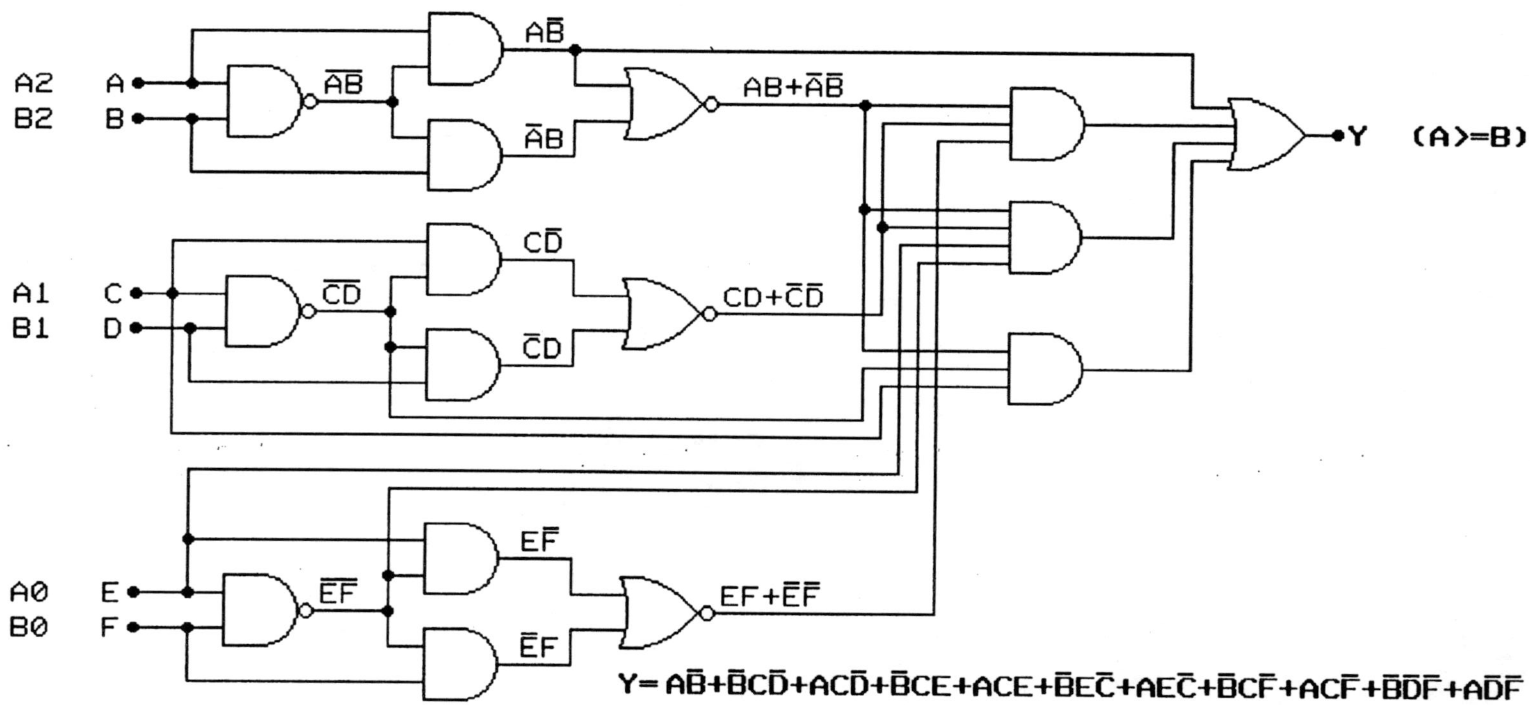

Figure 6.21 Three-bit magnitude comparitor

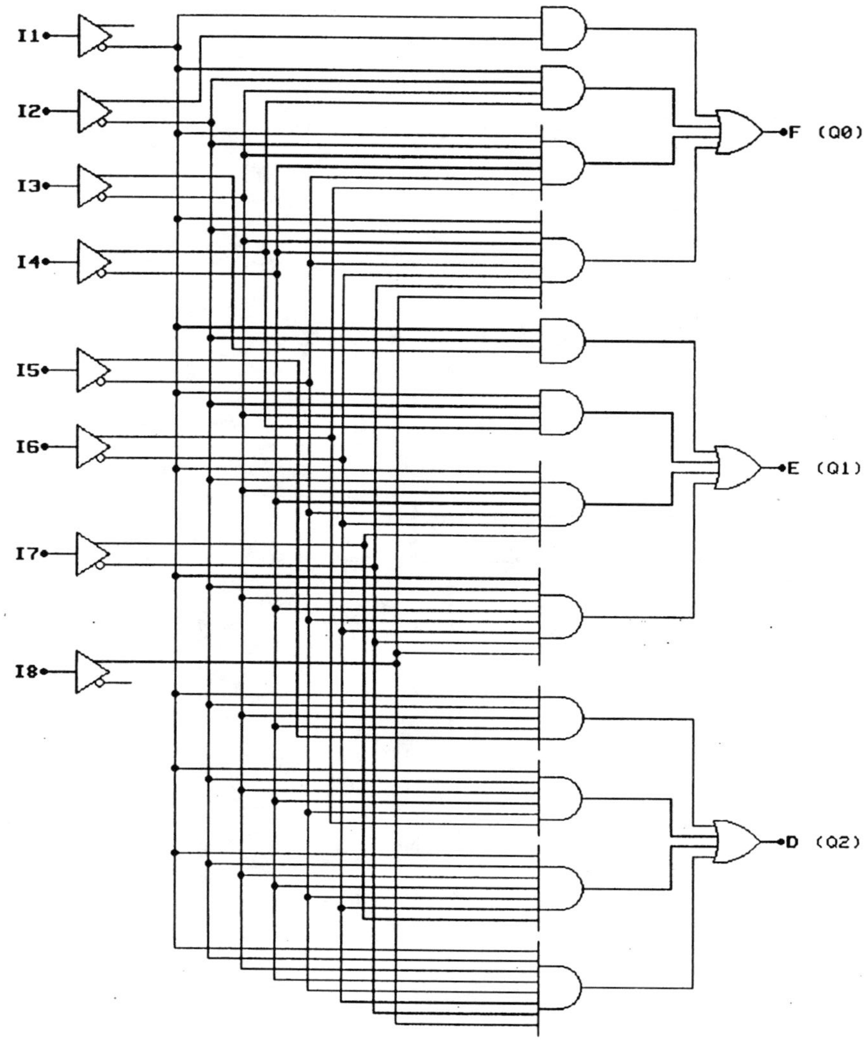

FIGURE 6.22 8-bit to 3-bit priority encoder circuit.

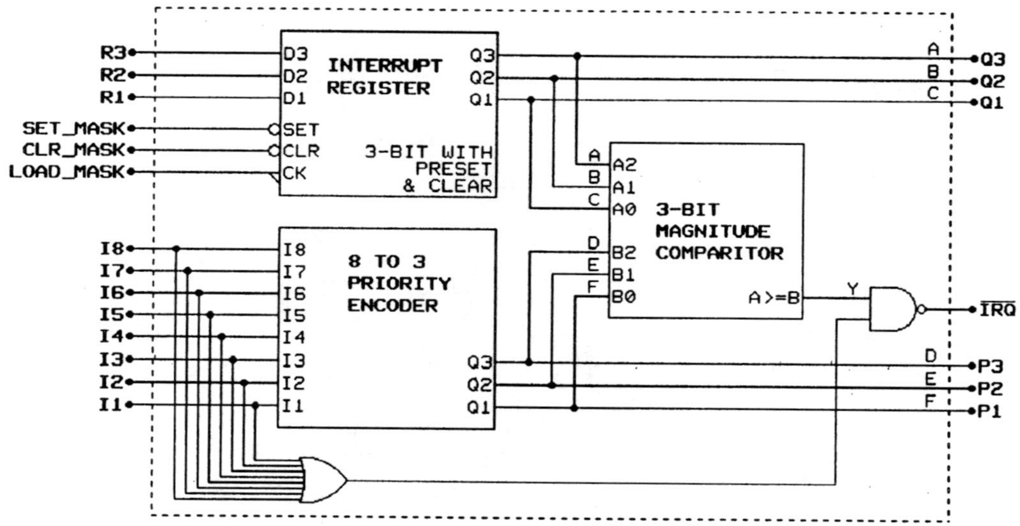

FIGURE 6.23 Priority Encoder with interrupt register, block diagram.

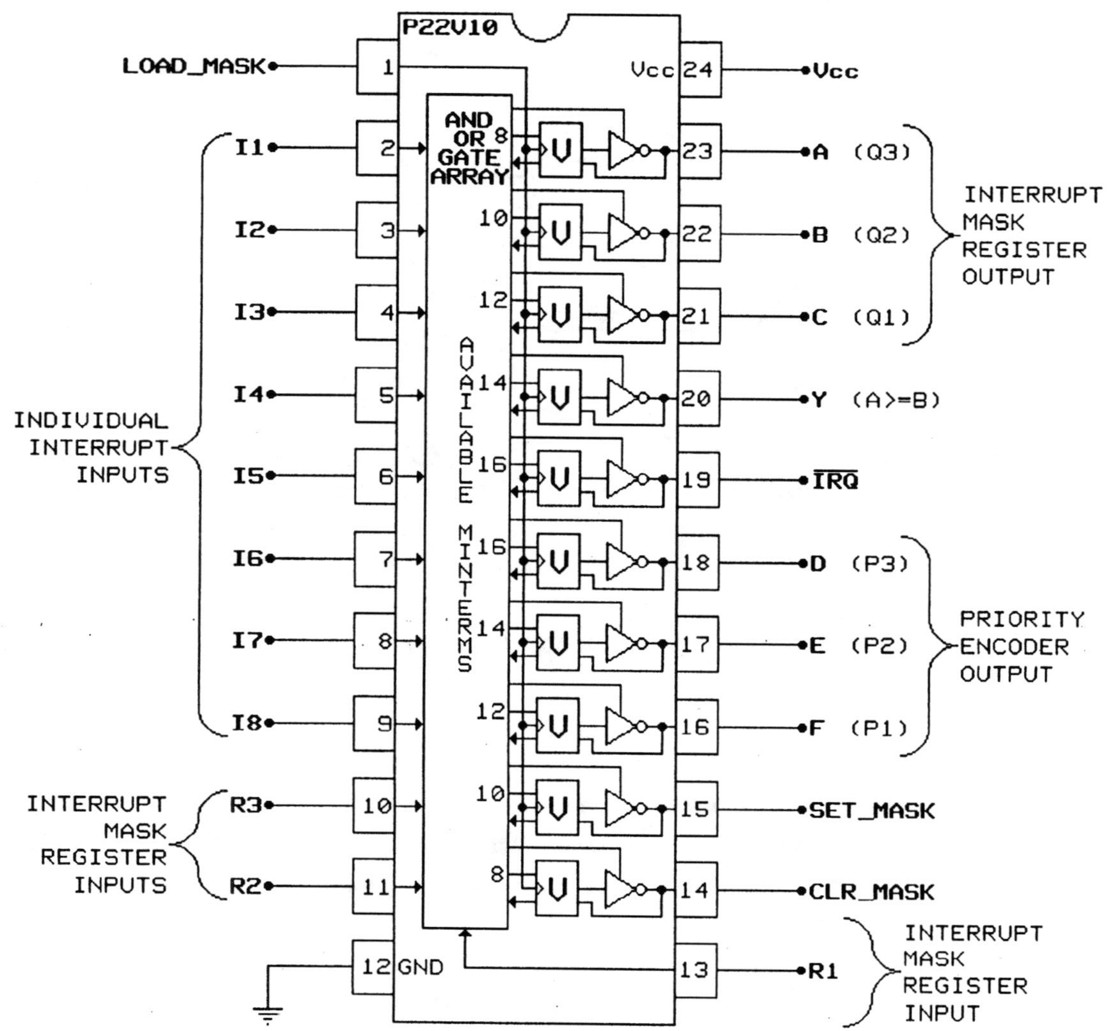

FIGURE 6.24 P22V10 configured as the priority encoder with mask register.

```
Name      EXP0606;
Date      01/01/96;
Revision  01;
Designer  J. Carter;
Device    G22V10;
Format    J;

/*****************************************************************/
/* Example 6.6  8 TO 3 Priority Encoder with Interrupt Masking  */
/*****************************************************************/

/* Inputs */
Pin 1 = LOAD_MASK;
Pin [2..9] = [I1..I8];       /* Individual IRQ inputs.                */
Pin 10 = R3;                 /* Interrupt Mask Register Inputs        */
Pin 11 = R2;
Pin 13 = R1;
Pin 14 = CLR_MASK;           /* Set Interrupt Mask Register to 000    */
Pin 15 = SET_MASK;           /* Set Interrupt Mask Register to 111    */

/* Outputs */
Pin 23 = A;                  /* Interrupt Mask Register  R3           */
Pin 22 = B;                  /*                          R2           */
Pin 21 = C;                  /*                          R1           */
Pin 20 = Y;
Pin 19 = IRQ;                /* !IRQ Output to Processor              */
Pin 18 = D;                  /* Priority Encoder Output  Q3           */
Pin 17 = E;                  /*                          Q2           */
Pin 16 = F;                  /*                          Q1           */
```

```
/* Three-Bit Comparitor */
Y = (A & !B) # (!B &  C & !D) # ( A &  C & !D) #
   (!B &  C &  E) # ( A &  C &  E) # (!B & !C &  E) # ( A & !C &  E) #
   (!B &  C & !F) # ( A &  C & !F) # (!B & !D & !F) # ( A & !D & !F);

/*  Priority Encoder */
D = (!I1 & !I2 & !I3 & !I4 & I5) #
    (!I1 & !I2 & !I3 & !I4 & !I5 & I6) #
    (!I1 & !I2 & !I3 & !I4 & !I5 & !I6 & I7) #
    (!I1 & !I2 & !I3 & !I4 & !I5 & !I6 & !I7 & I8);
E = (!I1 & !I2 & I3) # (!I1 & !I2 & !I3 & I4) #
    (!I1 & !I2 & !I3 & !I4 & !I5 & !I6 & I7) #
    (!I1 & !I2 & !I3 & !I4 & !I5 & !I6 & !I7 & I8);
F = (!I1 & I2) # (!I1 & !I2 & !I3 & I4) #
    (!I1 & !I2 & !I3 & !I4 & !I5 & I6) #
    (!I1 & !I2 & !I3 & !I4 & !I5 & !I6 & !I7 & I8);

/* Interrupt Mask Register */
A.D = R3;   A.AR = !CLR_MASK;  A.SP = !SET_MASK;
B.D = R2;   B.AR = !CLR_MASK;  B.SP = !SET_MASK;
C.D = R1;   C.AR = !CLR_MASK;  C.SP = !SET_MASK;

/*  IRQ Output */
IRQ  = !(I1 & I2 & I3 & I4 & I5 & I6 & I7 & I8) # !Y;
```

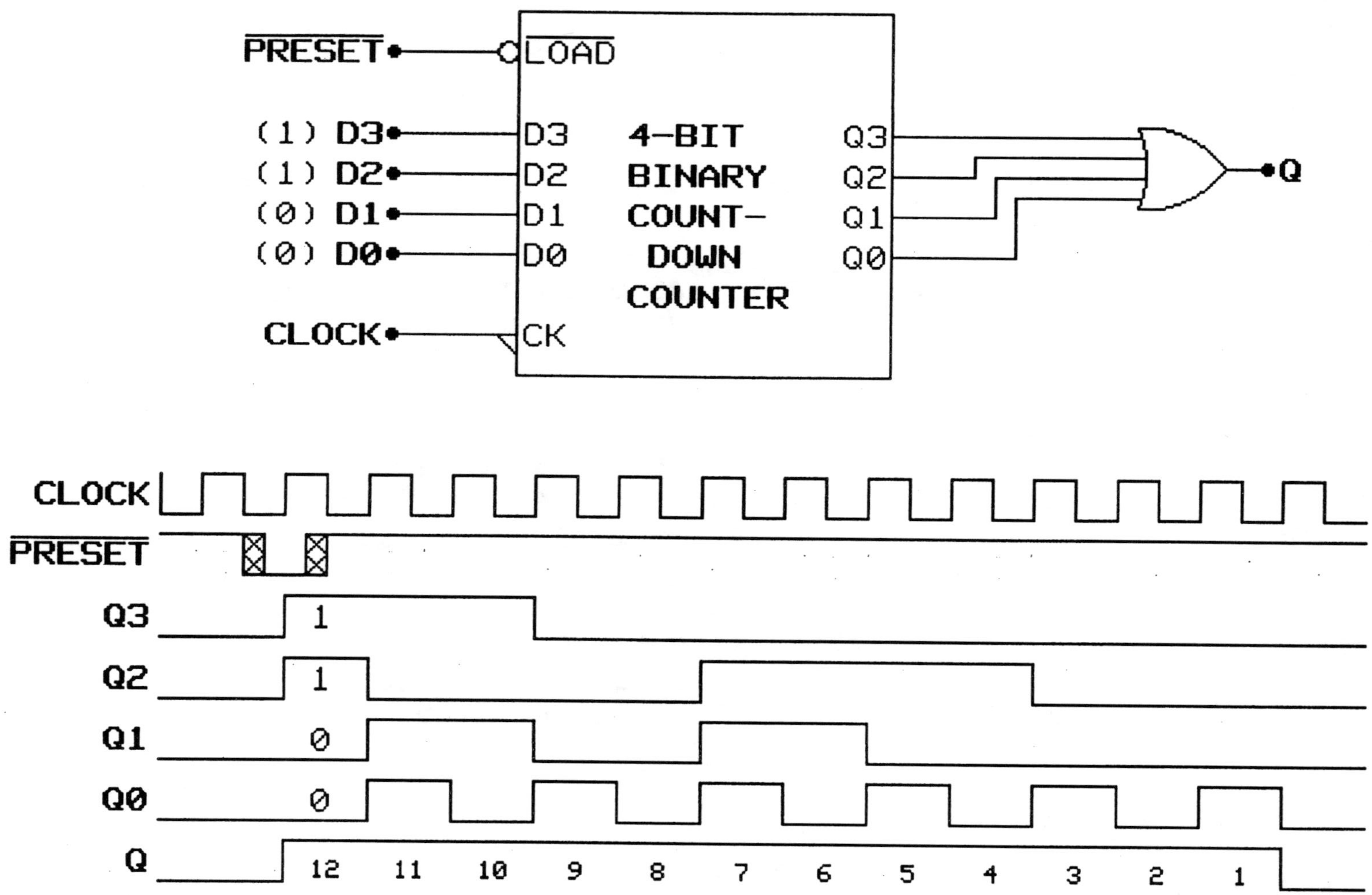

*FIGURE 6.25 **Programmable retriggerable monostable block diagram and timing.***

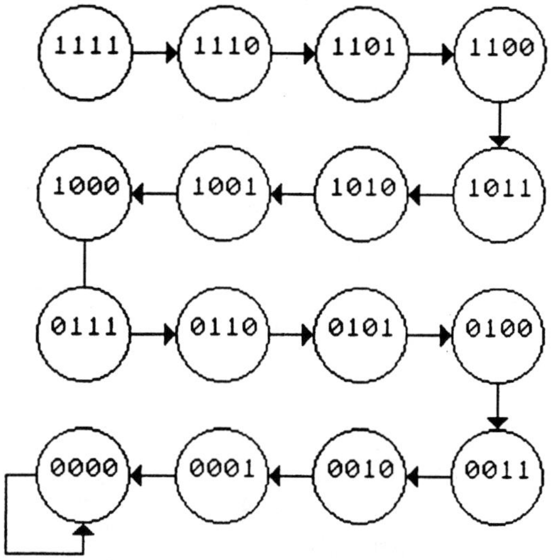

PRESENT ABCE	NEXT ABCE	INPUT DA	FORMING DB	DC	LOGIC DE
0000	0000	0	0	0	0
0001	0000	0	0	0	0
0010	0001	0	0	0	1
0011	0010	0	0	1	0
0100	0011	0	0	1	1
0101	0100	0	1	0	0
0110	0101	0	1	0	1
0111	0110	0	1	1	0
1000	0111	0	1	1	1
1001	1000	1	0	0	0
1010	1001	1	0	0	1
1011	1010	1	0	1	0
1100	1011	1	0	1	1
1101	1100	1	1	0	0
1110	1101	1	1	0	1
1111	1110	1	1	1	0

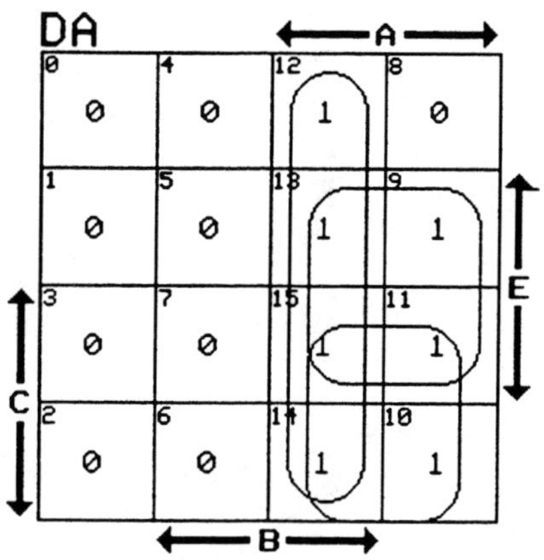

$DA = AB + AE + AC$

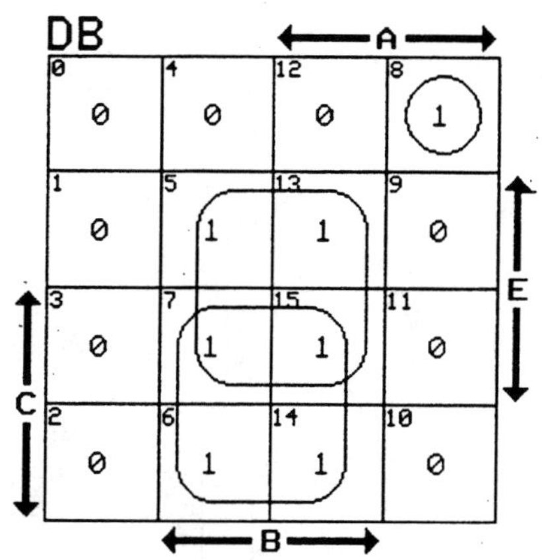

$DB = A\bar{B}\bar{C}\bar{E} + BC + BE$

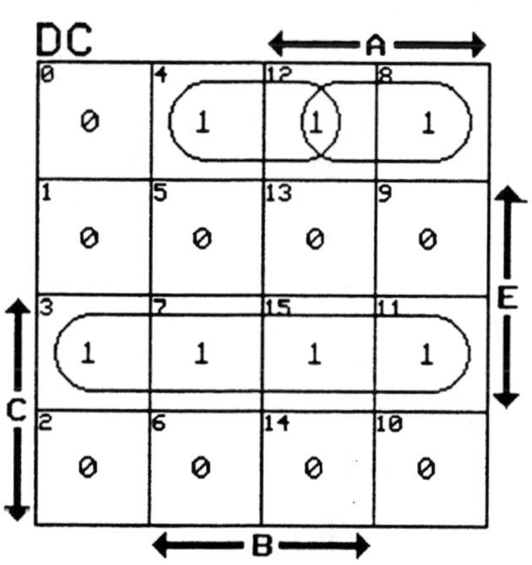

$DC = CE + A\bar{C}\bar{E} + B\bar{C}\bar{E}$

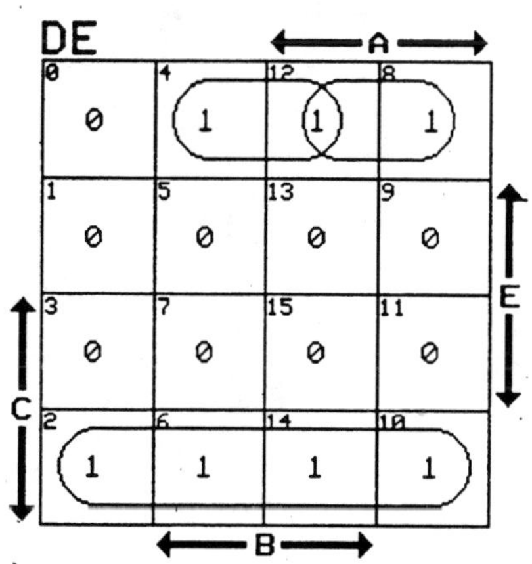

$DE = C\bar{E} + A\bar{C}\bar{E} + B\bar{C}\bar{E}$

FIGURE 6.26 State diagram, state table and Karnaugh map reduction.

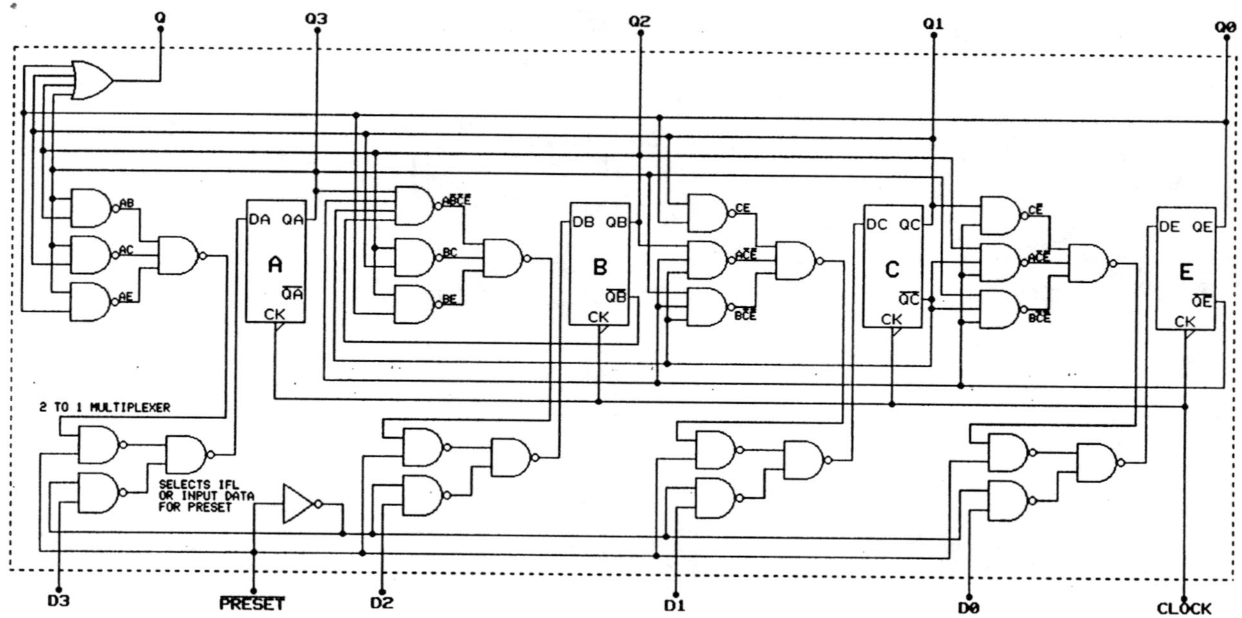

FIGURE 6.27 Detailed Schematic, programmable retriggerable monostable multivibrator.

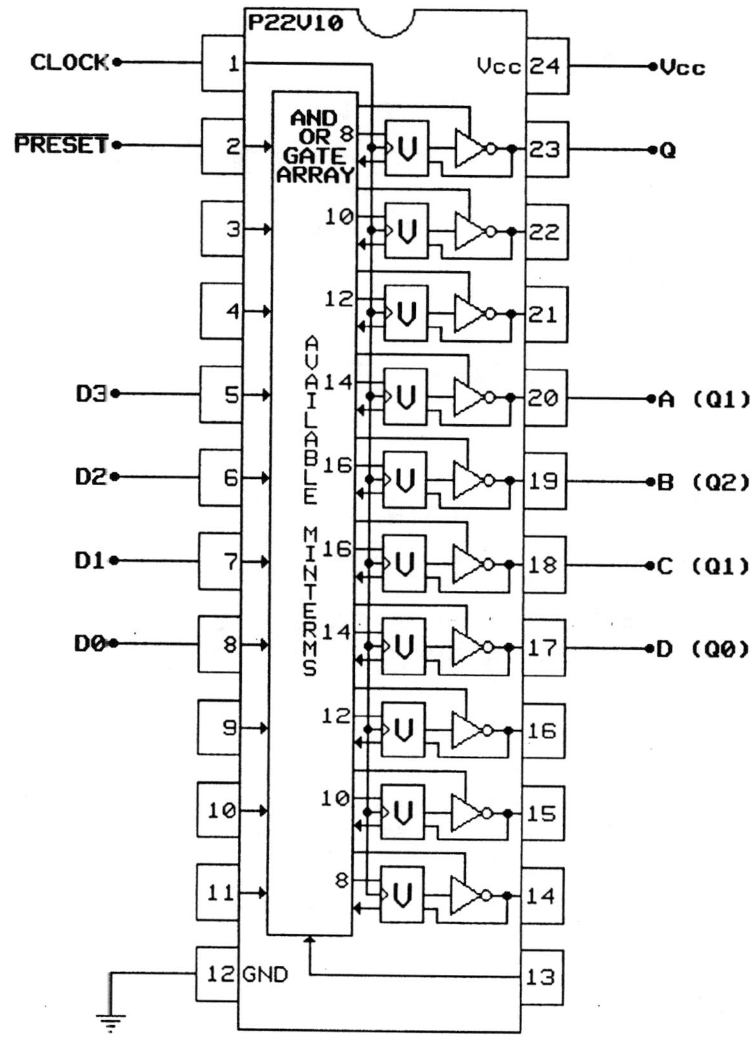

FIGURE 6.28 22V8 PAL configured as a monostable multivibrator.

```
Name        EXP0707;
Date        01/01/96;
Revision    01;
Designer    J. Carter;
Device      P22V10;
Format      J;

/*******************************************************************/
/*  Example 6.7                                                    */
/*  Programmable Retriggerable Monostable Multivibrator            */
/*******************************************************************/

/* Inputs */
Pin 1 = CLOCK;
Pin 2 = PRESET;             /* Trigger input Loads Count Register */
Pin [5..8] = [D3..0];       /* Input Data for Count Register      */

/* Outputs */
Pin 16 = E;                 /* Four-Bit Binary Counter  Q0         */
Pin 17 = C;                 /*                          Q1         */
Pin 18 = B;                 /*                          Q2         */
Pin 19 = A;                 /*                          Q3         */
Pin 23 = Q;                 /* Multivibrator output pulse          */

/* Four-Bit Binary Counter */
A.D = ((A & B) # (A & E) # (A & C)                 & PRESET) # (D3 & !PRESET);
B.D = ((A & !B & !C & !E) # (B & C) # (B & E)      & PRESET) # (D2 & !PRESET);
C.D = ((C & E) # (A & !C & !E) # (B & !C & !E)     & PRESET) # (D1 & !PRESET);
E.D = ((C & !E) # (A & !C & !E) # (B & !C & !E)    & PRESET) # (D0 & !PRESET);

/* Multivibrator Output */
Q = A # B # C # E;
```

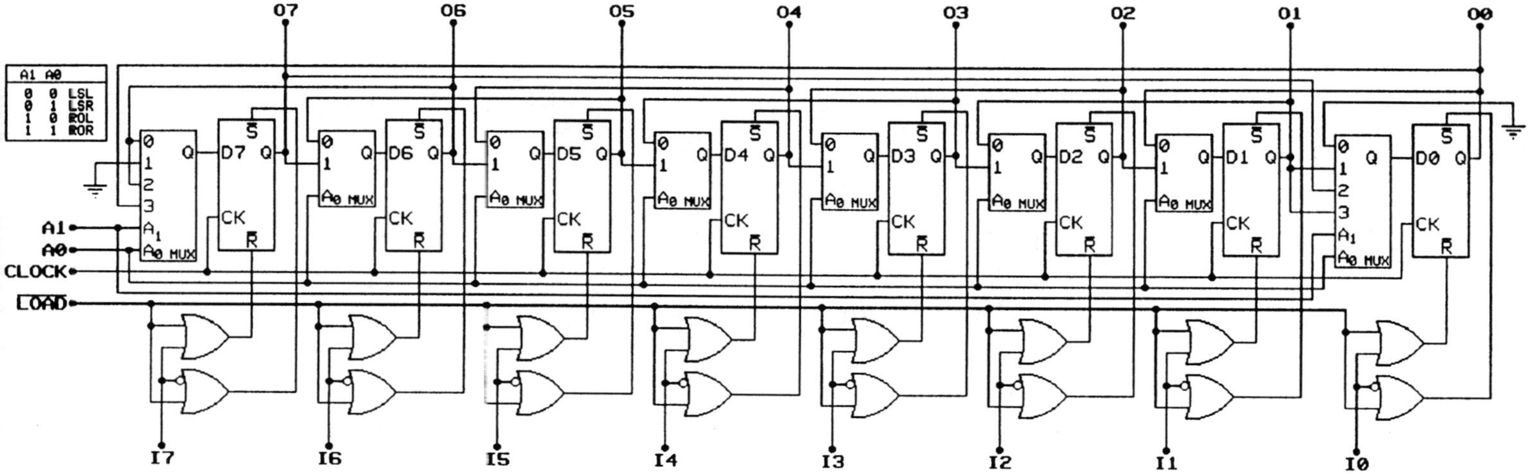

FIGURE 6.29 8-bit shift register with logical shift and rotate linkage.

A1	A0	MNEM	O7	O6	O5	O4	O3	O2	O1	O0
0	0	LSL	I6	I5	I4	I3	I2	I1	I0	0
0	1	LSR	0	I7	I6	I5	I4	I3	I2	I1
1	0	ROL	I6	I5	I4	I3	I2	I1	I0	I7
1	1	ROR	I0	I7	I6	I5	I4	I3	I2	I1

FIGURE 6.30 Combinational shifter output functions.

```
Name      EXP0608;
Date      01/01/96;
Revision  01;
Designer  J. Carter;
Device    P16L8;
Format    J;
/******************************************************************/
/* Example 6.8                                                    */
/* 8-Bit Combinational Shifter with Logical Shift & Rotate        */
/******************************************************************/

/* Inputs */
Pin [1..2] = [A1..0];        /* Function Control Lines          */
                             /* A1,A0 = 0,0 = Logical Shift Left  */
                             /* A1,A0 = 0,1 = Logical Shift Right */
                             /* A1,A0 = 1,0 = Rotate Left         */
                             /* A1,A0 = 1,1 = Rotate Right        */
Pin [3..9] = [I7..I1];       /* Input Data Byte                 */
Pin     11 = I0;

/* Outputs */
Pin [19..12] = [O7..O0];     /* Output Data Byte                */

/* Logical Output Functions */
O7 = (!A0 & I6) # (A1 & A0 & I0);        O7.OE = 'b'1;
O6 = (!A0 & I5) # (A0 & I7);             O6.OE = 'b'1;
O5 = (!A0 & I4) # (A0 & I6);             O5.OE = 'b'1;
O4 = (!A0 & I3) # (A0 & I5);             O4.OE = 'b'1;
O3 = (!A0 & I2) # (A0 & I4);             O3.OE = 'b'1;
O2 = (!A0 & I1) # (A0 & I3);             O2.OE = 'b'1;
O1 = (!A0 & I0) # (A0 & I2);             O1.OE = 'b'1;
O0 = ( A0 & I1) # (A1 & !A0 & I7);       O0.OE = 'b'1;
```

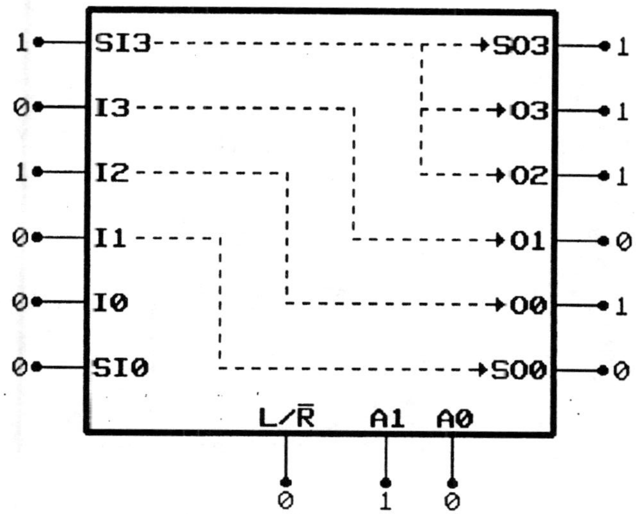

L/R̄	A1	A0	SO3	O3	O2	O1	O0	SO0
0	0	0	SI3	I3	I2	I1	I0	SI0
0	0	1	SI3	SI3	I3	I2	I1	I0
0	1	0	SI3	SI3	SI3	I3	I2	I1
0	1	1	SI3	SI3	SI3	SI3	I3	I2
1	0	0	SI3	I3	I2	I1	I0	SI0
1	0	1	I3	I2	I1	I0	SI0	SI0
1	1	0	I2	I1	I0	SI0	SI0	SI0
1	1	1	I1	I0	SI0	SI0	SI0	SI0

***FIGURE 6.31* 4-Bit combinational barrel shifter, block diagram and function table.**

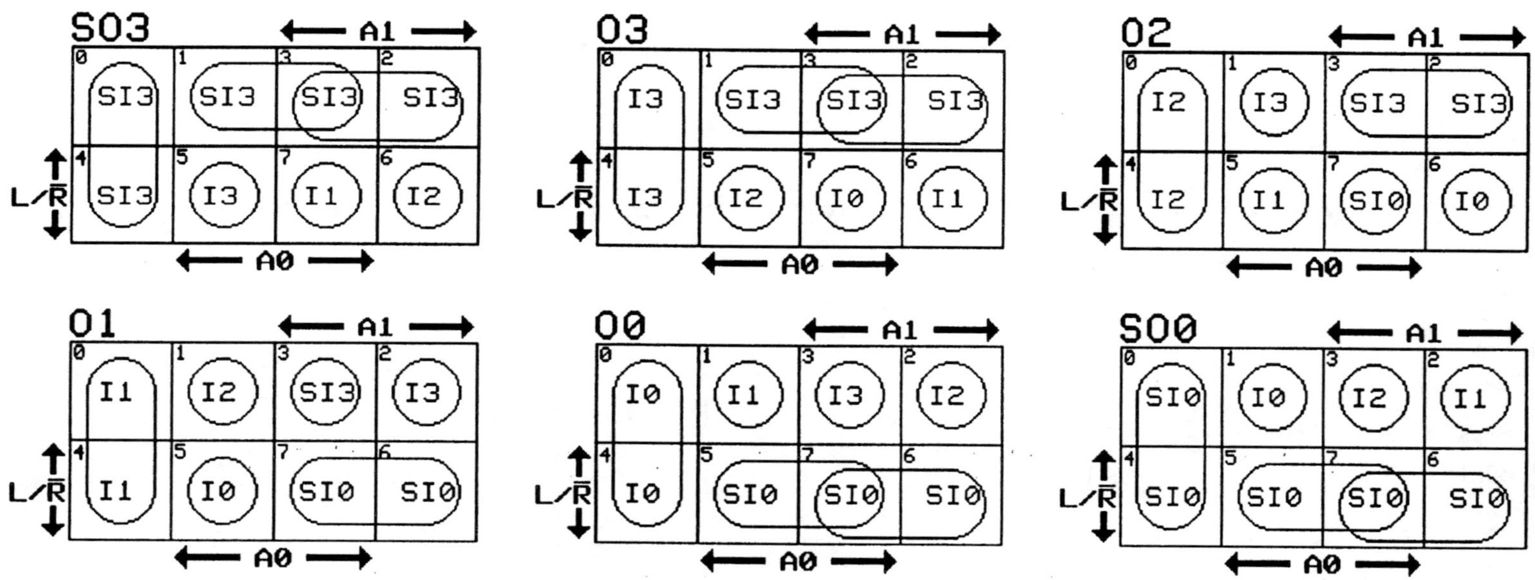

FIGURE 6.32 Karnaugh maps of barrel shifter functions.

$$
\begin{aligned}
SO3 &= (\overline{A0}\cdot\overline{A1}\cdot SI3)+(A0\cdot\overline{LR}\cdot SI3)+(A1\cdot\overline{LR}\cdot SI3)\\
&\quad+(A0\cdot\overline{A1}\cdot LR\cdot I3)+(\overline{A0}\cdot A1\cdot LR\cdot I2)+(A0\cdot A1\cdot LR\cdot I1)\\
O3 &= (\overline{A0}\cdot\overline{A1}\cdot I3)+(A0\cdot\overline{LR}\cdot SI3)+(A1\cdot\overline{LR}\cdot SI3)\\
&\quad+(A0\cdot\overline{A1}\cdot LR\cdot I2)+(\overline{A0}\cdot A1\cdot LR\cdot I1)+(A0\cdot A1\cdot LR\cdot I0)\\
O2 &= (\overline{A0}\cdot\overline{A1}\cdot I2)+(A0\cdot\overline{A1}\cdot\overline{LR}\cdot I3)+(A1\cdot\overline{LR}\cdot SI3)\\
&\quad+(A0\cdot\overline{A1}\cdot LR\cdot I0)+(\overline{A0}\cdot A1\cdot LR\cdot SI0)+(A0\cdot A1\cdot LR\cdot I1)\\
O1 &= (\overline{A0}\cdot\overline{A1}\cdot I1)+(A0\cdot\overline{A1}\cdot\overline{LR}\cdot I2)+(A0\cdot\overline{A1}\cdot\overline{LR}\cdot I3)\\
&\quad+(A0\cdot A1\cdot\overline{LR}\cdot SI3)+(A0\cdot\overline{A1}\cdot LR\cdot I0)+(A1\cdot LR\cdot SI0)\\
O0 &= (\overline{A0}\cdot\overline{A1}\cdot I0)+(A0\cdot\overline{A1}\cdot\overline{LR}\cdot I1)+(A0\cdot\overline{A1}\cdot\overline{LR}\cdot I2)\\
&\quad+(A0\cdot A1\cdot\overline{LR}\cdot I3)+(A0\cdot LR\cdot SI0)+(A1\cdot LR\cdot SI0)\\
SO0 &= (\overline{A0}\cdot\overline{A1}\cdot SI0)+(A0\cdot\overline{A1}\cdot\overline{LR}\cdot I0)+(A0\cdot\overline{A1}\cdot\overline{LR}\cdot I1)\\
&\quad+(A0\cdot A1\cdot\overline{LR}\cdot I2)+(A0\cdot LR\cdot SI0)+(A1\cdot LR\cdot SI0)
\end{aligned}
$$

```
Name        EXP0609;
Date        01/01/96;
Revision    01;
Designer    J. Carter;
Device      P16L8;
Format      J;

/*********************************************************************/
/* Practical Example 6.9                                             */
/* 4-Bit Combinational Barrel Shifter                                */
/*********************************************************************/

/* Inputs */
Pin 1 = SI3;                 /* Shift into Most Significant Bit    */
Pin [2..5] = [I3..I0];       /* Input Data Lines                   */
Pin 6 = SI0;                 /* Shift into Least Significant Bit   */
Pin 7 = LR;                  /* Left/!Right 1=Left, 0=Right        */
Pin [8..9] = [A1..A0];       /* Number of Shifts: 3, 2, 1 or 0     */

/* Outputs */
Pin 19 = SO3;                /* Data from Most Significant Bit     */
Pin [18..15] = [O3..O0];     /* Output Data Lines                  */
Pin 14 = SO0;                /* Data from Least Significant Bit    */

/* Logical Output Functions */
SO3 = (!A0 & !A1 &  SI3)      # ( A0 & !LR &  SI3)
    # ( A1 & !LR &  SI3)      # ( A0 & !A1 &  LR  & I3)
    # (!A0 &  A1 &  LR  & I2) # ( A0 &  A1 &  LR  & I1);
    SO3.OE = 'b'1;
```

```
O3 =   (!A0 & !A1 &  I3)          # ( A0 & !LR &  SI3)
     # ( A1 & !LR &  SI3)          # ( A0 & !A1 &  LR  & I2)
     # (!A0 &  A1 &  LR  & I1)    # ( A0 &  A1 &  LR  & I0);
     O3.OE = 'b'1;

O2 =   (!A0 & !A1 &  I2)          # ( A0 & !A1 & !LR  & I3)
     # ( A1 & !LR &  SI3)          # ( A0 & !A1 &  LR  & I0)
     # (!A0 &  A1 &  LR  & SI0)   # ( A0 &  A1 &  LR  & I1);
     O2.OE = 'b'1;

O1 =   (!A0 & !A1 &  I1)          # ( A0 & !A1 & !LR  & I2)
     # ( A0 & !A1 & !LR  & I3)    # ( A0 &  A1 & !LR  & SI3)
     # ( A0 & !A1 &  LR  & I0)    # ( A1 &  LR &  SI0);
     O1.OE = 'b'1;
O0 =   (!A0 & !A1 &  I0)          # ( A0 & !A1 & !LR  & I1)
     # ( A0 & !A1 & !LR  & I2)    # ( A0 &  A1 & !LR  & I3)
     # ( A0 &  LR &  SI0)          # ( A1 &  LR &  SI0);
     O0.OE = 'b'1;

SO0 =  (!A0 & !A1 &  SI0)         # ( A0 & !A1 & !LR  & I0)
     # ( A0 & !A1 & !LR  & I1)    # ( A0 &  A1 & !LR  & I2)
     # ( A0 &  LR &  SI0)          # ( A1 &  LR &  SI0);
     SO0.OE = 'b'1;
```

```
Name      Barrel22;
Date      05/11/89;
Designer  Kahl;
Company   Logical Devices, Inc.;
Device    P22V10;

/*****************************************************************/
/*                                                               */
/* 8-Bit Registered Barrel Shifter                               */
/*                                                               */
/* This 8-bit registered barrel shifter takes 8 data inputs      */
/* and cyclically rotates the data from 0 to 7 places under      */
/* control of the select ( S0, S1, S2 ) inputs.  A SET input     */
/* can be used to initialize the outputs to the all ones state   */
/*****************************************************************/

/**  Inputs  **/
PIN 1           = clock;        /* Register Clock                */
PIN [2..9]      = [D7..0];      /* Data Inputs                   */
PIN [10,11,14]  = [S2..0];      /* Shift Count Select Inputs     */
PIN 13          = !out_enable;  /* Register Output Enable        */
PIN 23          = SET;          /* Set to Ones Input             */

/**  Outputs  **/
PIN [15..22] = [Q7..0];         /* Register Outputs              */
```

```
/** Declarations and Intermediate Variable Definitions **/
field shift  = [S2..0];          /* Shift Width Field            */
field input  = [D7..0];          /* Inputs Field                 */
field output = [Q7..0];          /* Outputs Field                */

/** Logic Equations **/
output.d   =  [D7, D6, D5, D4, D3, D2, D1, D0]  &  shift:0
           #  [D0, D7, D6, D5, D4, D3, D2, D1]  &  shift:1
           #  [D1, D0, D7, D6, D5, D4, D3, D2]  &  shift:2
           #  [D2, D1, D0, D7, D6, D5, D4, D3]  &  shift:3
           #  [D3, D2, D1, D0, D7, D6, D5, D4]  &  shift:4
           #  [D4, D3, D2, D1, D0, D7, D6, D5]  &  shift:5
           #  [D5, D4, D3, D2, D1, D0, D7, D6]  &  shift:6
           #  [D6, D5, D4, D3, D2, D1, D0, D7]  &  shift:7;

output.sp = SET;                 /* synchronous preset for SET   */
output.oe = out_enable;          /* tri-state control            */
output.ar = 'b'0;                /* asynchronous reset not used  */
```

ANGLE RADIANS	(SIN(A)+1)*128 DECIMAL	HEX
0.000	128	80
0.196	152	98
0.393	176	B0
0.589	198	C6
0.785	218	DA
0.982	233	EA
1.178	245	F5
1.374	253	FD
1.571	255	FF
1.767	253	FD
1.963	245	F5
2.160	233	EA
2.356	218	DA
2.553	198	C6
2.749	176	B0
2.945	152	98
3.142	128	80
3.338	102	67
3.534	079	4F
3.731	057	39
3.927	037	25
4.123	021	15
4.320	010	0A
4.516	002	02
4.712	000	00
4.909	002	02
5.105	010	0A
5.301	021	15
5.498	037	25
5.694	057	39
5.890	079	4F
6.087	102	67
6.283	128	80

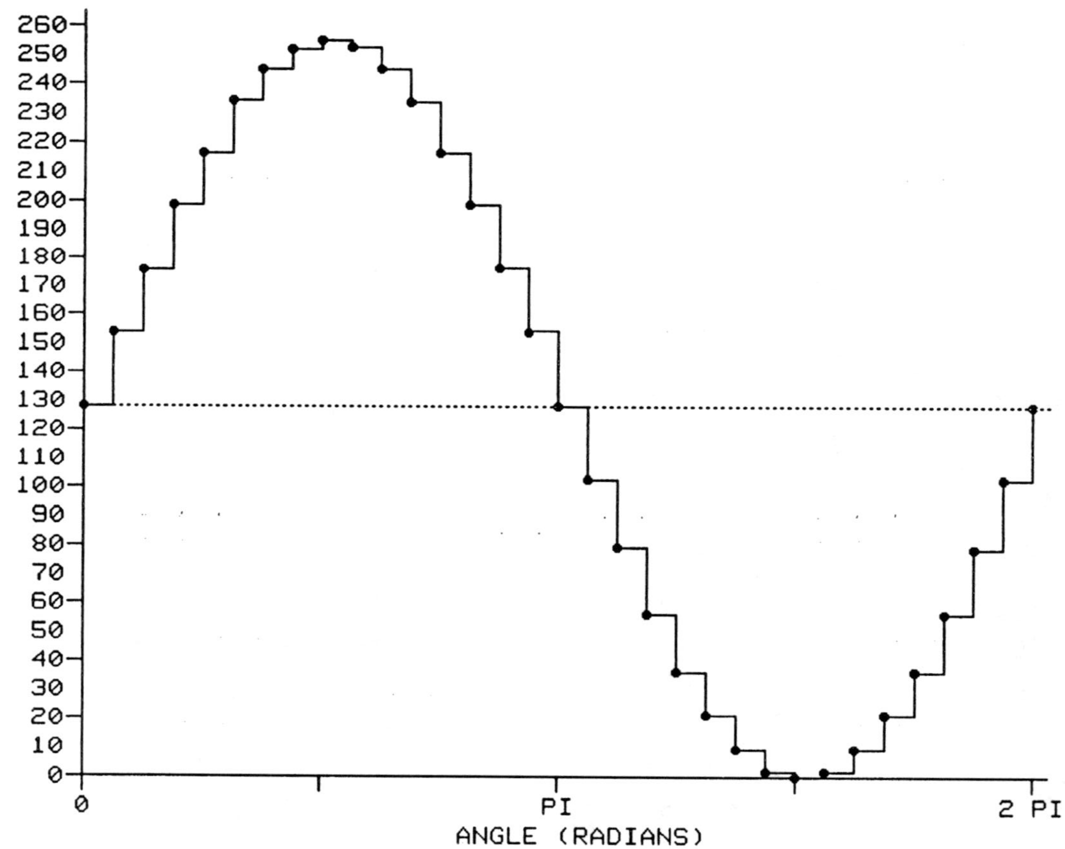

FIGURE 6.33 Sine wave envelope.

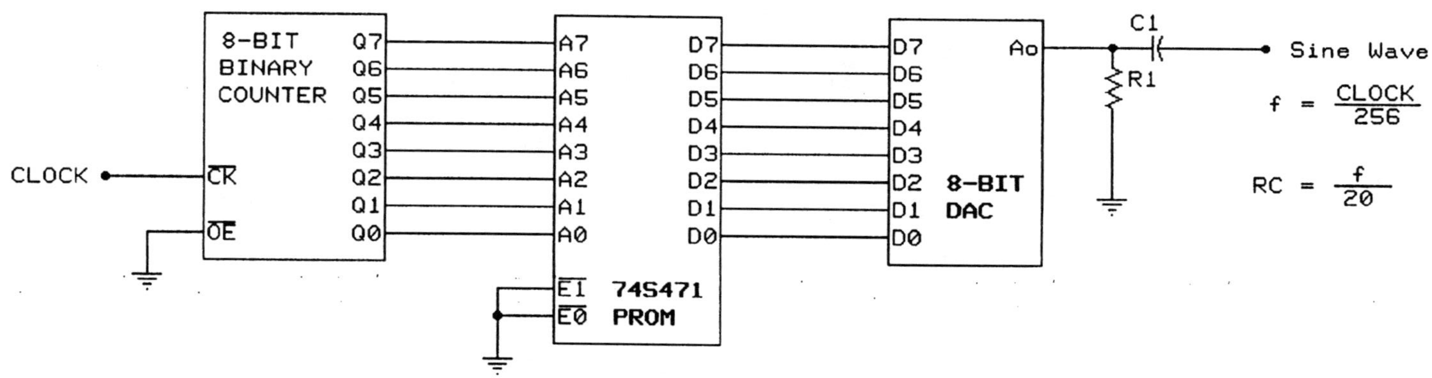

FIGURE 6.34 ***Sine wave synthesizer schematic diagram.***

```
Name       SINEWAVE;
Date       01/01/96;
Designer   J.W. Carter;
Company    Logical Devices, Inc.;
Device     RA8P8;

/******************************************************************/
/* 8-Bit Sine Wave Synthesis                                      */
/*                                                                */
/* This programs a DM74LS471 PROM such that the output follows   */
/* a sine amplitude of 0 to 255 as the angle varies in 256       */
/* steps from 0 to 2 pi radians.  If driven by an 8-bit binary   */
/* clock, the output will generate a sine wave when using a DAC */
/******************************************************************/

/**  Inputs  **/
PIN [1..5]   = [I0..4];          /* Angle Inputs                 */
PIN [17..19] = [I5..7];

/**  Outputs  **/
PIN [6..9]   = [O0..3];          /* Waveform Outputs             */
PIN [11..14] = [O4..7];

FIELD input =  [I7..I0];
FIELD output = [O7..O0];
TABLE input =>output  {

 'H'0 =>'H'80;  'H'2B=>'H'EE;  'H'56=>'H'ED;  'H'81=>'H'7C;  'H'AC=>'H'F ;  'H'D7=>'H'14;
 'H'1 =>'H'83;  'H'2C=>'H'F0;  'H'57=>'H'EB;  'H'82=>'H'79;  'H'AD=>'H'E ;  'H'D8=>'H'15;
 'H'2 =>'H'86;  'H'2D=>'H'F1;  'H'58=>'H'EA;  'H'83=>'H'76;  'H'AE=>'H'C ;  'H'D9=>'H'17;
 'H'3 =>'H'89;  'H'2E=>'H'F3;  'H'59=>'H'E8;  'H'84=>'H'73;  'H'AF=>'H'B ;  'H'DA=>'H'19;
 'H'4 =>'H'8C;  'H'2F=>'H'F4;  'H'5A=>'H'E6;  'H'85=>'H'70;  'H'B0=>'H'A ;  'H'DB=>'H'1B;
 'H'5 =>'H'8F;  'H'30=>'H'F5;  'H'5B=>'H'E4;  'H'86=>'H'6D;  'H'B1=>'H'9 ;  'H'DC=>'H'1D;
 'H'6 =>'H'92;  'H'31=>'H'F6;  'H'5C=>'H'E2;  'H'87=>'H'6A;  'H'B2=>'H'7 ;  'H'DD=>'H'1F;
 'H'7 =>'H'95;  'H'32=>'H'F8;  'H'5D=>'H'E0;  'H'88=>'H'67;  'H'B3=>'H'6 ;  'H'DE=>'H'21;
 'H'8 =>'H'98;  'H'33=>'H'F9;  'H'5E=>'H'DE;  'H'89=>'H'64;  'H'B4=>'H'5 ;  'H'DF=>'H'23;
 'H'9 =>'H'9B;  'H'34=>'H'FA;  'H'5F=>'H'DC;  'H'8A=>'H'61;  'H'B5=>'H'5 ;  'H'E0=>'H'25;
```

```
'H'A =>'H'9E;  'H'35=>'H'FA;  'H'60=>'H'DA;  'H'8B=>'H'5D;  'H'B6=>'H'4 ;  'H'E1=>'H'28;
'H'B =>'H'A2;  'H'36=>'H'FB;  'H'61=>'H'D7;  'H'8C=>'H'5A;  'H'B7=>'H'3 ;  'H'E2=>'H'2A;
'H'C =>'H'A5;  'H'37=>'H'FC;  'H'62=>'H'D5;  'H'8D=>'H'58;  'H'B8=>'H'2 ;  'H'E3=>'H'2C;
'H'D =>'H'A7;  'H'38=>'H'FD;  'H'63=>'H'D3;  'H'8E=>'H'55;  'H'B9=>'H'2 ;  'H'E4=>'H'2F;
'H'E =>'H'AA;  'H'39=>'H'FD;  'H'64=>'H'D0;  'H'8F=>'H'52;  'H'BA=>'H'1 ;  'H'E5=>'H'31;
'H'F =>'H'AD;  'H'3A=>'H'FE;  'H'65=>'H'CE;  'H'90=>'H'4F;  'H'BB=>'H'1 ;  'H'E6=>'H'34;
'H'10=>'H'B0;  'H'3B=>'H'FE;  'H'66=>'H'CB;  'H'91=>'H'4C;  'H'BC=>'H'1 ;  'H'E7=>'H'36;
'H'11=>'H'B3;  'H'3C=>'H'FE;  'H'67=>'H'C9;  'H'92=>'H'49;  'H'BD=>'H'0 ;  'H'E8=>'H'39;
'H'12=>'H'B6;  'H'3D=>'H'FF;  'H'68=>'H'C6;  'H'93=>'H'46;  'H'BE=>'H'0 ;  'H'E9=>'H'3B;
'H'13=>'H'B9;  'H'3E=>'H'FF;  'H'69=>'H'C4;  'H'94=>'H'43;  'H'BF=>'H'0 ;  'H'EA=>'H'3E;
'H'14=>'H'BC;  'H'3F=>'H'FF;  'H'6A=>'H'C1;  'H'95=>'H'41;  'H'C0=>'H'0 ;  'H'EB=>'H'41;
'H'15=>'H'BE;  'H'40=>'H'FF;  'H'6B=>'H'BE;  'H'96=>'H'3E;  'H'C1=>'H'0 ;  'H'EC=>'H'43;
'H'16=>'H'C1;  'H'41=>'H'FF;  'H'6C=>'H'BC;  'H'97=>'H'3B;  'H'C2=>'H'0 ;  'H'ED=>'H'46;
'H'17=>'H'C4;  'H'42=>'H'FF;  'H'6D=>'H'B9;  'H'98=>'H'39;  'H'C3=>'H'0 ;  'H'EE=>'H'49;
'H'18=>'H'C6;  'H'43=>'H'FF;  'H'6E=>'H'B6;  'H'99=>'H'36;  'H'C4=>'H'1 ;  'H'EF=>'H'4C;
'H'19=>'H'C9;  'H'44=>'H'FE;  'H'6F=>'H'B3;  'H'9A=>'H'34;  'H'C5=>'H'1 ;  'H'F0=>'H'4F;
'H'1A=>'H'CB;  'H'45=>'H'FE;  'H'70=>'H'B0;  'H'9B=>'H'31;  'H'C6=>'H'1 ;  'H'F1=>'H'52;
'H'1B=>'H'CE;  'H'46=>'H'FE;  'H'71=>'H'AD;  'H'9C=>'H'2F;  'H'C7=>'H'2 ;  'H'F2=>'H'55;
'H'1C=>'H'D0;  'H'47=>'H'FD;  'H'72=>'H'AA;  'H'9D=>'H'2C;  'H'C8=>'H'2 ;  'H'F3=>'H'58;
'H'1D=>'H'D3;  'H'48=>'H'FD;  'H'73=>'H'A7;  'H'9E=>'H'2A;  'H'C9=>'H'3 ;  'H'F4=>'H'5A;
'H'1E=>'H'D5;  'H'49=>'H'FC;  'H'74=>'H'A5;  'H'9F=>'H'28;  'H'CA=>'H'4 ;  'H'F5=>'H'5D;
'H'1F=>'H'D7;  'H'4A=>'H'FB;  'H'75=>'H'A2;  'H'A0=>'H'25;  'H'CB=>'H'5 ;  'H'F6=>'H'61;
'H'20=>'H'DA;  'H'4B=>'H'FA;  'H'76=>'H'9E;  'H'A1=>'H'23;  'H'CC=>'H'5 ;  'H'F7=>'H'64;
'H'21=>'H'DC;  'H'4C=>'H'FA;  'H'77=>'H'9B;  'H'A2=>'H'21;  'H'CD=>'H'6 ;  'H'F8=>'H'67;
'H'22=>'H'DE;  'H'4D=>'H'F9;  'H'78=>'H'98;  'H'A3=>'H'1F;  'H'CE=>'H'7 ;  'H'F9=>'H'6A;
'H'23=>'H'E0;  'H'4E=>'H'F8;  'H'79=>'H'95;  'H'A4=>'H'1D;  'H'CF=>'H'9 ;  'H'FA=>'H'6D;
'H'24=>'H'E2;  'H'4F=>'H'F6;  'H'7A=>'H'92;  'H'A5=>'H'1B;  'H'D0=>'H'A ;  'H'FB=>'H'70;
'H'25=>'H'E4;  'H'50=>'H'F5;  'H'7B=>'H'8F;  'H'A6=>'H'19;  'H'D1=>'H'B ;  'H'FC=>'H'73;
'H'26=>'H'E6;  'H'51=>'H'F4;  'H'7C=>'H'8C;  'H'A7=>'H'17;  'H'D2=>'H'C ;  'H'FD=>'H'76;
'H'27=>'H'E8;  'H'52=>'H'F3;  'H'7D=>'H'89;  'H'A8=>'H'15;  'H'D3=>'H'E ;  'H'FE=>'H'79;
'H'28=>'H'EA;  'H'53=>'H'F1;  'H'7E=>'H'86;  'H'A9=>'H'14;  'H'D4=>'H'F ;  'H'FF=>'H'7C;
'H'29=>'H'EB;  'H'54=>'H'F0;  'H'7F=>'H'83;  'H'AA=>'H'12;  'H'D5=>'H'11;
'H'2A=>'H'ED;  'H'55=>'H'EE;  'H'80=>'H'7F;  'H'AB=>'H'11;  'H'D6=>'H'12;
}
```

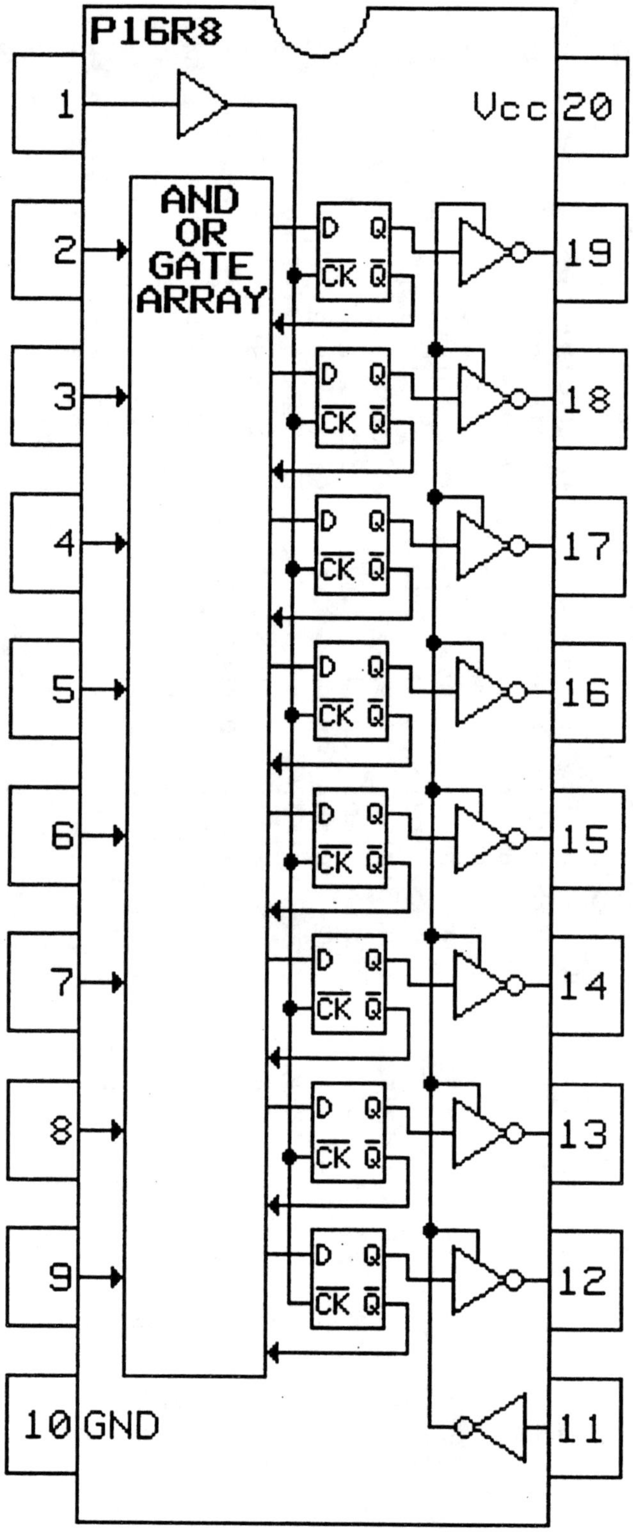

FIGURE 7.1 P16R8 programmable logic array.

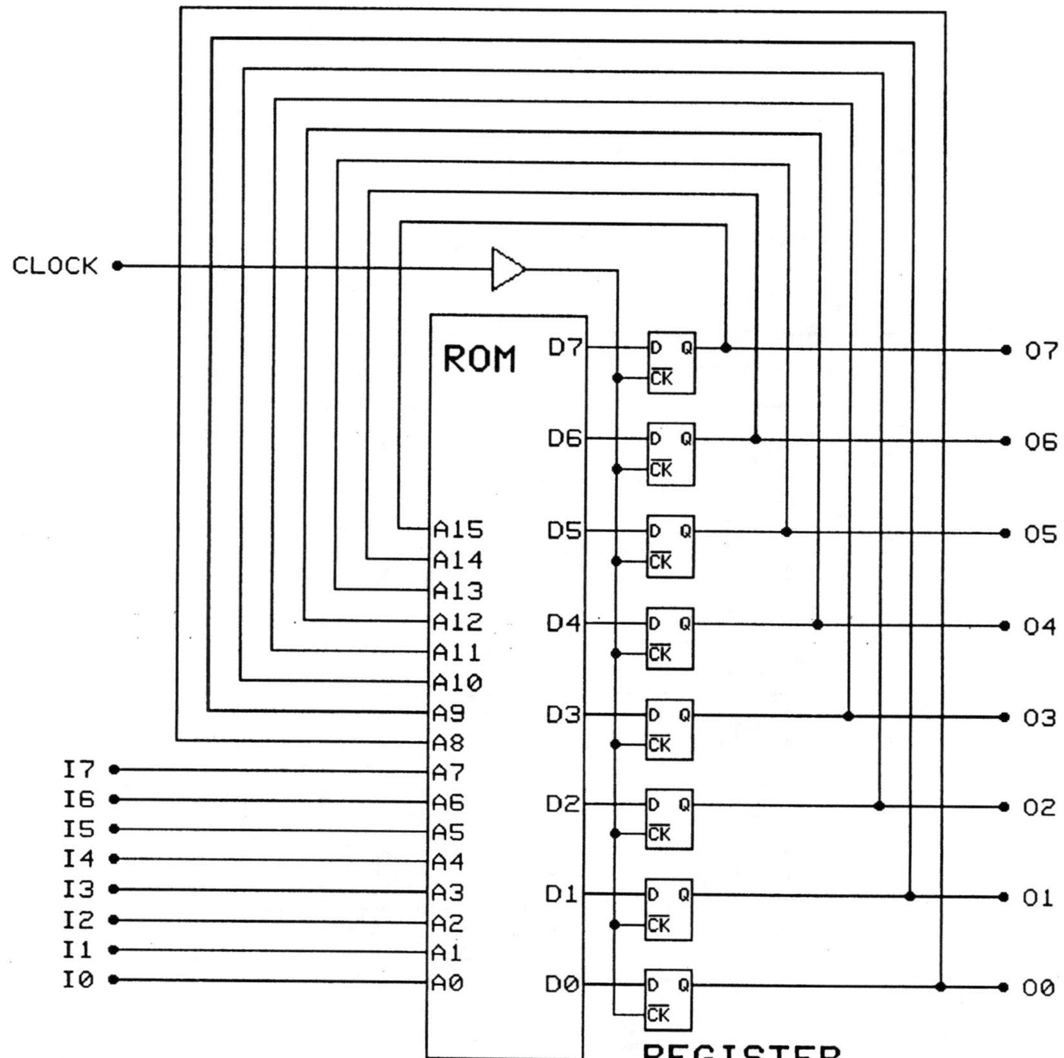

FIGURE 7.2 ROM/Register emulation of a 16R8 PAL device.

ADDRESS	DATA
000	001
001	010
010	011
011	100
100	101
101	110
110	111
111	000

TABLE 7.1 Simple ROM program.

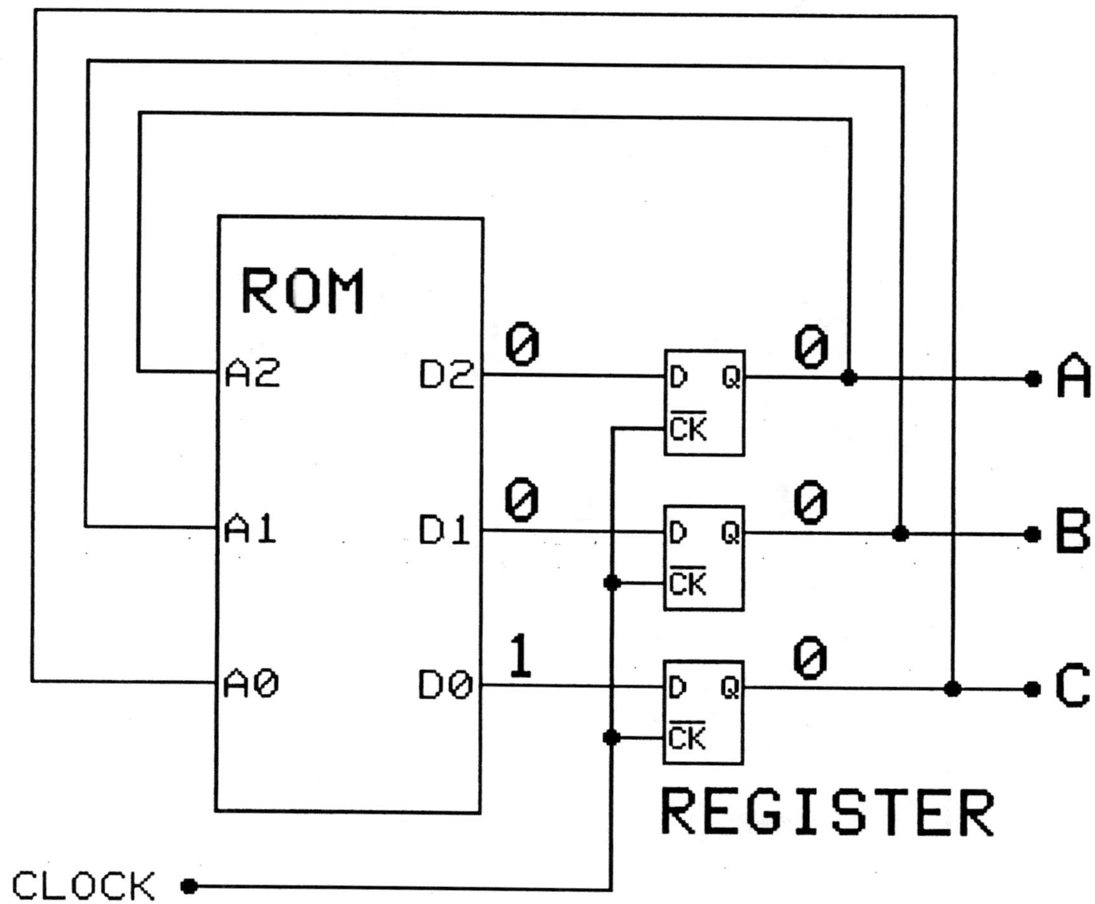

FIGURE 7.3 Three-bit ROM-based binary counter.

PRESENT STATE (Address) A B C	NEXT STATE (Data) A B C
0 0 0	0 0 1
0 0 1	0 1 0
0 1 0	0 1 1
0 1 1	1 0 0
1 0 0	1 0 1
1 0 1	1 1 0
1 1 0	1 1 1
1 1 1	0 0 0

TABLE 7.2 Simple ROM program represented as a state table.

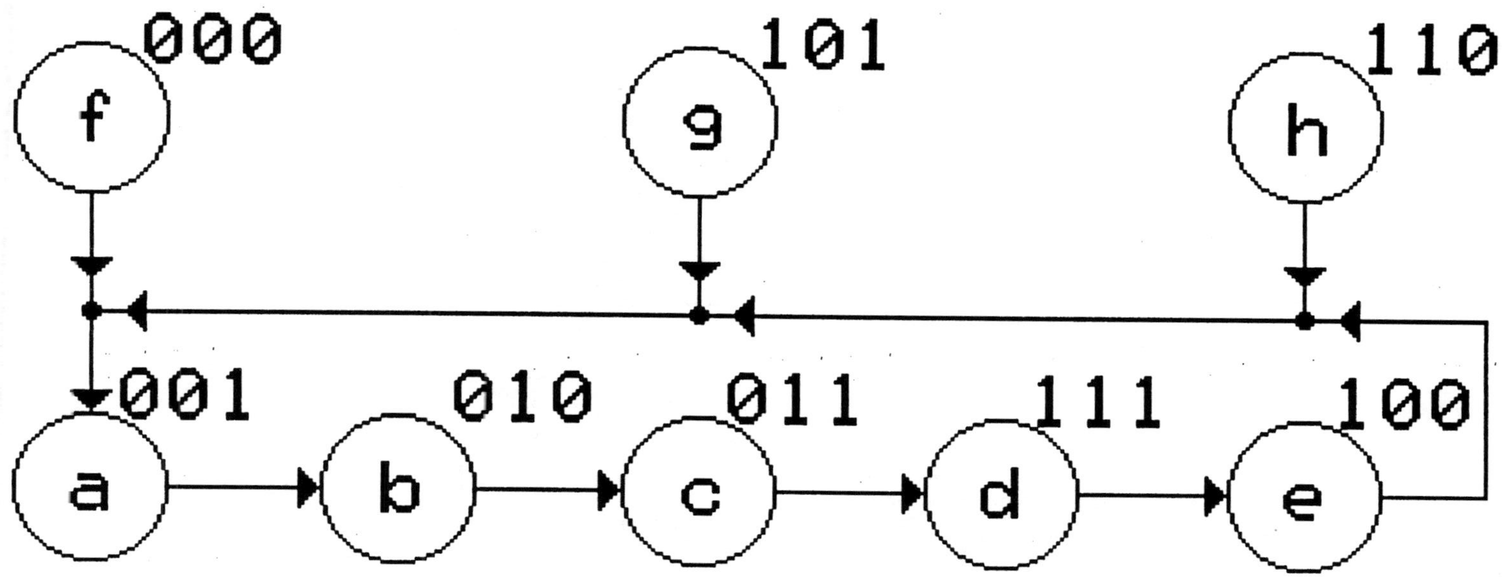

FIGURE 7.4 Practical Example 7.1, state diagram.

PRESENT STATE ABC	NEXT STATE ABC
0 0 0	0 0 1
0 0 1	0 1 0
0 1 0	0 1 1
0 1 1	1 1 1
1 0 0	0 0 1
1 0 1	0 0 0
1 1 0	0 0 1
1 1 1	1 0 0

TABLE 7.3 Practical Example 7.1, state table.

ADDRESS	DATA
0 0 0	0 0 1
0 0 1	0 1 0
0 1 0	0 1 1
0 1 1	1 1 1
1 0 0	0 0 1
1 0 1	0 0 0
1 1 0	0 0 1
1 1 1	1 0 0

TABLE 7.4 Practical Example 7.1, ROM program.

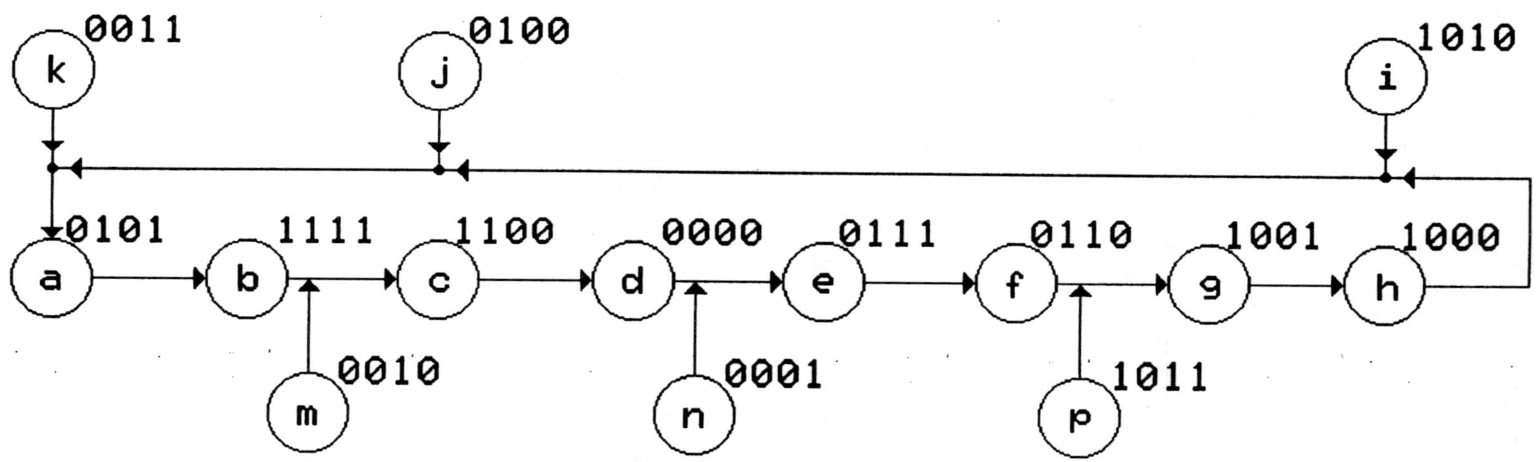

FIGURE 7.5 Practical Exercise 7.1, state diagram.

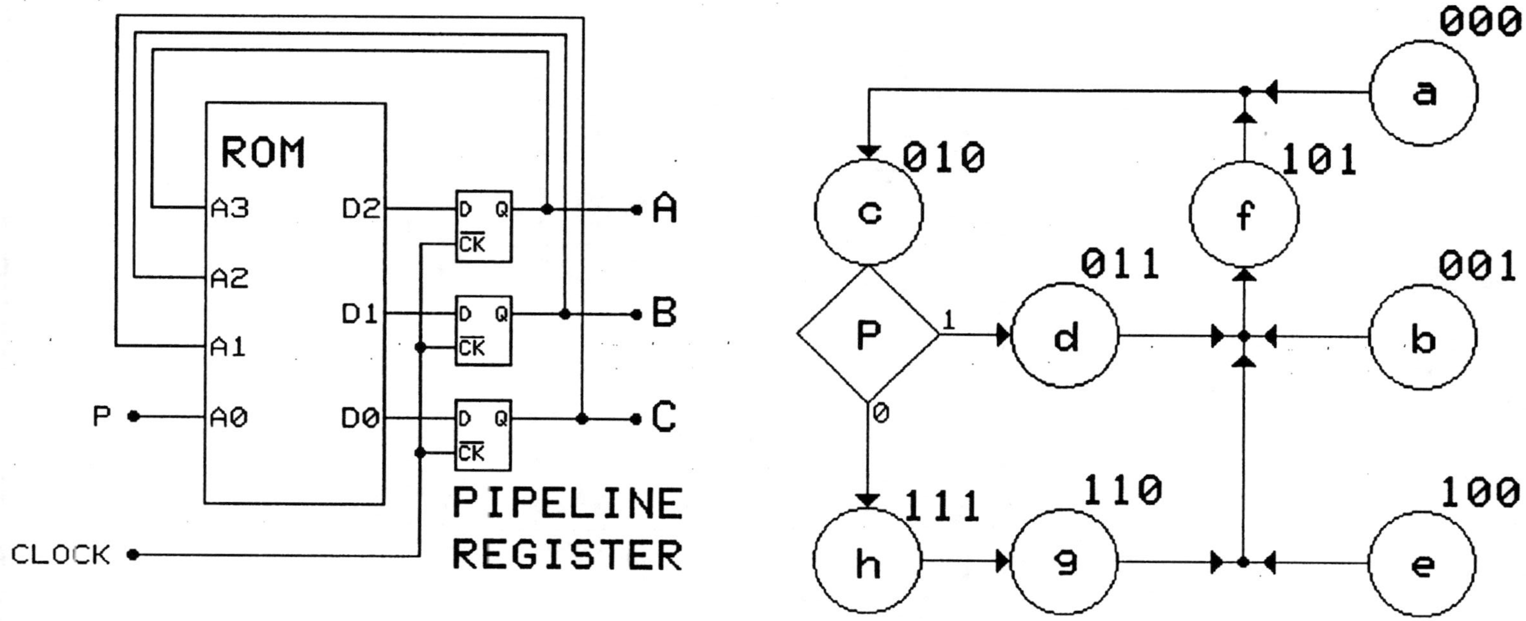

FIGURE 7.6 Practical Example 7.2, schematic and state diagrams.

	Present ABC	Input P	NExt ABC
a	0 0 0	X	0 1 0
b	0 0 1	X	1 0 1
c	0 1 0	0	1 1 1
		1	0 1 1
d	0 1 1	X	1 0 1
e	1 0 0	X	1 0 1
f	1 0 1	X	0 1 0
g	1 1 0	X	1 0 1
h	1 1 1	X	1 1 0

TABLE 7.5 Practical Example 7.2, State Table.

	ADDRESS $A_3A_2A_1A_0$	DATA $D_2D_1D_0$
a	0 0 0 0	0 1 0
	0 0 0 1	0 1 0
b	0 0 1 0	1 0 1
	0 0 1 1	1 0 1
c	0 1 0 0	1 1 1
	0 1 0 0	0 1 1
d	0 1 1 0	1 0 1
	0 1 1 1	1 0 1
e	1 0 0 0	1 0 1
	1 0 0 1	1 0 1
f	1 0 1 0	0 1 0
	1 0 1 1	0 1 0
g	1 1 0 0	1 0 1
	1 1 0 0	1 0 1
h	1 1 1 0	1 1 0
	1 1 1 1	1 1 0

TABLE 7.6 Practical Example 7.2, ROM Program.

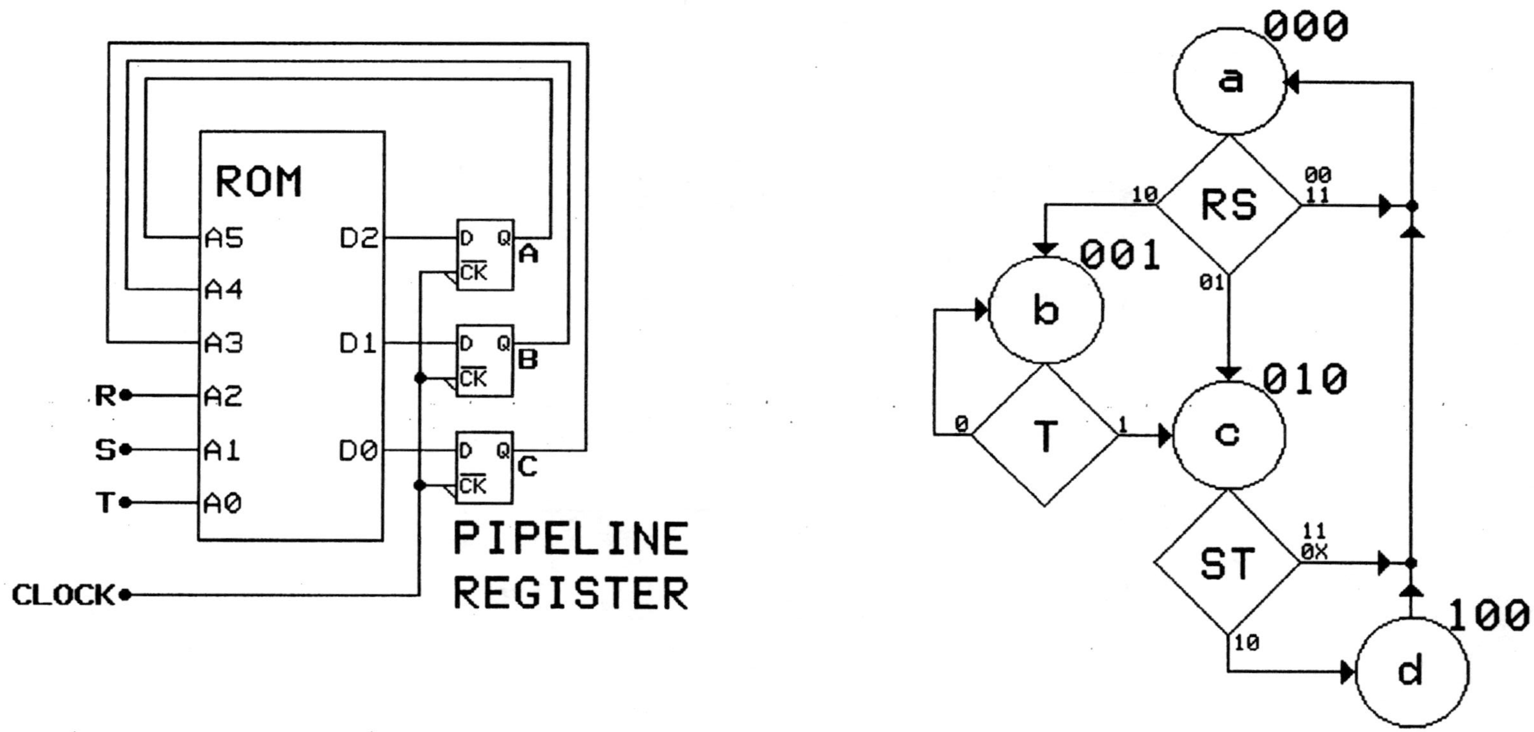

FIGURE 7.7 Practical Example 7.3, schematic and state diagrams.

STATE	N	PRES ABC	INPUT RST	NEXT ABC
a	0	000	00*	000
			01*	010
			10*	001
			11*	000
b	1	001	**0	001
			**1	010
c	2	010	*00	000
			*01	000
			*10	100
			*11	000
	3	011	***	***
d	4	100	***	000
	5	101	***	***
	6	110	***	***
	7	111	***	***

* = Don't Care

FIGURE 7.8 Practical Example 7.3, state table.

	ADDRESS $A_5A_4A_3$	$A_2A_1A_0$	DATA $D_2D_1D_0$
a	0 0 0	0 0 0	0 0 0
	0 0 0	0 0 1	0 0 0
	0 0 0	0 1 0	0 1 0
	0 0 0	0 1 1	0 1 0
	0 0 0	1 0 0	0 0 1
	0 0 0	1 0 1	0 0 1
	0 0 0	1 1 0	0 0 0
	0 0 0	1 1 1	0 0 0
b	0 0 1	0 0 0	0 0 1
	0 0 1	0 0 1	0 1 0
	0 0 1	0 1 0	0 0 1
	0 0 1	0 1 1	0 1 0
	0 0 1	1 0 0	0 0 1
	0 0 1	1 0 1	0 1 0
	0 0 1	1 1 0	0 0 1
	0 0 1	1 1 1	0 1 0
c	0 1 0	0 0 0	0 0 0
	0 1 0	0 0 1	0 0 0
	0 1 0	0 1 0	1 0 0
	0 1 0	0 1 1	0 0 0
	0 1 0	1 0 0	0 0 0
	0 1 0	1 0 1	0 0 0
	0 1 0	1 1 0	1 0 0
	0 1 0	1 1 1	0 0 0
d	1 0 0	0 0 0	0 0 0
	1 0 0	0 0 1	0 0 0
	1 0 0	0 1 0	0 0 0
	1 0 0	0 1 1	0 0 0
	1 0 0	1 0 0	0 0 0
	1 0 0	1 0 1	0 0 0
	1 0 0	1 1 0	0 0 0
	1 0 0	1 1 1	0 0 0

TABLE 7.7 Practical Example 7.3, ROM program.

```
Name      EXP0703a;
Date      01/01/96;
Designer  J.W. Carter;
Device    RA8P4;

/*****************************************************************/
/* Practical Example 7.3                                        */
/*    State machine with three inputs, uses DM74S387 256x4 PROM */
/*    All ROM data is manually included in the TABLE construct  */
/*****************************************************************/

/**  Inputs  **/
PIN [3..1]   = [A2..A0];          /* Inputs                     */
PIN [6..4]   = [A5..3];           /* Present State              */

/**  Outputs  **/
PIN [11..9]  = [D2..D0];          /* Next State                 */

FIELD input  = [A5..A0];
FIELD output = [D2..D0];

TABLE input =>output  {
   /* State a */          /* State b */          /* State c */          /* State d */
'B'000000 => 'B'000; 'B'001000 => 'B'001; 'B'010000 => 'B'000; 'B'100000 => 'B'000;
'B'000001 => 'B'000; 'B'001001 => 'B'010; 'B'010001 => 'B'000; 'B'100001 => 'B'000;
'B'000010 => 'B'010; 'B'001010 => 'B'001; 'B'010010 => 'B'100; 'B'100010 => 'B'000;
'B'000011 => 'B'010; 'B'001011 => 'B'010; 'B'010011 => 'B'000; 'B'100011 => 'B'000;
'B'000100 => 'B'001; 'B'001100 => 'B'001; 'B'010100 => 'B'000; 'B'100100 => 'B'000;
'B'000101 => 'B'001; 'B'001101 => 'B'010; 'B'010101 => 'B'000; 'B'100101 => 'B'000;
'B'000110 => 'B'000; 'B'001110 => 'B'001; 'B'010110 => 'B'100; 'B'100110 => 'B'000;
'B'000111 => 'B'000; 'B'001111 => 'B'010; 'B'010111 => 'B'000; 'B'100111 => 'B'000;
}
```

```
Name      EXP0703b;
Date      01/01/96;
Designer  J.W. Carter;
Device    RA8P4;

/******************************************************************/
/* Practical Example 7.3                                          */
/*     State machine with three inputs, uses DM74S387 256x4 PROM */
/*     TABLE construct is built from the state table directly.   */
/******************************************************************/

/**  Inputs   **/
PIN [3..1]    = [A2..A0];        /* Inputs                        */
PIN [6..4]    = [A5..3];         /* Present State                 */

/**  Outputs   **/
PIN [11..9]   = [D2..D0];        /* Next State                    */

FIELD input  = [A5..A0];
FIELD output = [D2..D0];

TABLE input =>output  {
/* State a */   'b'00000X => 'b'000;
                'b'00001X => 'b'010;
                'b'00010X => 'b'001;
                'b'00011X => 'b'000;
/* State b */   'b'001XX0 => 'b'001;
                'b'001XX1 => 'b'010;
/* State c */   'b'010X0X => 'b'000;
                'b'010X10 => 'b'100;
                'b'010X11 => 'b'000;
/* State d */   'b'100XXX => 'b'000;
}
```

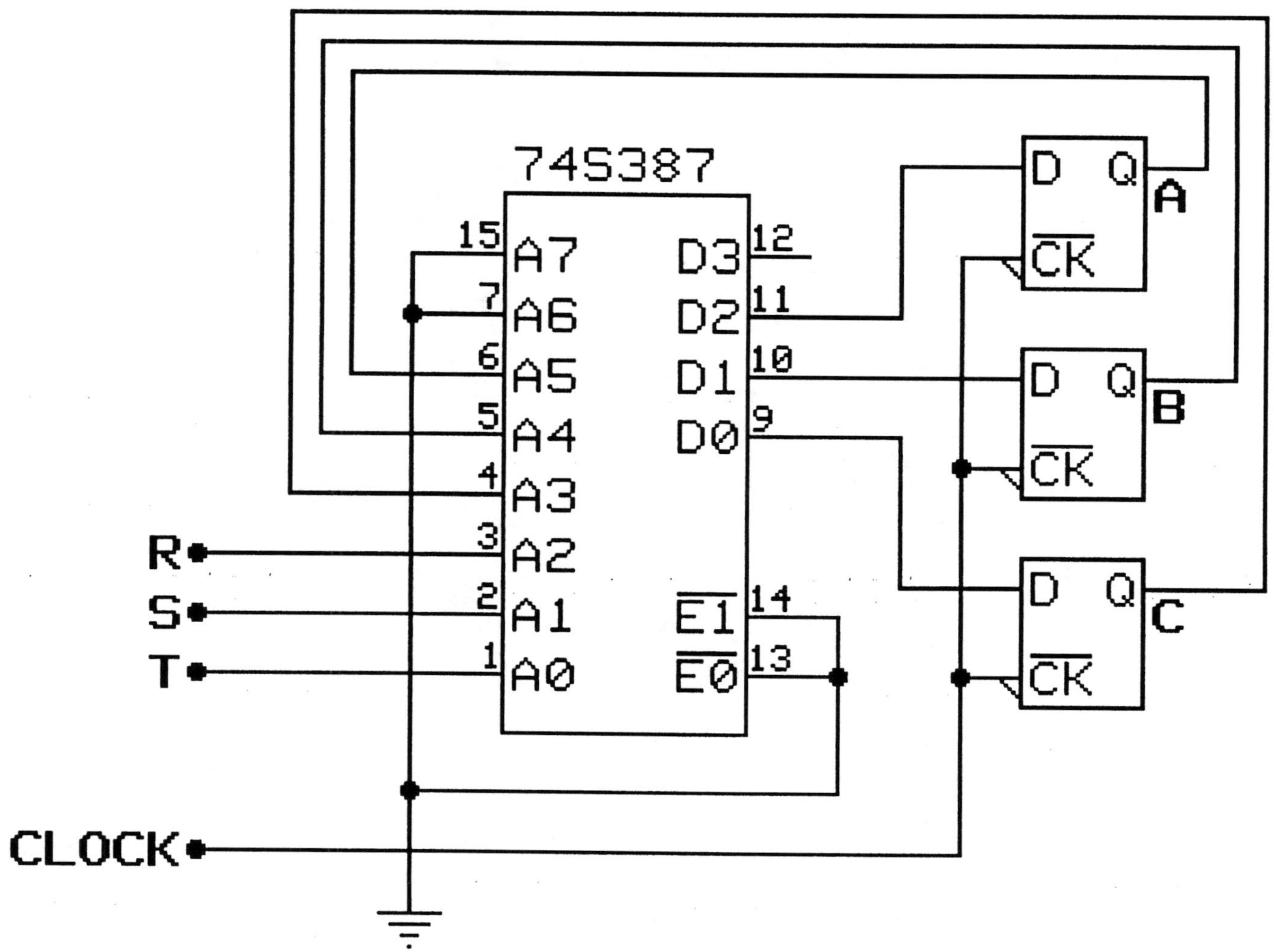

FIGURE 7.9 ***Practical Example 7.3, final schematic diagram.***

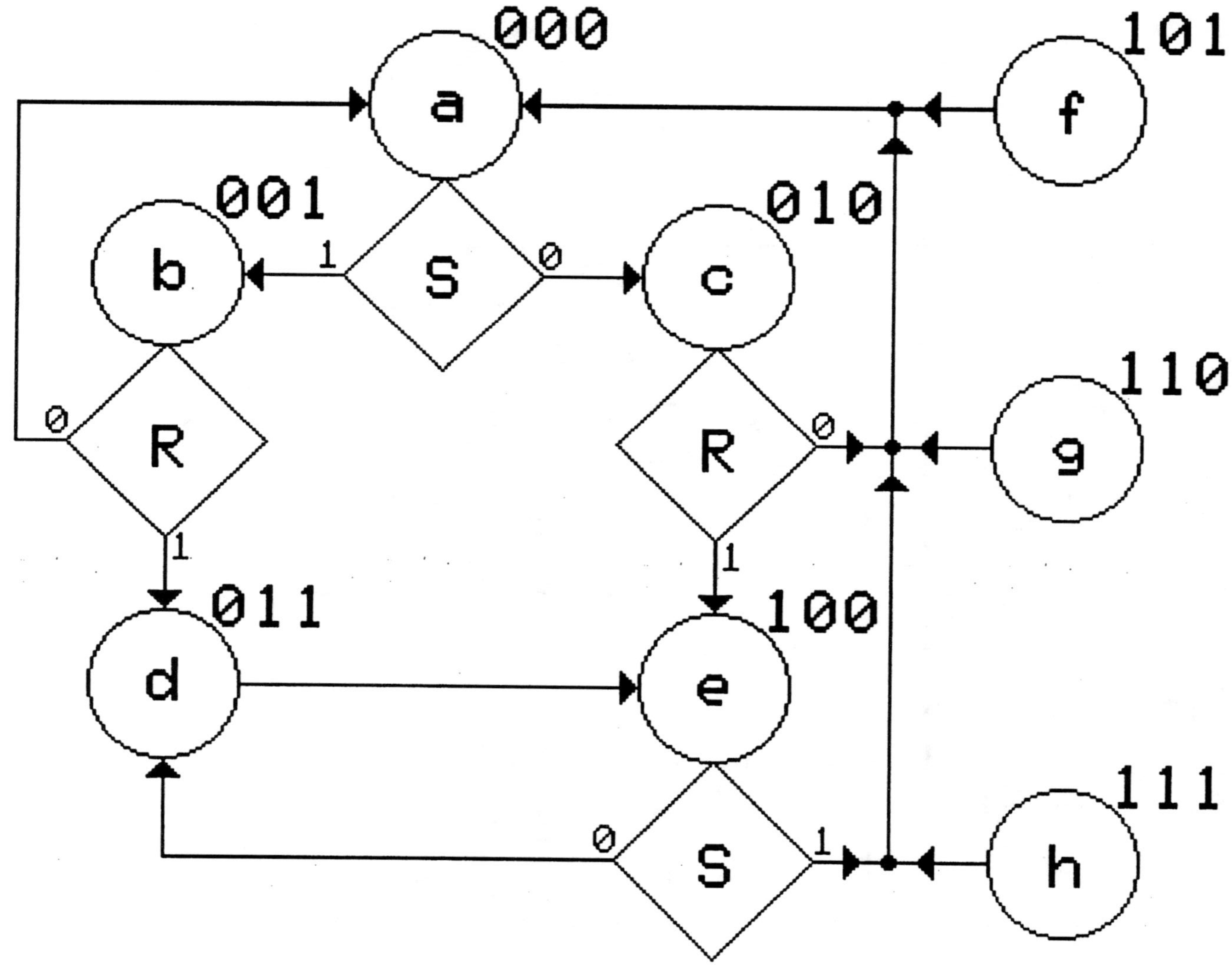

FIGURE 7.10 Practical Exercise 7.2, state diagram.

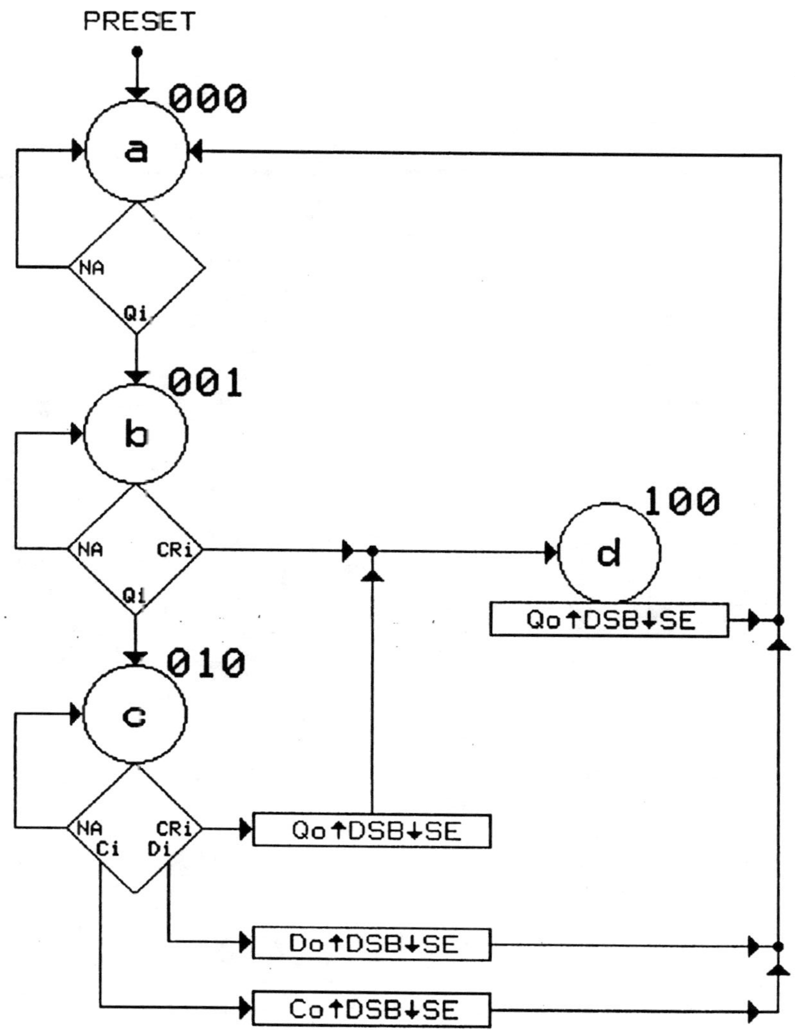

FIGURE 7.11 Practical Example 7.4, state diagram.

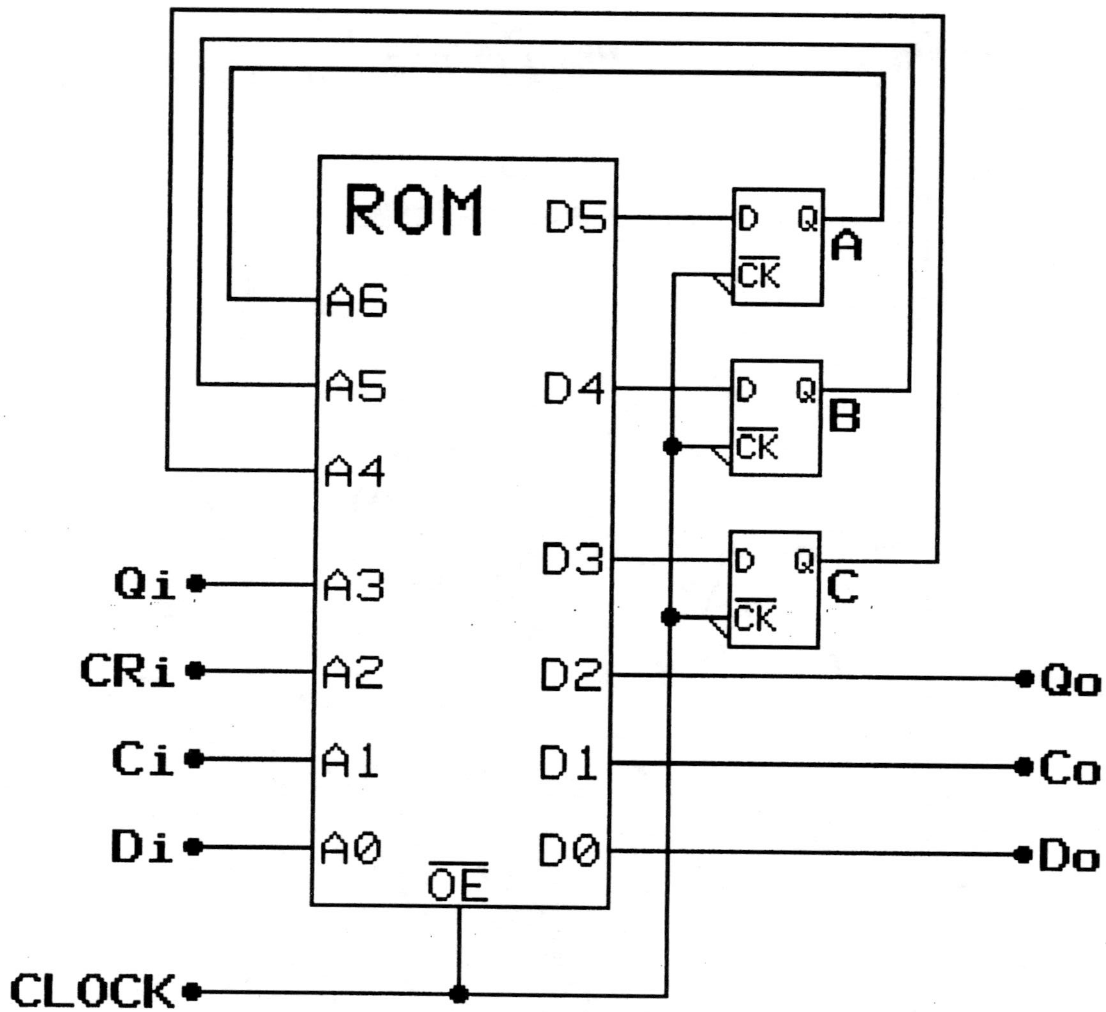

FIGURE 7.12 Practical Example 7.4, block diagram.

	PRES ABC	INPUTS Qi	Cri	Ci	Di	NEXT ABC	Qo	Co	Do
a	000	0	X	X	X	000	0	0	0
		1	X	X	X	001			
b	001	0	0	X	X	001	0	0	0
		1	X	X	X	010			
		X	1	X	X	100			
c	010	X	0	0	0	010	0	0	0
		X	1	X	X	100	↑DSB↓SE	0	0
		X	X	1	X	000	0	↑DSB↓SE	0
		X	X	X	1	000	0	0	↑DSB↓SE
	011	X	X	X	X	XXX	X	X	X
d	100	X	X	X	X	000	↑DSB↓SE	0	0
	101	X	X	X	X	XXX	X	X	X
	110	X	X	X	X	XXX	X	X	X
	111	X	X	X	X	XXX	X	X	X

FIGURE 7.13 Practical Exercise 7.4, state table.

	ADDRESS		DATA			ADDRESS		DATA	
	A B C	Q_i Cr_i C_i D_i	D_A D_B D_C	Q_o C_o D_o		A B C	Q_i Cr_i C_i D_i	D_A D_B D_C	Q_o C_o D_o
	$A_6A_5A_4$	$A_3A_2A_1A_0$	$D_5D_4D_3$	$D_2D_1D_0$		$A_6A_5A_4$	$A_3A_2A_1A_0$	$D_5D_4D_3$	$D_2D_1D_0$
a	0 0 0	0 0 0 0	0 0 0	0 0 0	c	0 1 0	0 0 0 0	0 1 0	0 0 0
	0 0 0	0 0 0 1	0 0 0	0 0 0		0 1 0	0 0 0 1	0 0 0	0 0 1
	0 0 0	0 0 1 0	0 0 0	0 0 0		0 1 0	0 0 1 0	0 0 0	0 1 0
	0 0 0	0 0 1 1	0 0 0	0 0 0		0 1 0	0 0 1 1	X X X	X X X
	0 0 0	0 1 0 0	0 0 0	0 0 0		0 1 0	0 1 0 0	1 0 0	1 0 0
	0 0 0	0 1 0 1	0 0 0	0 0 0		0 1 0	0 1 0 1	X X X	X X X
	0 0 0	0 1 1 0	0 0 0	0 0 0		0 1 0	0 1 1 0	X X X	X X X
	0 0 0	0 1 1 1	0 0 0	0 0 0		0 1 0	0 1 1 1	X X X	X X X
	0 0 0	1 0 0 0	0 0 1	0 0 0		0 1 0	1 0 0 0	0 1 0	0 0 0
	0 0 0	1 0 0 1	0 0 1	0 0 0		0 1 0	1 0 0 1	0 0 0	0 0 1
	0 0 0	1 0 1 0	0 0 1	0 0 0		0 1 0	1 0 1 0	0 0 0	0 1 0
	0 0 0	1 0 1 1	0 0 1	0 0 0		0 1 0	1 0 1 1	X X X	X X X
	0 0 0	1 1 0 0	0 0 1	0 0 0		0 1 0	1 1 0 0	1 0 0	1 0 0
	0 0 0	1 1 0 1	0 0 1	0 0 0		0 1 0	1 1 0 1	X X X	X X X
	0 0 0	1 1 1 0	0 0 1	0 0 0		0 1 0	1 1 1 0	X X X	X X X
	0 0 0	1 1 1 1	0 0 1	0 0 0		0 1 0	1 1 1 1	X X X	X X X
b	0 0 1	0 0 0 0	0 0 1	0 0 0	d	1 0 0	0 0 0 0	0 0 0	1 0 0
	0 0 1	0 0 0 1	0 0 1	0 0 0		1 0 0	0 0 0 1	0 0 0	1 0 0
	0 0 1	0 0 1 0	0 0 1	0 0 0		1 0 0	0 0 1 0	0 0 0	1 0 0
	0 0 1	0 0 1 1	0 0 1	0 0 0		1 0 0	0 0 1 1	0 0 0	1 0 0
	0 0 1	0 1 0 0	1 0 0	0 0 0		1 0 0	0 1 0 0	0 0 0	1 0 0
	0 0 1	0 1 0 1	1 0 0	0 0 0		1 0 0	0 1 0 1	0 0 0	1 0 0
	0 0 1	0 1 1 0	1 0 0	0 0 0		1 0 0	0 1 1 0	0 0 0	1 0 0
	0 0 1	0 1 1 1	1 0 0	0 0 0		1 0 0	0 1 1 1	0 0 0	1 0 0
	0 0 1	1 0 0 0	0 1 0	0 0 0		1 0 0	1 0 0 0	0 0 0	1 0 0
	0 0 1	1 0 0 1	0 1 0	0 0 0		1 0 0	1 0 0 1	0 0 0	1 0 0
	0 0 1	1 0 1 0	0 1 0	0 0 0		1 0 0	1 0 1 0	0 0 0	1 0 0
	0 0 1	1 0 1 1	0 1 0	0 0 0		1 0 0	1 0 1 1	0 0 0	1 0 0
	0 0 1	1 1 0 0	X X X	X X X		1 0 0	1 1 0 0	0 0 0	1 0 0
	0 0 1	1 1 0 1	X X X	X X X		1 0 0	1 1 0 1	0 0 0	1 0 0
	0 0 1	1 1 1 0	X X X	X X X		1 0 0	1 1 1 0	0 0 0	1 0 0
	0 0 1	1 1 1 1	X X X	X X X		1 0 0	1 1 1 1	0 0 0	1 0 0

TABLE 7.8 Practical Example 7.4, ROM program.

```
Name      EXP0704;
Date      01/01/96;
Designer  J.W. Carter;
Device    RA8P8;

/******************************************************************/
/* Practical Example 7.4                                          */
/*    Soda Machine Controller using a DM74S471 256 x 8 PROM       */
/******************************************************************/

/**  Inputs  **/
PIN [5..1]   = [A4..A0];        /* Present State & Inputs       */
PIN [18..17] = [A6..A5];

/**  Outputs  **/
PIN [9..6]   = [D3..D0];        /* Next State                   */
PIN [12..11] = [D5..D4];   /* Outputs                       */

FIELD input =  [A6..A0];
FIELD output = [D5..D0];

TABLE input =>output  {
/* State a */ 'b'0000XXX => 'b'000000;
              'b'0001XXX => 'b'001000;
/* State b */ 'b'00100XX => 'b'001000;
              'b'0011XXX => 'b'010000;
              'b'001X1XX => 'b'100000;
/* State c */ 'b'010X000 => 'b'010000;
              'b'010X1XX => 'b'100100;
              'b'010XX1X => 'b'000010;
              'b'010XXX1 => 'b'000001;
/* State d */ 'b'100XXXX => 'b'000100;
}
```

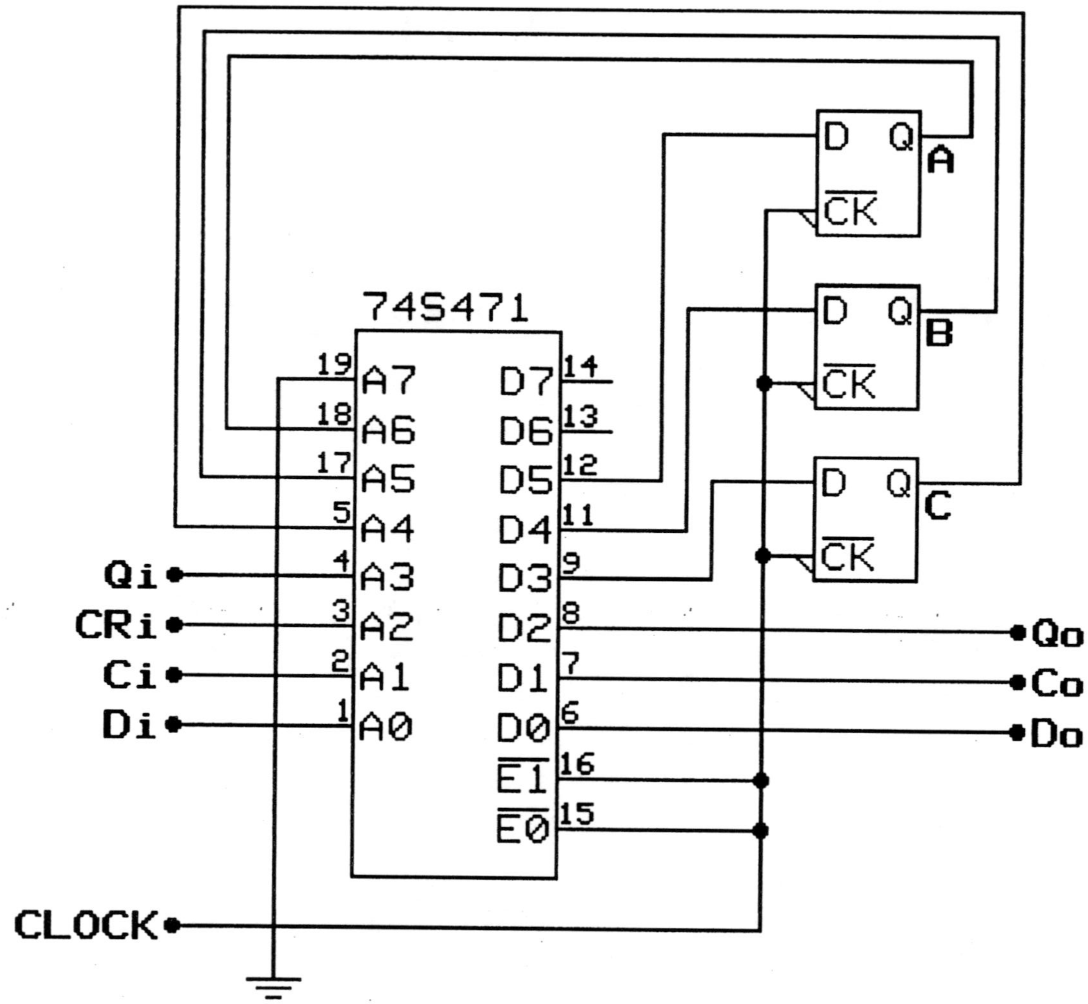

FIGURE 7.14 Practical Example 7.4, final schematic diagram.

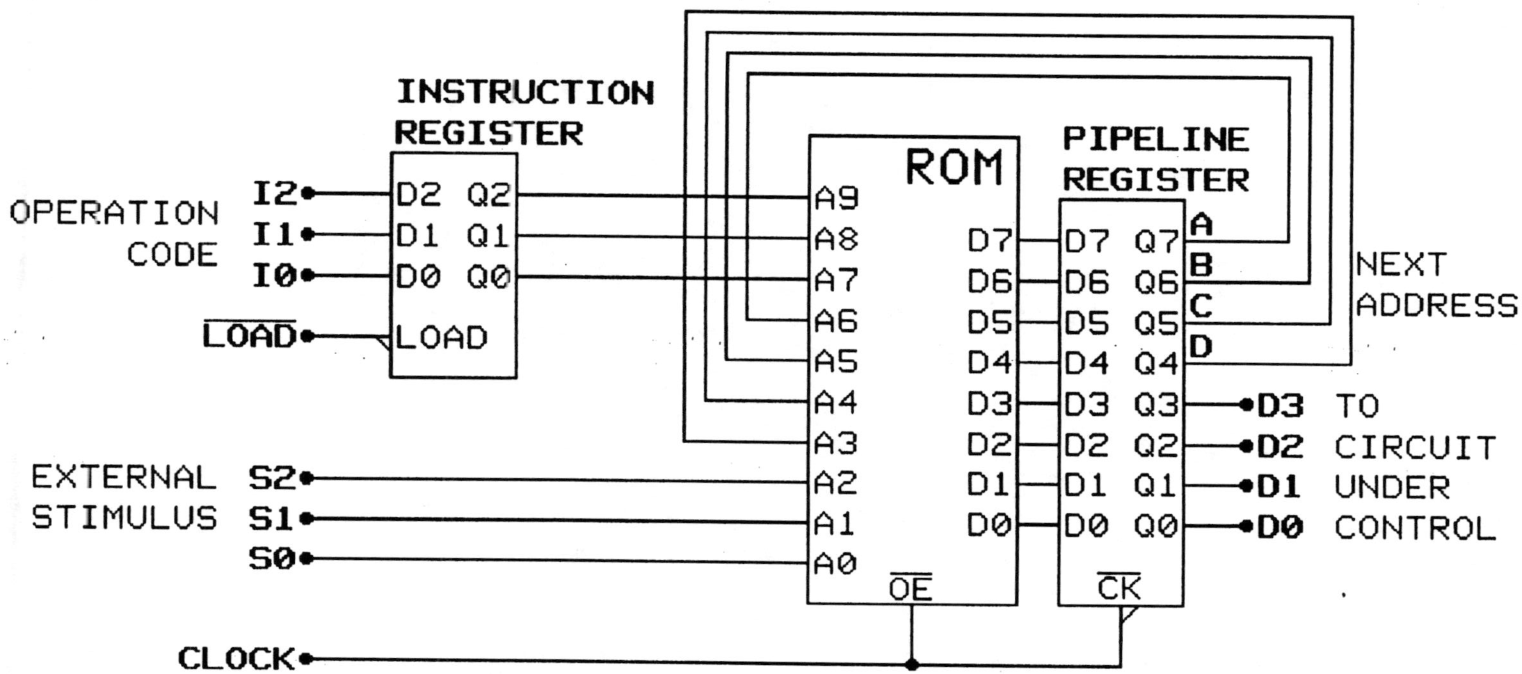

FIGURE 7.15 Microprogrammed control unit.

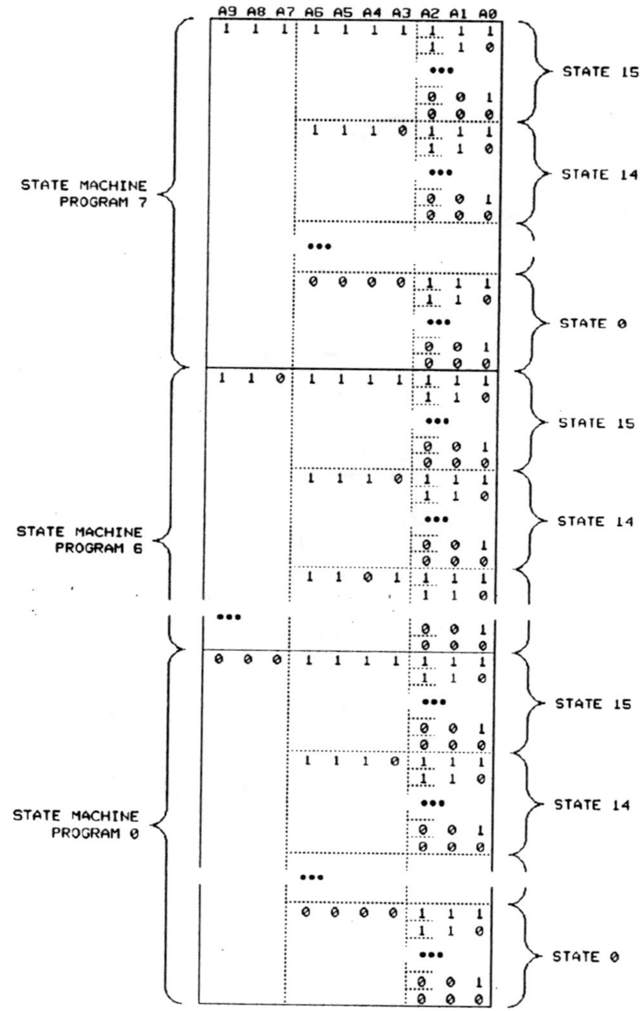

FIGURE 7.16 State machine memory map

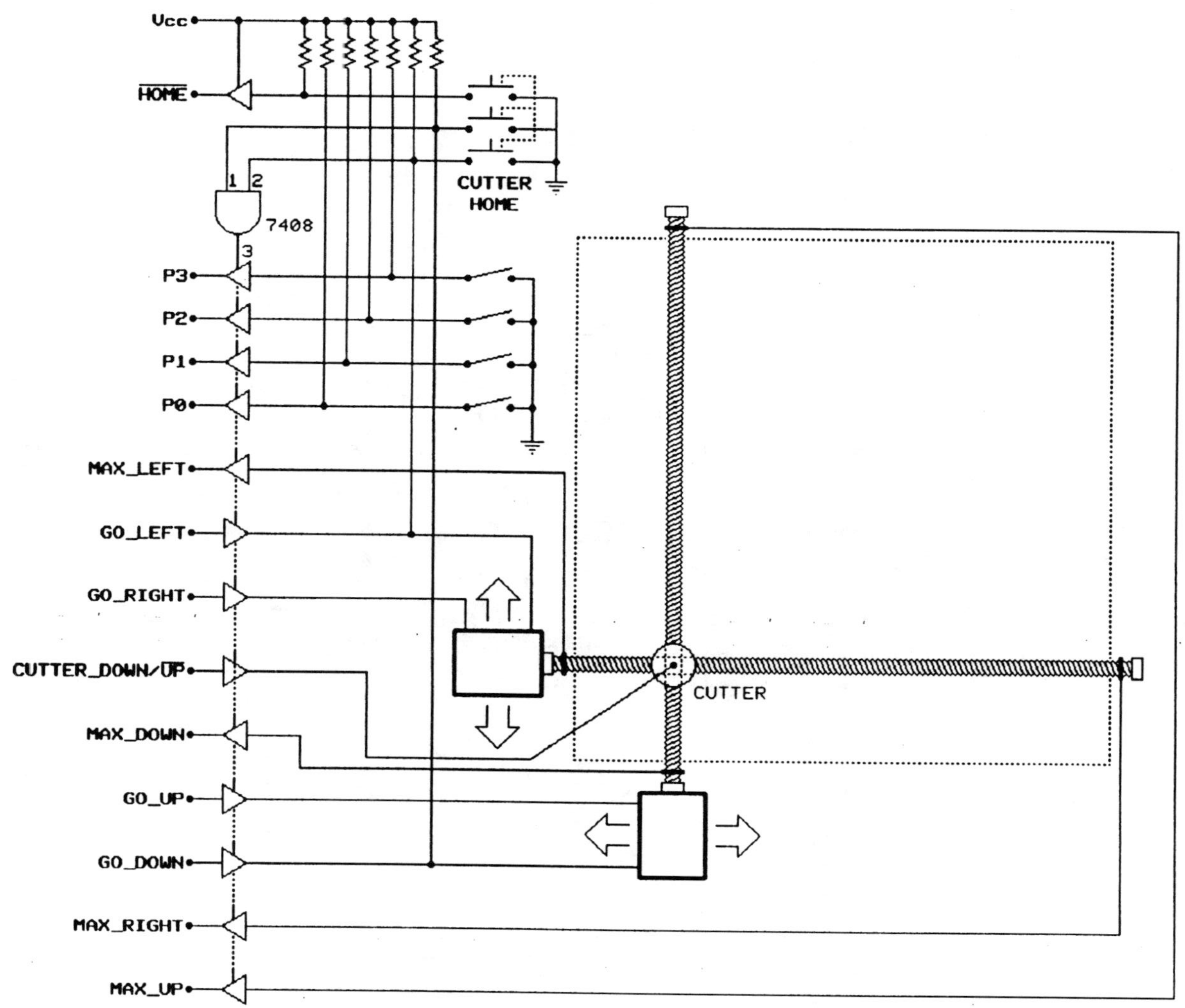

FIGURE 7.17 Part pattern cutter schematic diagram.

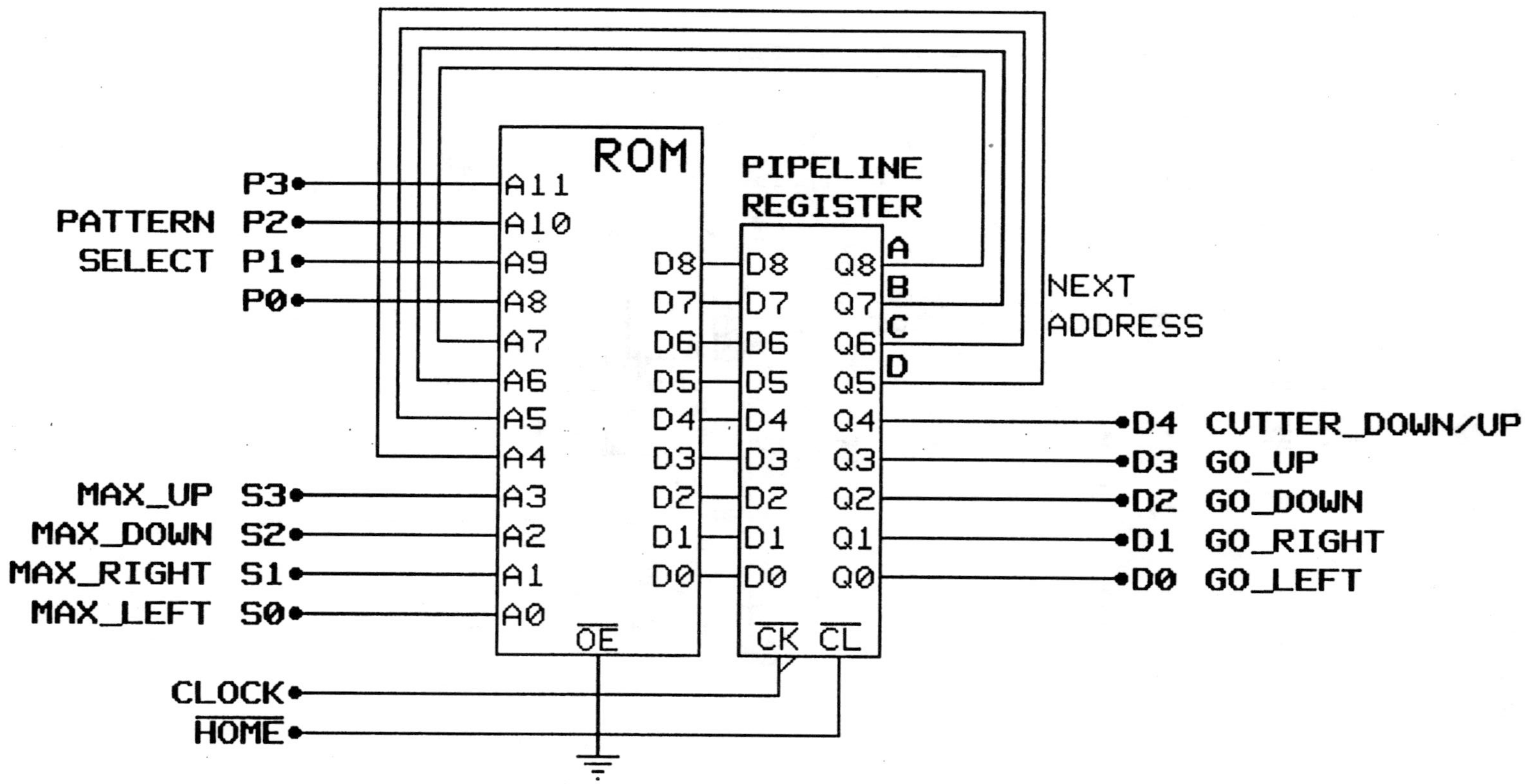

FIGURE 7.18 Practical Example 7.5 State Machine

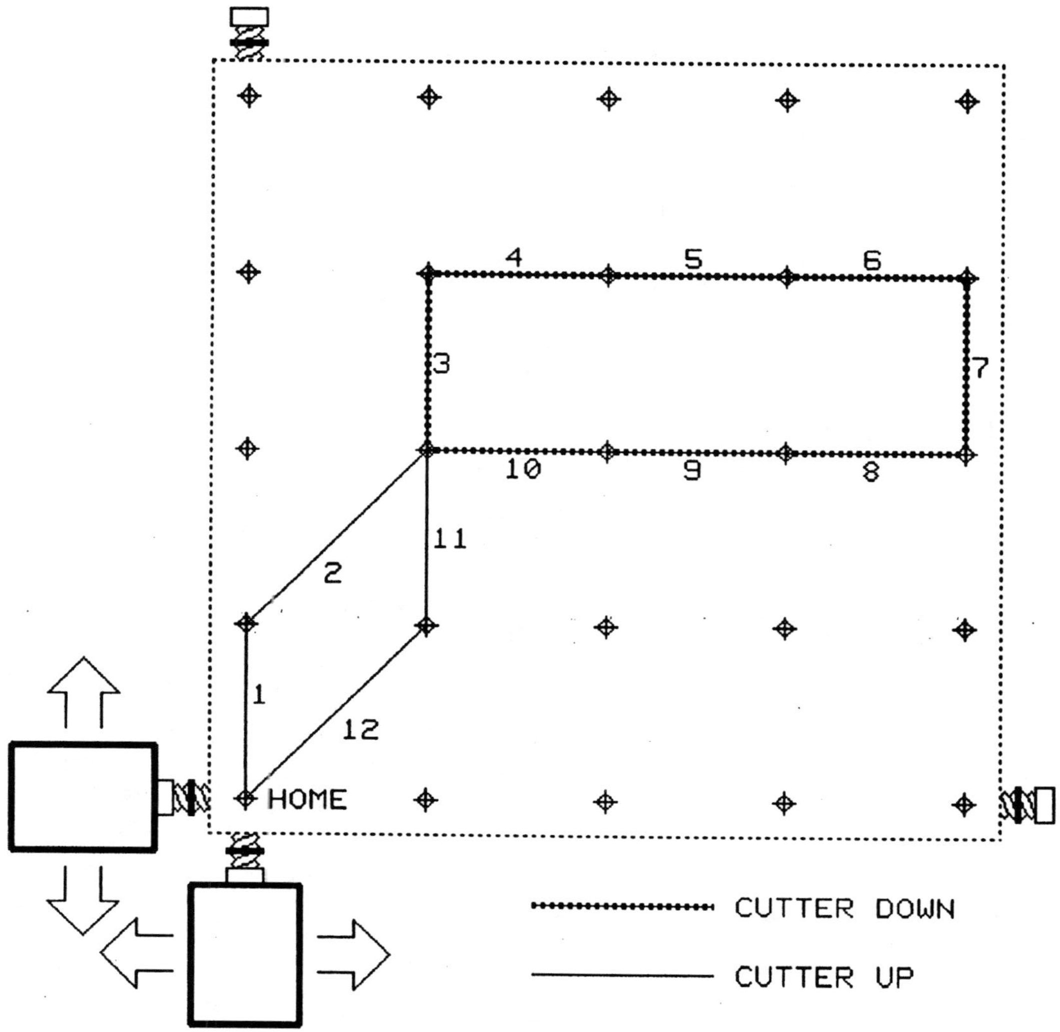

FIGURE 7.19 Pattern 0, cutter path.

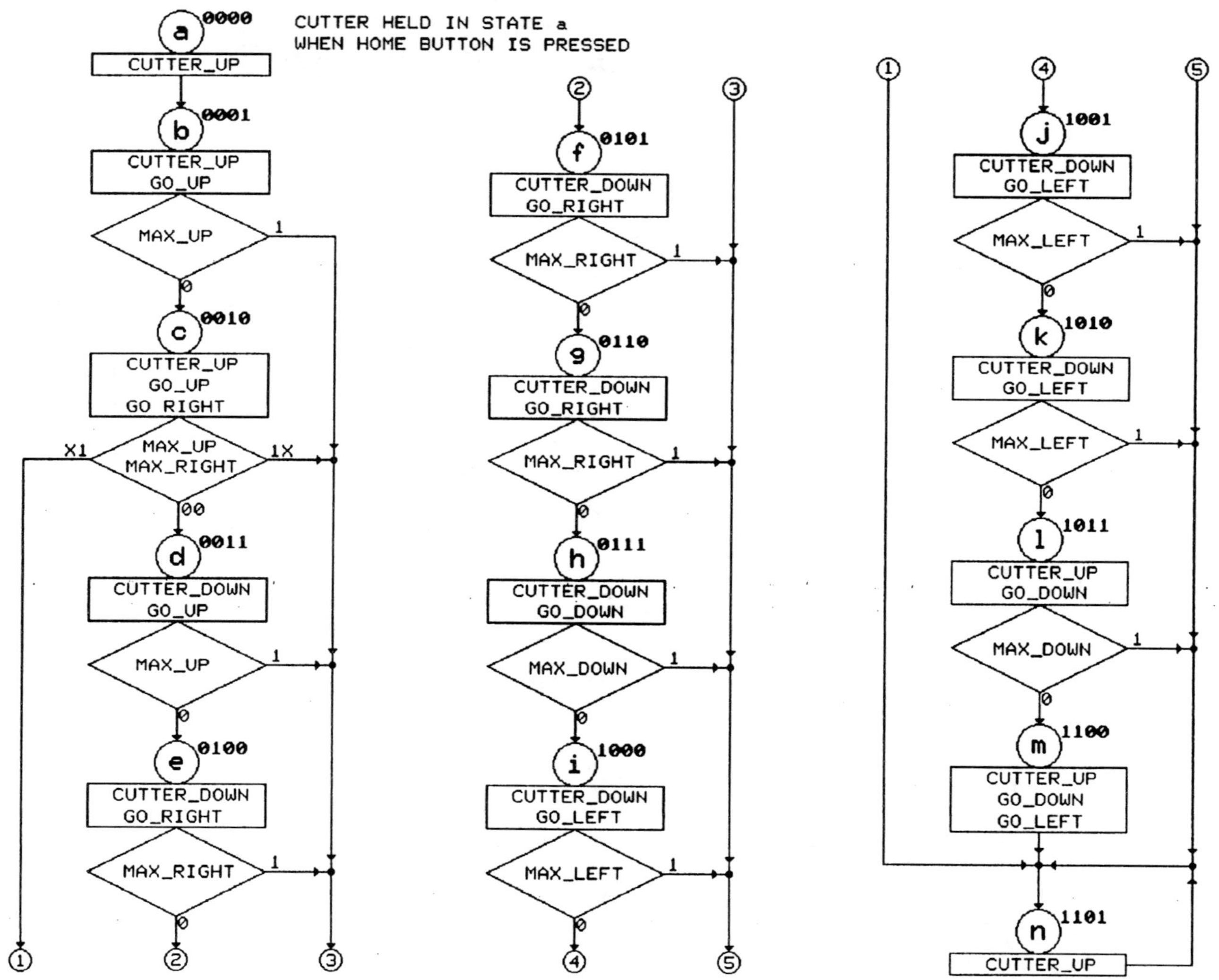

FIGURE 7.20 Practical Example 7.5, pattern 0, state diagram.

	PARTITION P3210	PRESENT ABCD	INPUTS S3210	NEXT ABCD	CUTTER C_DN	MOTORS G_UP	 G_DN	 G_RT	 G_LF
a	0000	0000	XXXX	0001	0	0	0	0	0
b	0000	0001	0XXX 1XXX	0010 1101	0	1	0	0	0
c	0000	0010	0X0X 1XXX XX1X	0011 1101 1101	0	1	0	1	0
d	0000	0011	0XXX 1XXX	0100 1101	1	1	0	0	0
e	0000	0100	XX0X XX1X	0101 1101	1	0	0	1	0
f	0000	0101	XX0X XX1X	0110 1101	1	0	0	1	0
g	0000	0110	XX0X XX1X	0111 1101	1	0	0	1	0
h	0000	0111	X0XX X1XX	1000 1101	1	0	1	0	0
i	0000	1000	XXX0 XXX1	1001 1101	1	0	0	0	1
j	0000	1001	XXX0 XXX1	1010 1101	1	0	0	0	1
k	0000	1010	XXX0 XXX1	1011 1101	1	0	0	0	1
l	0000	1011	X0XX X1XX	1100 1101	0	0	1	0	0
m	0000	1100	XXXX	1101	0	0	1	0	1
n	0000	1101	XXXX	1101	0	0	0	0	0

FIGURE 7.21 Pattern 0, state table.

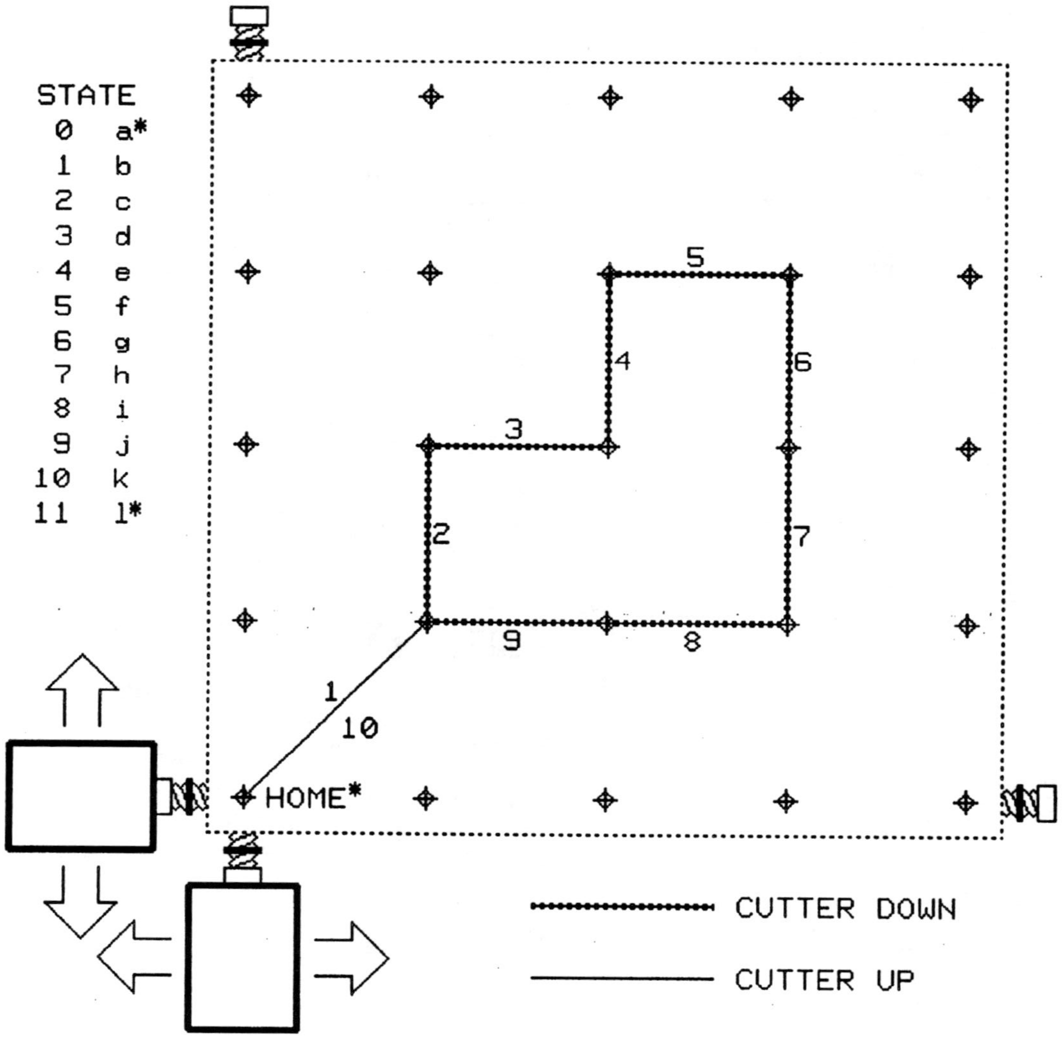

FIGURE 7.22 ***Pattern 1, cutter path.***

	PARTITION P3210	PRESENT ABCD	INPUTS S3210	NEXT ABCD	CUTTER C_DN	MOTORS G_UP	G_DN	G_RT	G_LF
a	ØØØ1	ØØØØ	XXXX	ØØØ1	Ø	Ø	Ø	Ø	Ø
b	ØØØ1	ØØØ1	ØXØX 1XXX XX1X	ØØ1Ø 1Ø11 1Ø11	Ø	1	Ø	1	Ø
c	ØØØ1	ØØ1Ø	ØXXX 1XXX	ØØ11 1Ø11	1	1	Ø	Ø	Ø
d	ØØØ1	ØØ11	XXØX XX1X	Ø1ØØ 1Ø11	1	Ø	Ø	1	Ø
e	ØØØ1	Ø1ØØ	ØXXX 1XXX	Ø1Ø1 1Ø11	1	1	Ø	Ø	Ø
f	ØØØ1	Ø1Ø1	XXØX XX1X	Ø11Ø 1Ø11	1	Ø	Ø	1	Ø
g	ØØØ1	Ø11Ø	XØXX X1XX	Ø111 1Ø11	1	Ø	1	Ø	Ø
h	ØØØ1	Ø111	XØXX X1XX	1ØØØ 1Ø11	1	Ø	1	Ø	Ø
i	ØØØ1	1ØØØ	XXXØ XXX1	1ØØ1 1Ø11	1	Ø	Ø	Ø	1
j	ØØØ1	1ØØ1	XXXØ XXX1	1Ø1Ø 1Ø11	1	Ø	Ø	Ø	1
k	ØØØ1	1Ø1Ø	XXXX	1Ø11	Ø	Ø	1	Ø	1
l	ØØØ1	1Ø11	XXXX	1Ø11	Ø	Ø	Ø	Ø	Ø

FIGURE 7.23 Pattern 1, state table.

```
Name      EXP0704;
Date      01/01/96;
Designer  J.W. Carter;
Device    RA12P9;
/*****************************************************************/
/* Practical Example 7.5:Pattern Cutter using a 4096 x 9 PROM   */
/*****************************************************************/
/**  Inputs  **/
PIN [01..04] = [A11..A08];      /* Partition Inputs P3..P0      */
PIN [05..08] = [A07..A04];      /* Counter Bits A, B, C, D      */
PIN [09..10] = [A03..A02];      /* Input Stimulus S3..S0        */
PIN [13..14] = [A01..A00];
/**  Outputs  **/
PIN [15..19] = [D00..D04];      /* Output Forming Logic         */
PIN [20..23] = [D05..D08];       /* Next Address Bits            */
FIELD input =  [A11..A00];
FIELD output = [D08..D00];
TABLE input =>output  {
*/ PATTERN 0 */
     /*              PPPP    SSSS          DDDDD  */
     /*              3210ABCD3210      ABCD43210  */
     /* State a */ 'b'00000000XXXX => 'b'000100000;
     /* State b */ 'b'000000010XXX => 'b'001001000;
                   'b'000000011XXX => 'b'110101000;
     /* State c */ 'b'000000100X0X => 'b'001101010;
                   'b'000000101XXX => 'b'110101010;
                   'b'00000010XX1X => 'b'110101010;
     /* State d */ 'b'000000110XXX => 'b'010011000;
                   'b'000000111XXX => 'b'110111000;
     /* State e */ 'b'00000100XX0X => 'b'010110010;
                   'b'00000100XX1X => 'b'110110010;
     /* State f */ 'b'00000101XX0X => 'b'011010010;
                   'b'00000101XX1X => 'b'110110010;
     /* State g */ 'b'00000110XX0X => 'b'011110010;
                   'b'00000110XX1X => 'b'110110010;
     /* State h */ 'b'00000111X0XX => 'b'100010100;
```

```
                    'b'00000111X1XX => 'b'110110100;
    /* State i */   'b'00001000XXX0 => 'b'100110001;
                    'b'00001000XXX1 => 'b'110110001;
    /* State j */   'b'00001001XXX0 => 'b'101010001;
                    'b'00001001XXX1 => 'b'110110001;
    /* State k */   'b'00001010XXX0 => 'b'101110001;
                    'b'00001010XXX1 => 'b'110110001;
    /* State l */   'b'00001011X0XX => 'b'111000100;
                    'b'00001011X1XX => 'b'110100100;
    /* State m */   'b'00001100XXXX => 'b'110100101;
    /* State n */   'b'00001101XXXX => 'b'110100000;
*/ PATTERN 1 */
    /* State a */   'b'00010000XXXX => 'b'000100000;
    /* State b */   'b'000100010X0X => 'b'001001010;
                    'b'000100011XXX => 'b'101101010;
                    'b'00010001XX1X => 'b'101101010;
    /* State c */   'b'000100100XXX => 'b'001111000;
                    'b'000100101XXX => 'b'101111000;
    /* State d */   'b'00010011XX0X => 'b'010010010;
                    'b'00010011XX1X => 'b'101110010;
    /* State e */   'b'000101000XXX => 'b'010111000;
                    'b'000101001XXX => 'b'101111000;
    /* State f */   'b'00010101XX0X => 'b'011010010;
                    'b'00010101XX1X => 'b'101110010;
    /* State g */   'b'00010110X0XX => 'b'011110100;
                    'b'00010110X1XX => 'b'101110100;
    /* State h */   'b'00010111X0XX => 'b'100010100;
                    'b'00010111X1XX => 'b'101110100;
    /* State i */   'b'00011000XXX0 => 'b'100110001;
                    'b'00011000XXX1 => 'b'101110001;
    /* State j */   'b'00011001XXX0 => 'b'101010001;
                    'b'00011001XXX1 => 'b'101110001;
    /* State k */   'b'00011010XXXX => 'b'101110101;
    /* State l */   'b'00011011XXXX => 'b'101100000;
}
```

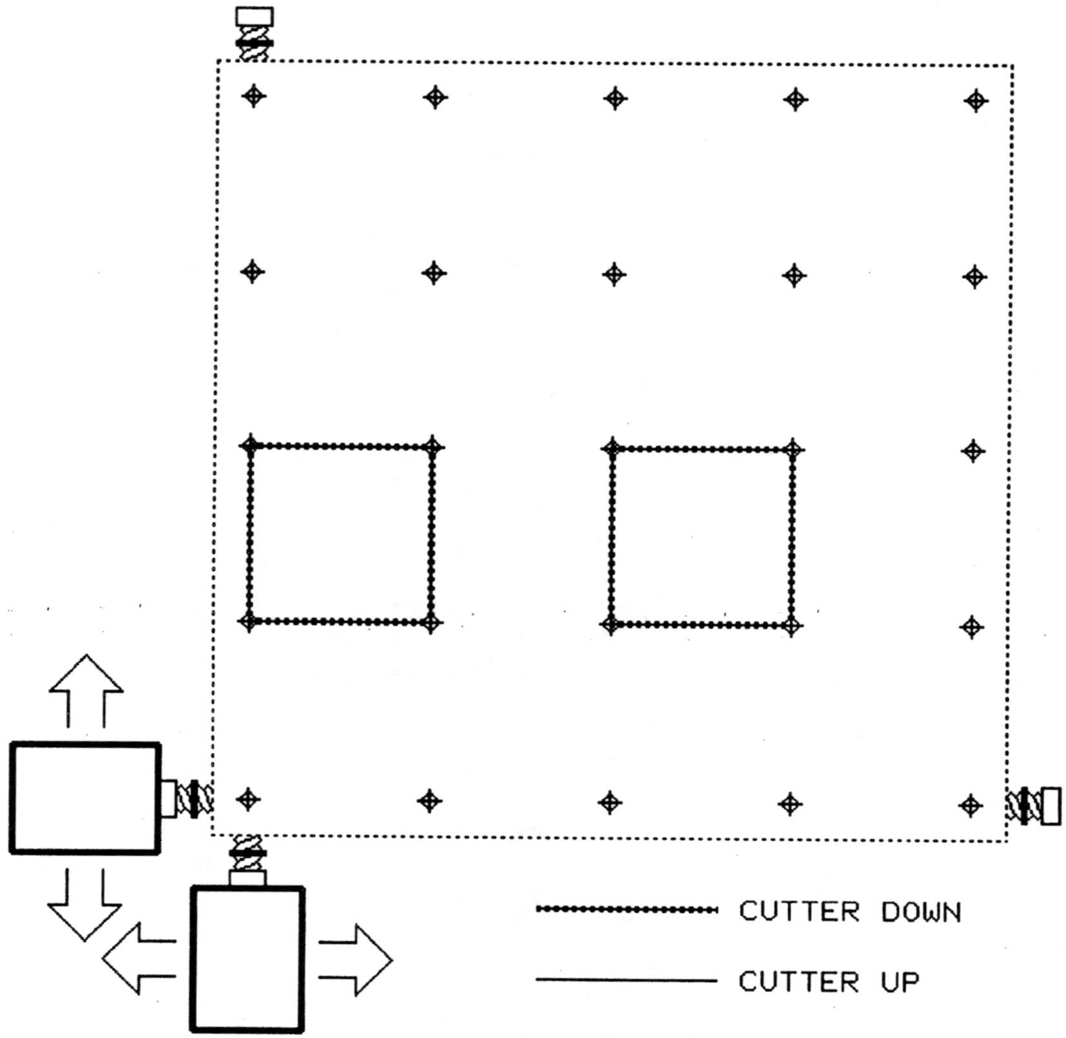

FIGURE 7.24 Practical Exercise 7.8, Cutter Pattern 2.

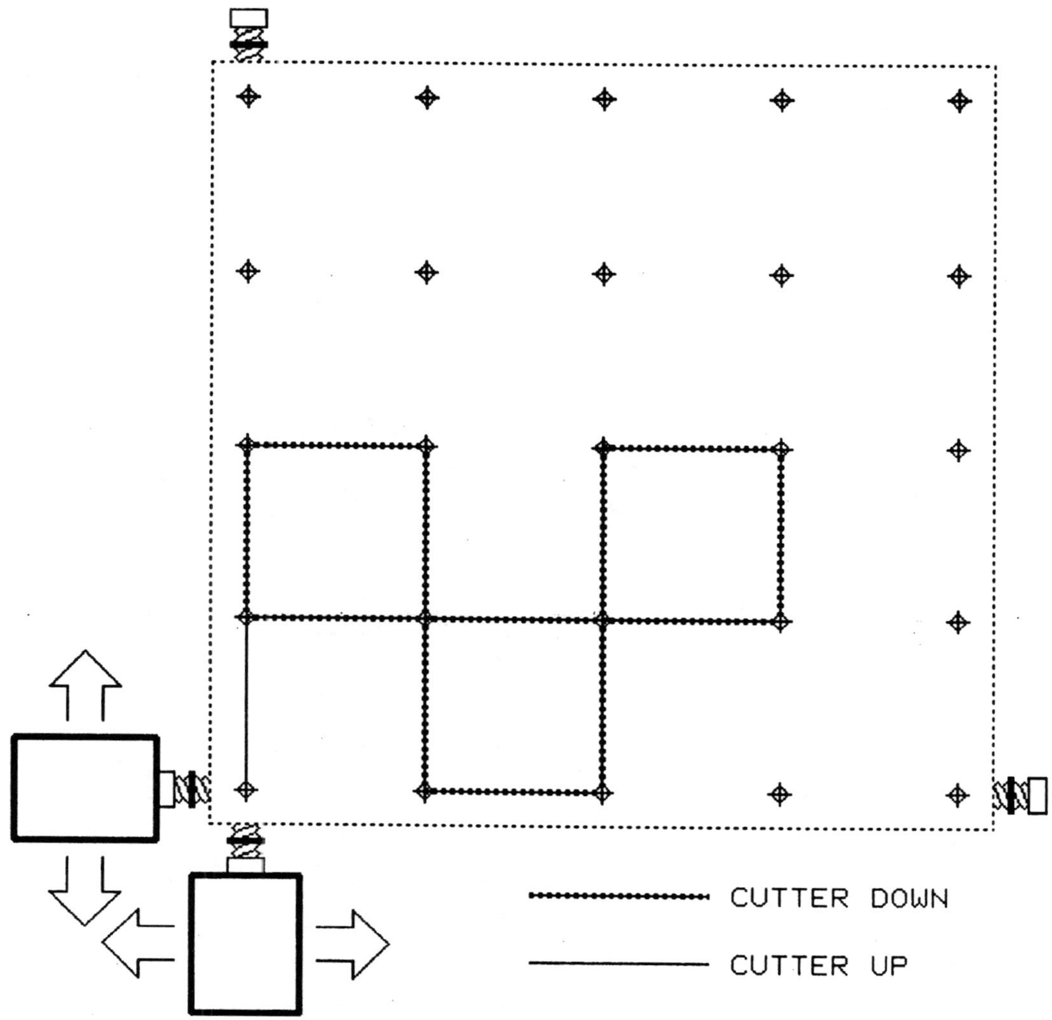

FIGURE 7.25 Practical Exercise 7.9, Cutter Pattern 3.

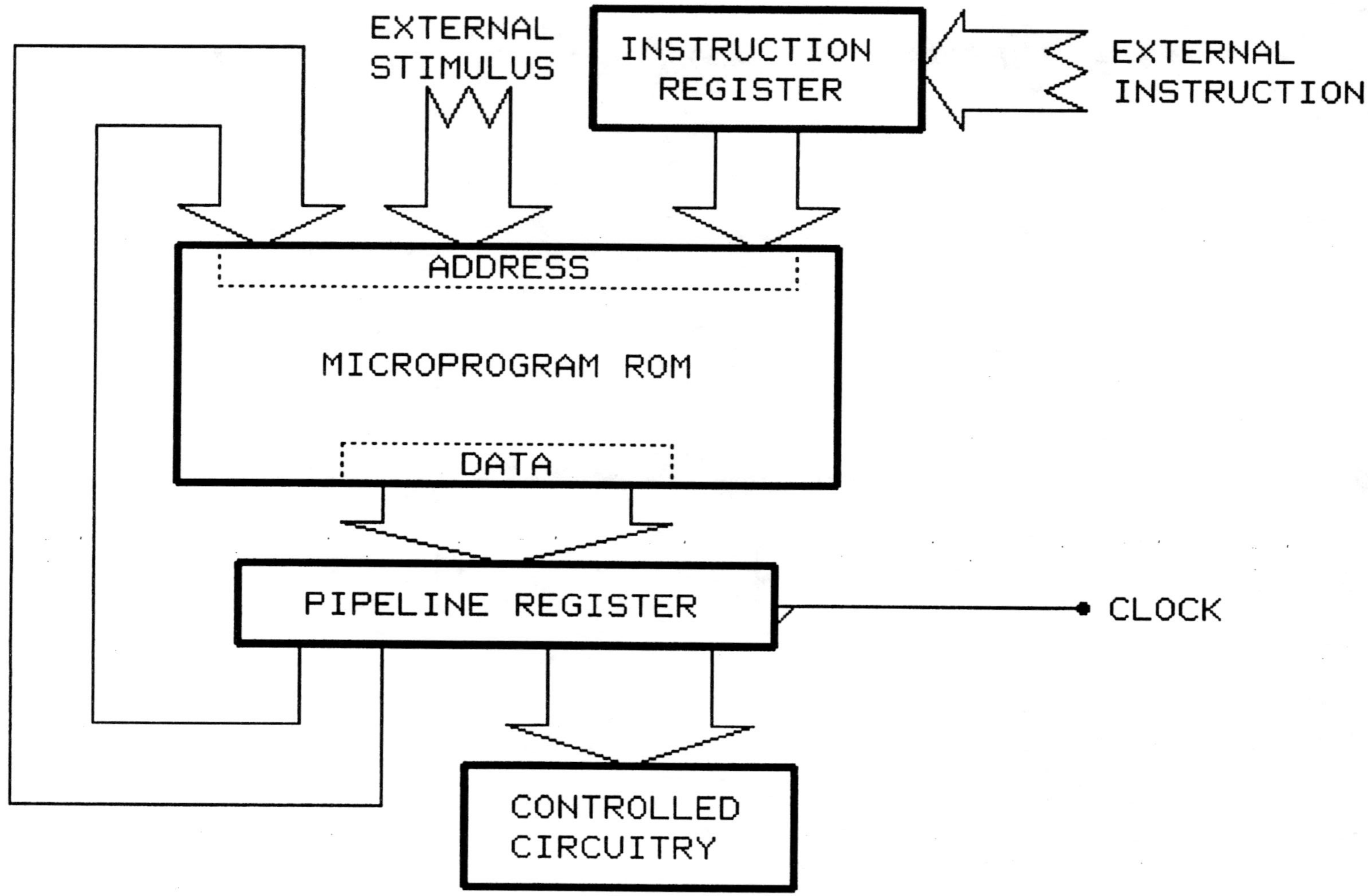

FIGURE 8.1 ROM-based state machine block diagram

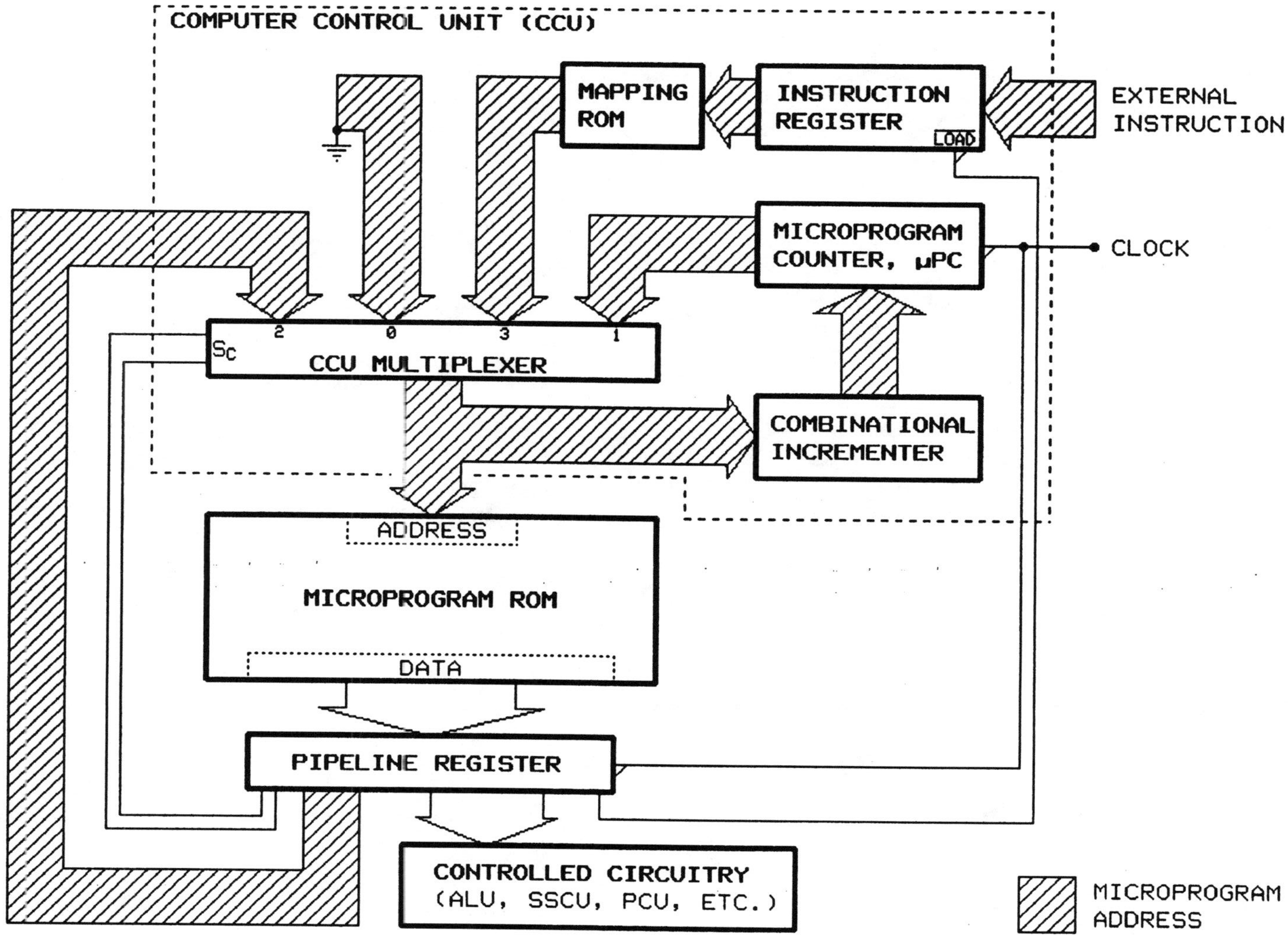

FIGURE 8.2 *A simple microprocessor as a state machine.*

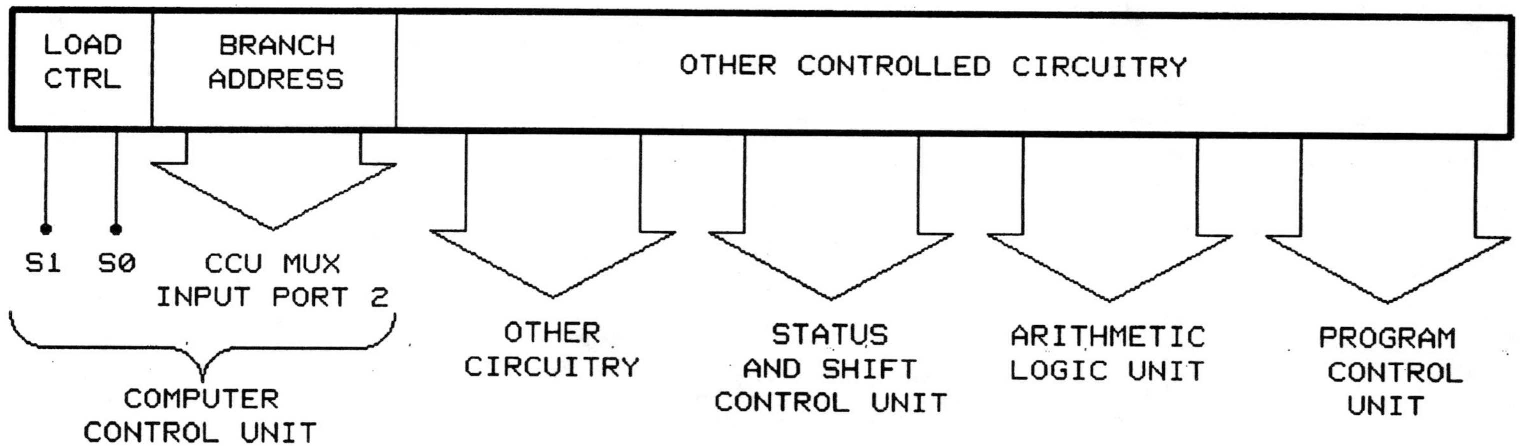

FIGURE 8.3   Microprocessor pipeline register.

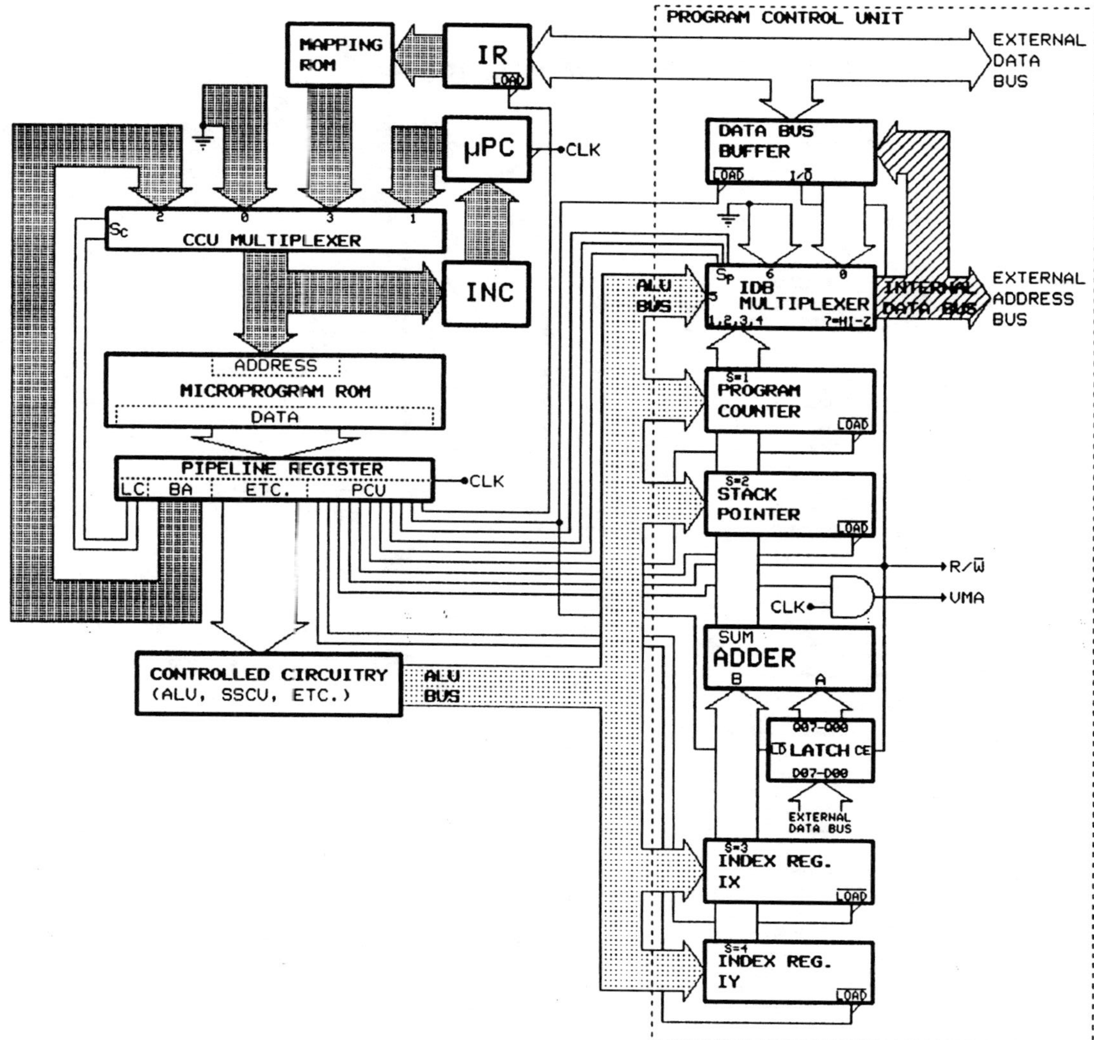

FIGURE 8.4 Addition of the program control unit to the state machine.

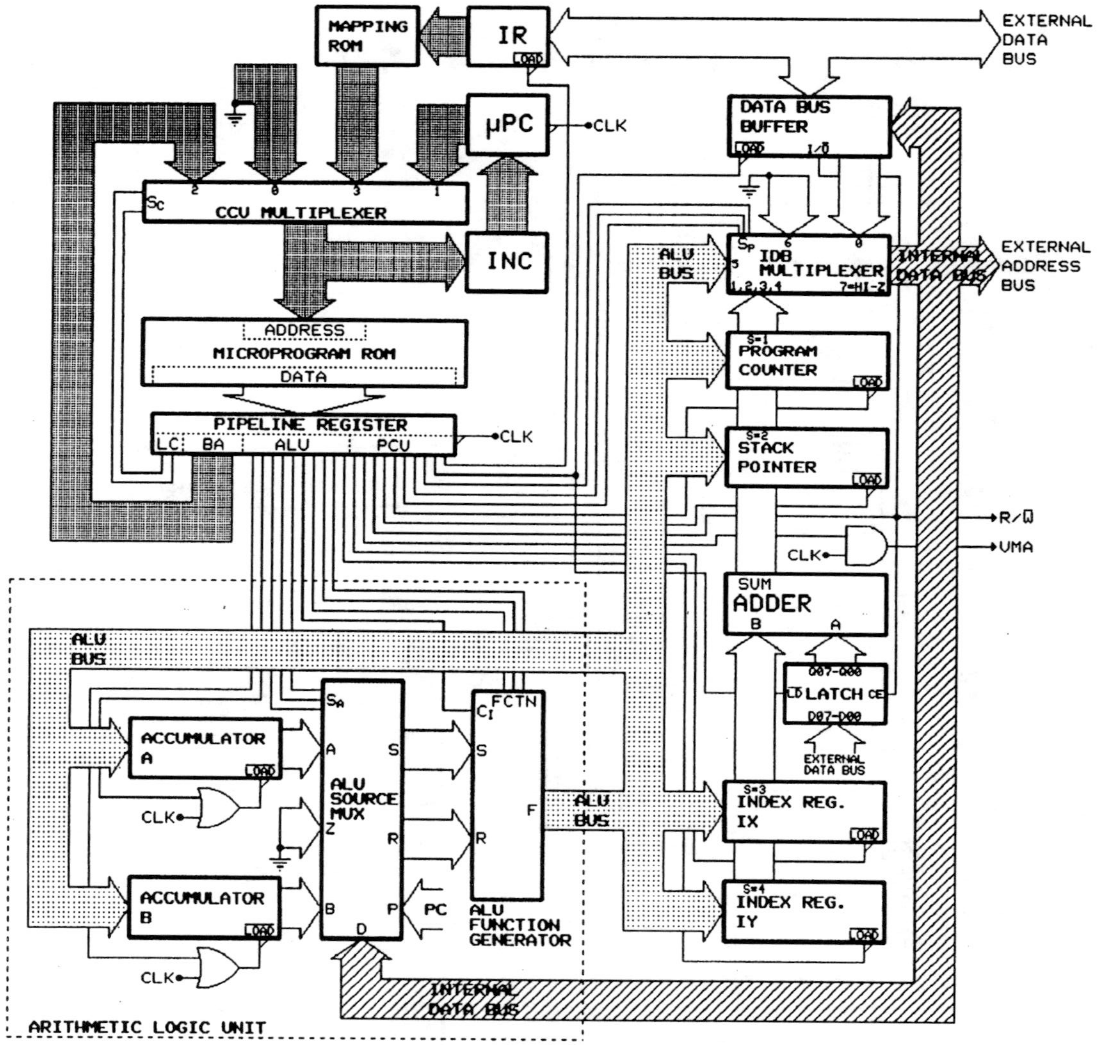

FIGURE 8.5 Addition of the arithmetic logic unit to the state machine.

ALU SOURCE MULTIPLEXER			
I2 1 0	MNEM	R	S
0 0 0	DZ	D	Z
0 0 1	AZ	A	Z
0 1 0	AB	A	B
0 1 1	AD	A	D
1 0 0	DP	D	PC
1 0 1	BZ	B	Z
1 1 0	BA	B	A
1 1 1	BD	B	D

ALU FUNCTION GENERATOR		
I2 1 0	MNEM	FUNCTION
0 0 0	PAR	R + CI
0 0 1	PAS	S + CI
0 1 0	ADD	S + R + 1 − CI
0 1 1	SUB	S − R − 1 + CI
1 0 0	AND	$S \wedge R$
1 0 1	OR	$S \vee R$
1 1 0	XOR	$S \veebar R$
1 1 1	NOT	$\bar{R}$

FIGURE 8.6 ***ALU sources and functions.***

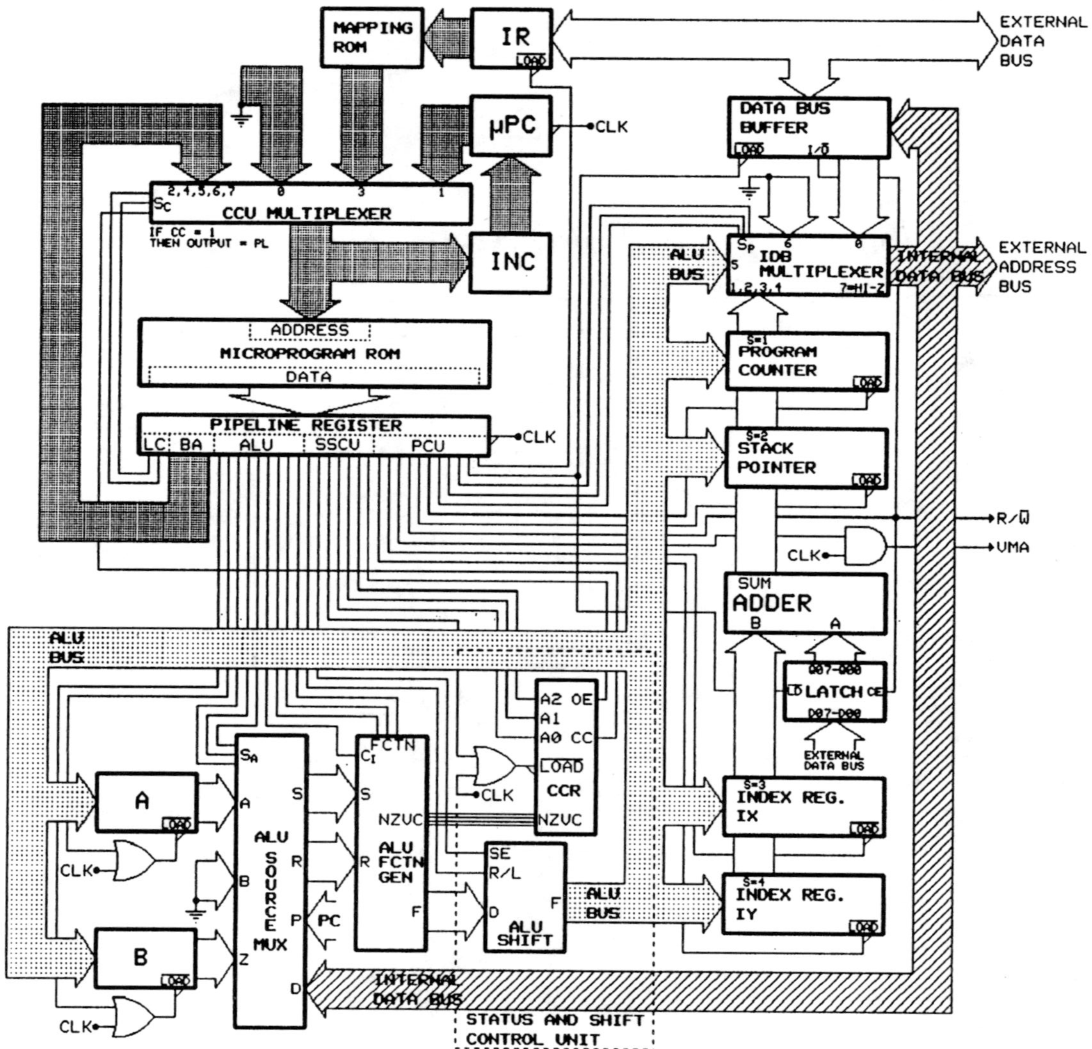

FIGURE 8.7 *Addition of the status and shift control unit to the state machine.*

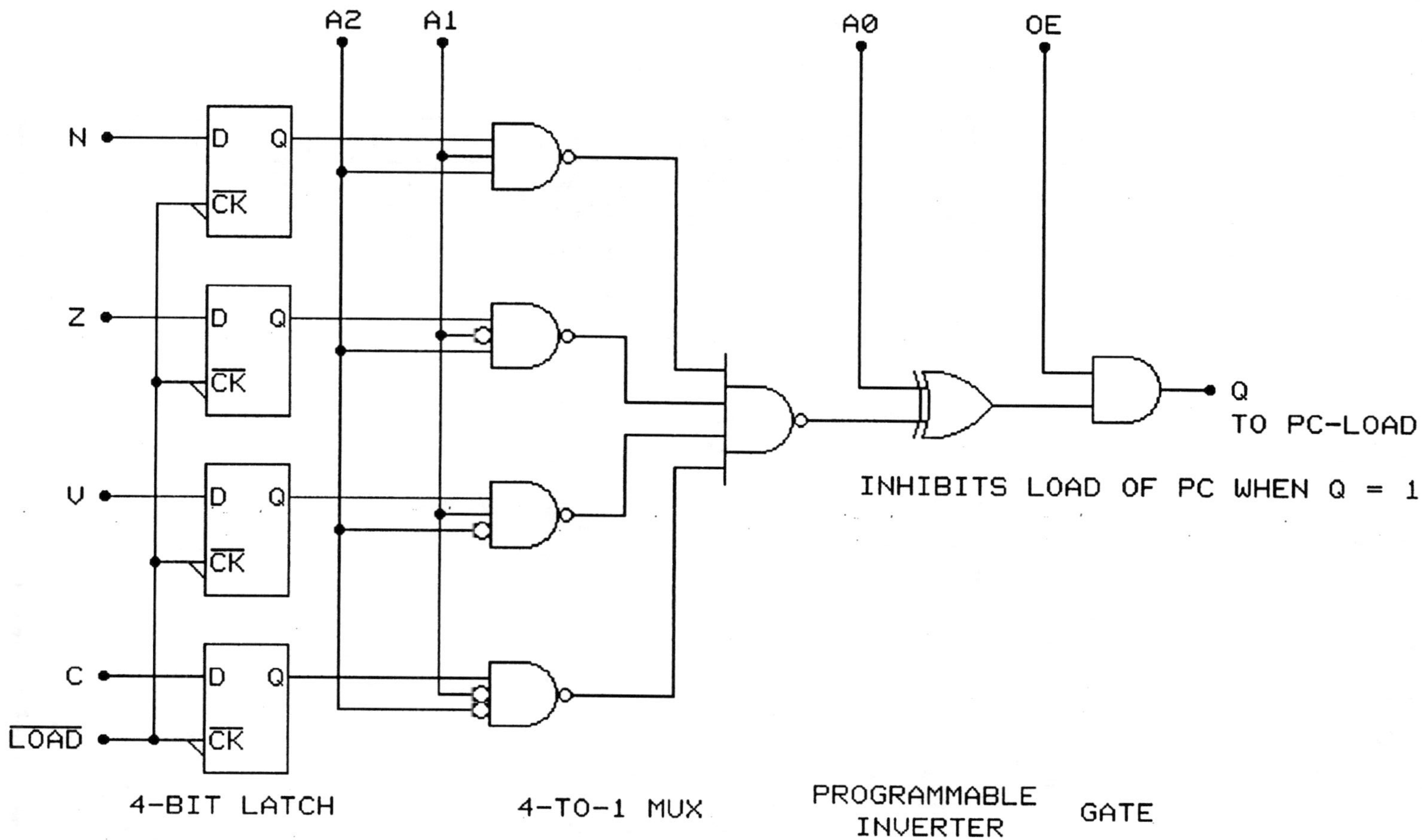

FIGURE 8.8 Condition Code Register Architecture

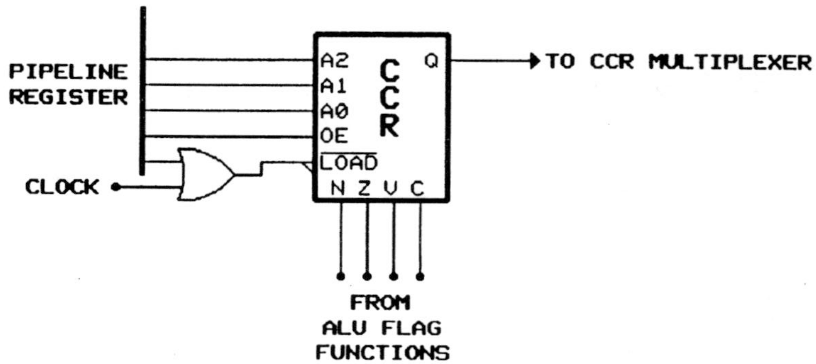

A2	1	0	OE	Q	USE	
×	×	×	0	0	--	OUTPUT = 0
0	0	0	1	C	CS	CARRY SET
0	0	1	1	$\bar{C}$	CC	CARRY CLEAR
0	1	0	1	V	VS	OVERFLOW SET
0	1	1	1	$\bar{V}$	VC	OVERFLOW CLEAR
1	0	0	1	Z	EQ	EQUAL
1	0	1	1	$\bar{Z}$	NE	NOT EQUAL
1	1	0	1	N	MI	MINUS
1	1	1	1	$\bar{N}$	PL	PLUS

FIGURE 8.9 Condition code register use.

UNIT	COMPUTER CONTROL UNIT									
DESCRIPTION	LOAD CTRL		BRANCH ADDRESS							
	I1	I0	A7	A6	A5	A4	A3	A2	A1	A0
REGISTER BIT	36	35	34	33	32	31	30	29	28	27
DEFAULT VALUE	0	1	X	X	X	X	X	X	X	X

UNIT	ARITHMETIC LOGIC UNIT								
DESCRIPTION	LDB	LDA	SOURCE			CI	FCTN		
			I2	I1	I0		I2	I1	I0
REGISTER BIT	26	25	24	23	22	21	20	19	18
DEFAULT VALUE	1	1	0	0	0	0	0	0	0

UNIT	STATUS AND SHIFT						
DESCRIPTION	R/L	SE	LD	FLAG SELECT			OE
				A2	A1	A0	
REGISTER BIT	17	16	15	14	13	12	11
DEFAULT VALUE	X	0	1	X	X	X	0

UNIT	PROGRAM CONTROL UNIT										
DESCRIPTION	IY	IX	VMA	R/W	SP	PC	IDB			DBB	IR
							S2	S1	S0		
REGISTER BIT	10	09	08	07	06	05	04	03	02	01	00
DEFAULT VALUE	1	1	0	1	1	1	0	0	0	1	1

FIGURE 8.10 Pipeline register default values.

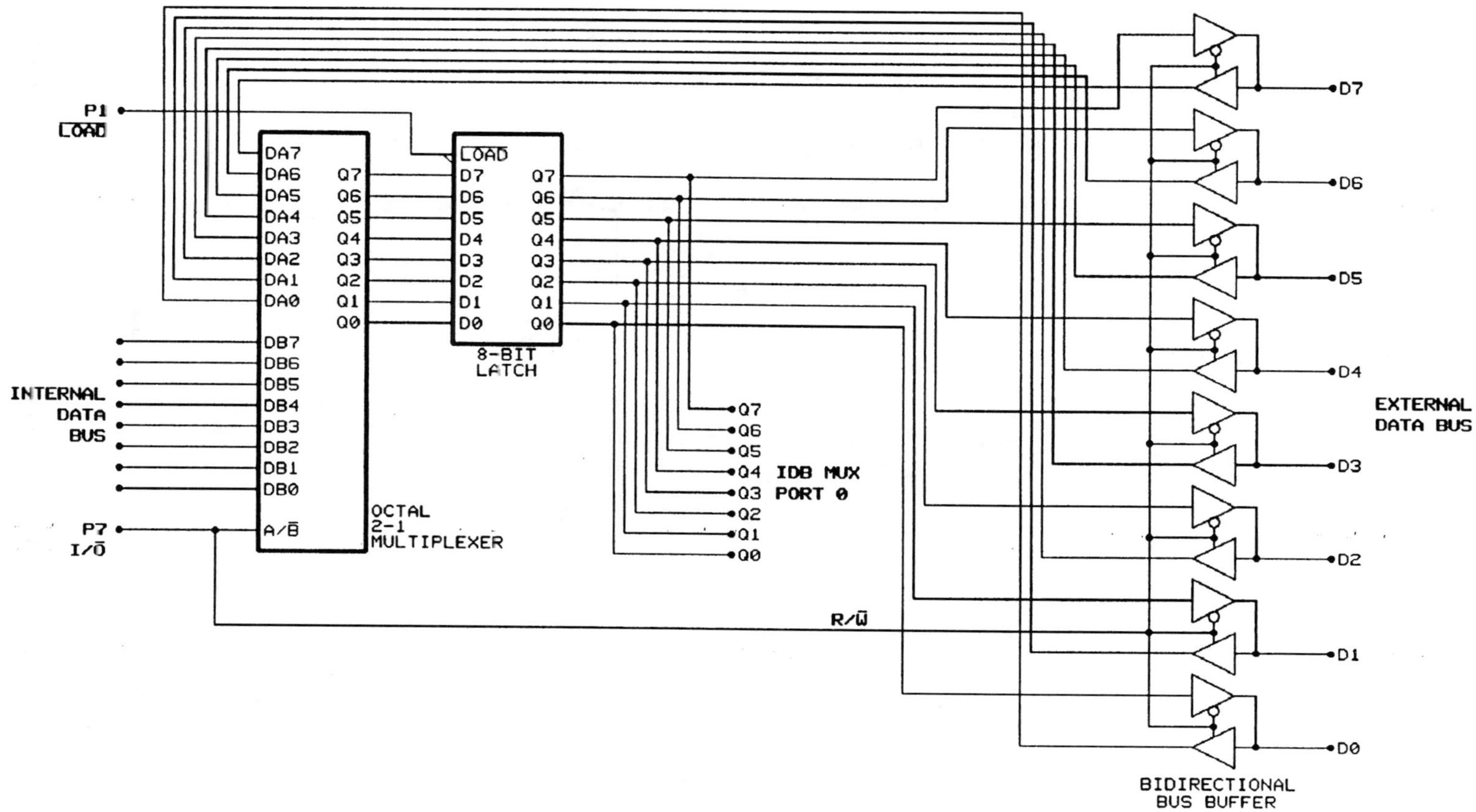

FIGURE 8.11 Data bus buffer.

S2	1	0	IDB MUX OUTPUT
0	0	0	DBB
0	0	1	PC
0	1	0	SP
0	1	1	IX + OPERAND
1	0	0	IY + OPERAND
1	0	1	ALU BUS
1	1	0	0
1	1	1	HI-Z

FIGURE 8.12 IDB multiplexer selection.

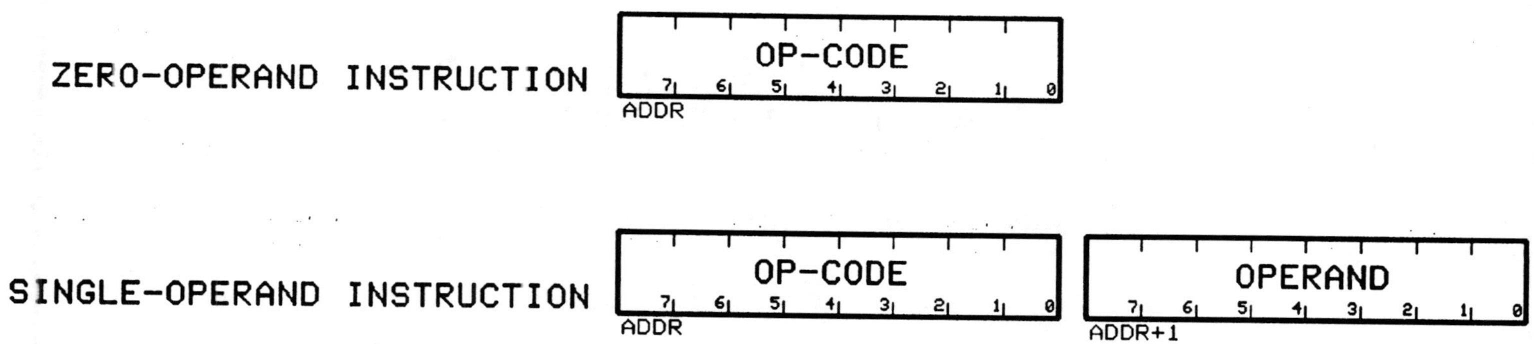

FIGURE 8.13 Example instruction formats.

Addressing Mode	Description	Assembly Language	Effective Address
Inherent (or implied)	The opcode is self-explanatory. No operand is required.	No operand	None
Immediate	The operand is an eight-bit data constant	#$NN	PC
Direct	The operand is an eight-bit address.	$NN	Operand
Indexed	The address is calculated by adding the index register contents to the operand.	$NN,X $NN,Y	Operand + X
Relative	The address is calculated by adding the program counter contents to the operand.	*+$NN	Operand + PC

TABLE 8.1 Example addressing modes.

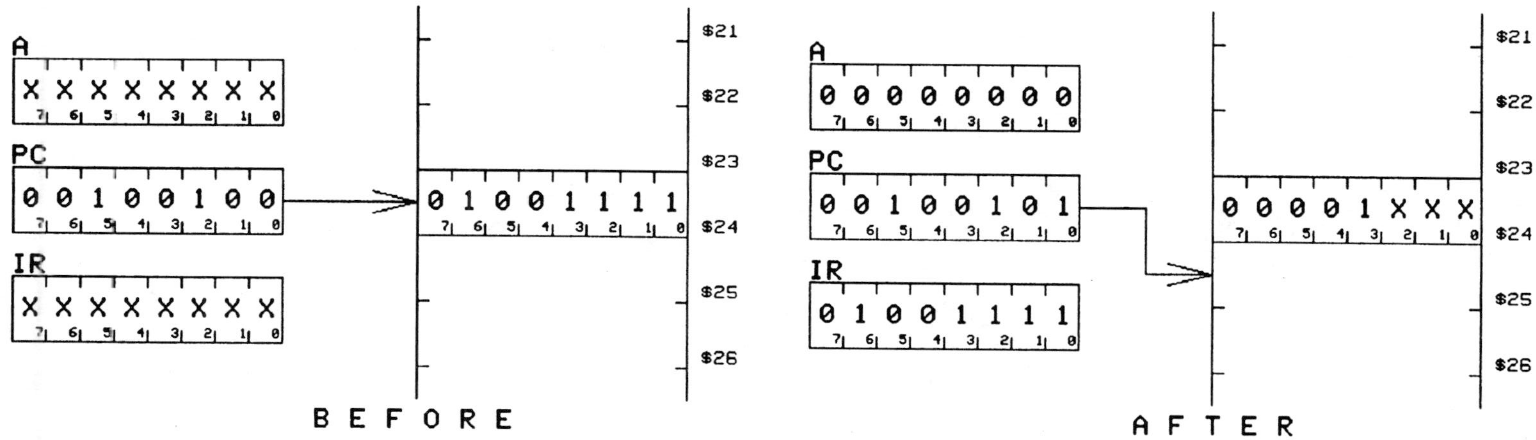

FIGURE 8.14 Implied addressing instruction: CLRA.

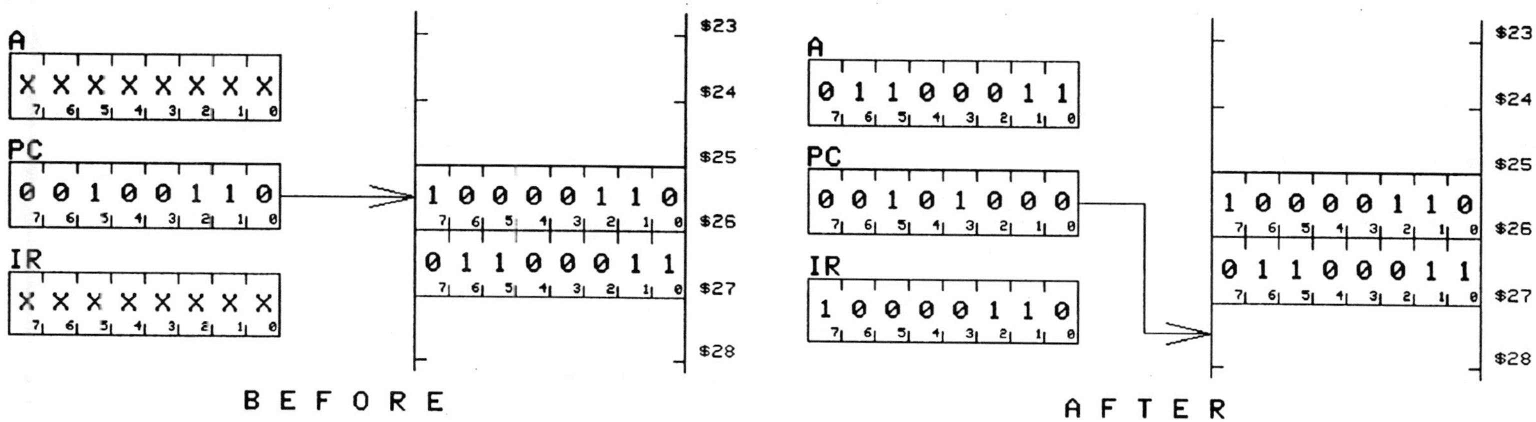

FIGURE 8.15 Immediate addressing instruction: MOVE #$63,A.

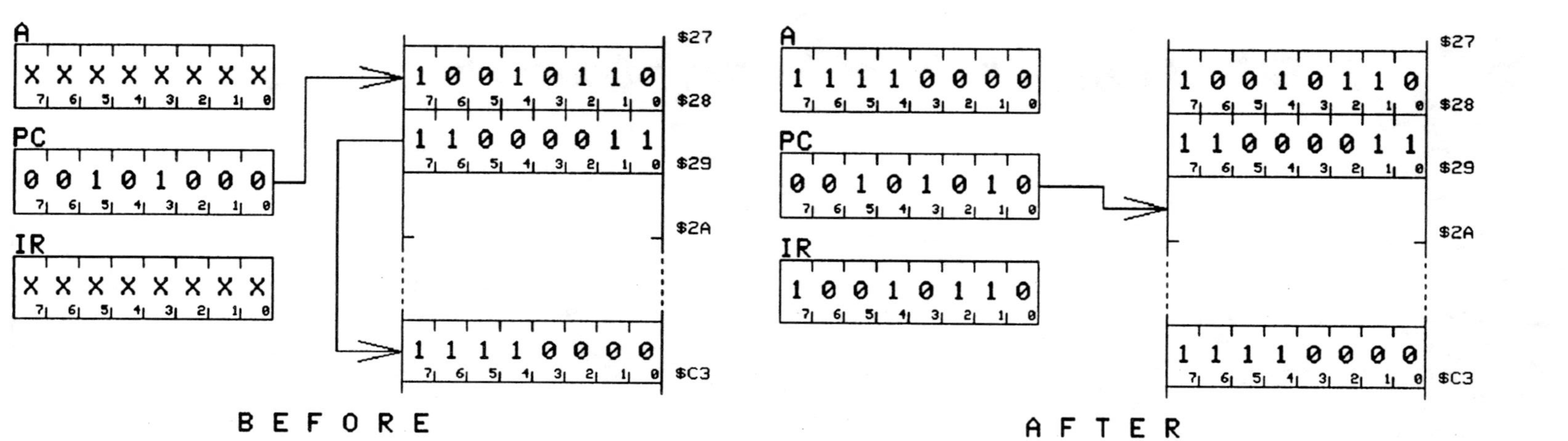

FIGURE 8.16 Direct addressing instruction: LDAA $C3.

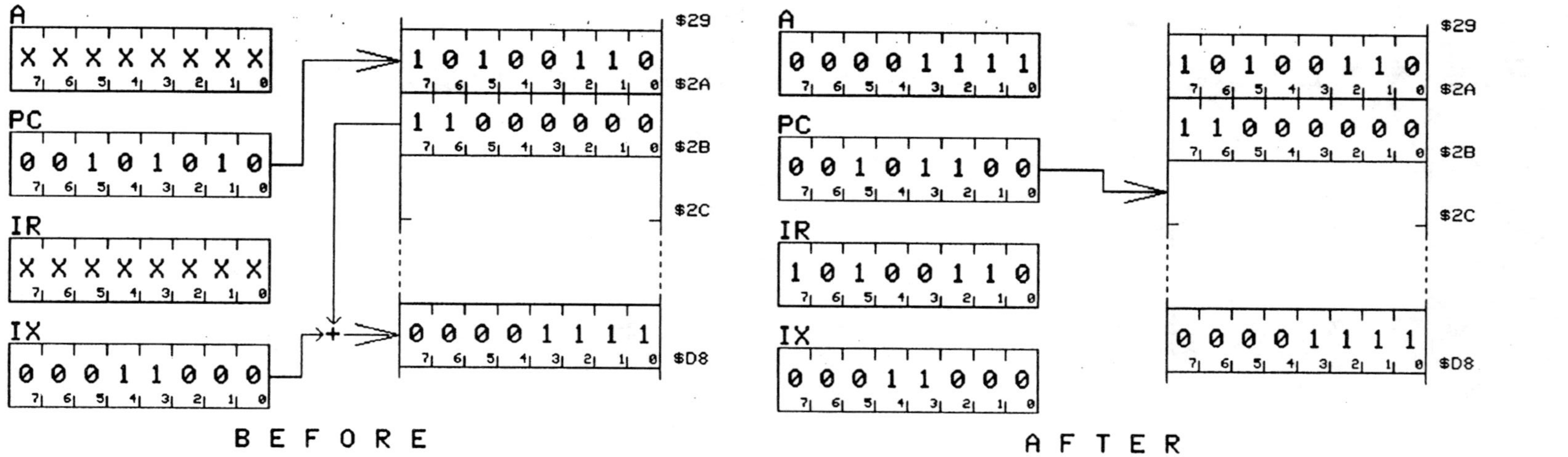

FIGURE 8.17 Direct indexed addressing instruction: LDAA $C0,X.

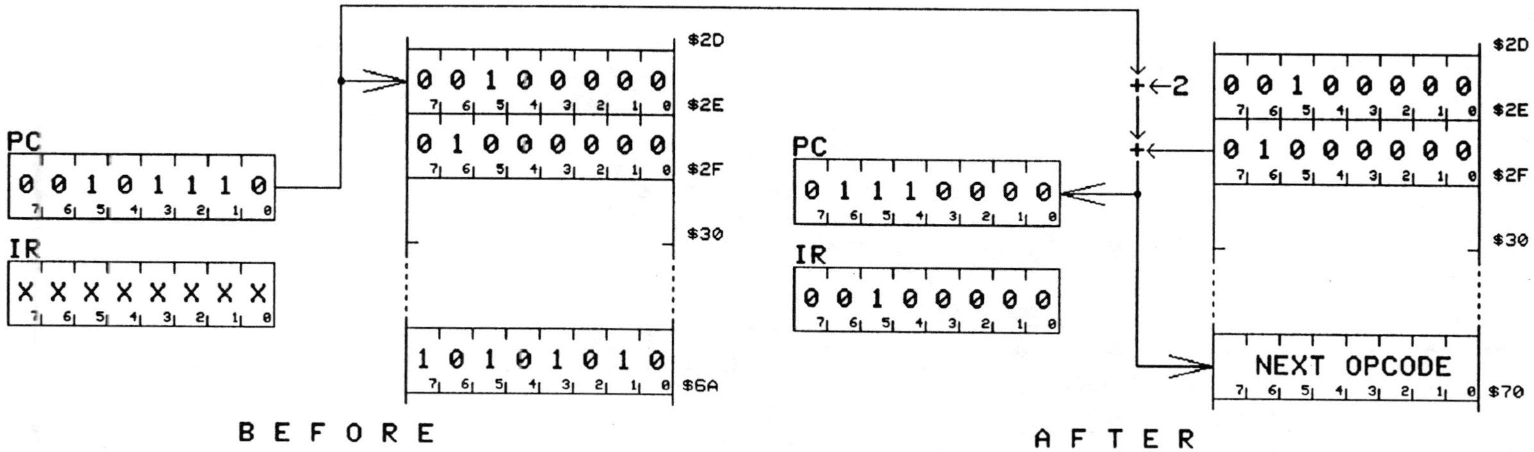

FIGURE 8.18 Relative addressing instruction: BRA *+$40.

NAME: DATE:	μP	COMPUTER CONTROL UNIT										ARITHMETIC LOGIC UNIT									STATUS AND SHIFT							PROGRAM CONTROL UNIT										
MNEMONIC: OPCODE:	ROM	LOAD CTRL		BRANCH ADDRESS								LDB	LDA	SOURCE			CI	FCTN			R/L	SE	LD	FLAG SELECT			OE	IY	IX	VMA	R/W	SP	PC	IDB			DBB	IR
	ADDRESS	I1	I0	A7	A6	A5	A4	A3	A2	A1	A0			I2	I1	I0		I2	I1	I0				A2	A1	A0								S2	S1	S0		
		36	35	34	33	32	31	30	29	28	27	26	25	24	23	22	21	20	19	18	17	16	15	14	13	12	11	10	09	08	07	06	05	04	03	02	01	00
T0																																						
T1																																						
T2																																						
T3																																						
T4																																						
T5																																						

FIGURE 8.19 Microprogramming worksheet.

Pipeline	Example RTL	Description
P0 = 0	(IR) ← (EDB)	Load the IR with the macroinstruction opcode on the external data bus.
P1 = 0, P7 = 1	(DBB) ← (EDB)	Load the DBB with the information in the EDB following a READ cycle.
P1 = 0, P7 = 0	(DBB) ← (IDB)	Load the DBB with the information on the IDB prior to a WRITE cycle.
P4,3,2 = 0,0,0	DBB → IDB Path	Place the contents of the data bus buffer on the IDB
P4,3,2 = 0,0,1	PC → IDB Path	Place the contents of the program counter on the IDB
P4,3,2 = 0,1,0	SP → IDB Path	Place the contents of the stack pointer on the IDB
P4,3,2 = 0,1,1	IX → IDB Path	Place the sum of the IX register and the instruction operand on the IDB
P4,3,2 = 1,0,0	IY → IDB Path	Place the sum of the IY register and the instruction operand on the IDB
P4,3,2 = 1,0,1	ALU bus → IDB Path	Place the contents of the ALU bus on the IDB
P4,3,2 = 1,1,0	Gnd → IDB Path	Set the internal data bus to a logical zero.
P4,3,2 = 1,1,1	Hi-Z → IDB Path	Set the internal data bus to a high impedance state.
P5 = 0	(PC) ← (ALU Bus)	Load the PC with the data on the ALU bus.
P6 = 0	(SP) ← (ALU Bus)	Load the SP with the data on the ALU bus.
P8,7 = 1,1	Memory READ	Enable the external memory system to execute a READ cycle.
P8,7 = 1,0	Memory WRITE	Enable the external memory system to execute a WRITE cycle.
P9 = 0	(IX) ← (IDB)	Load the index register IX with the data on the IDB.
P10 = 0	(IY) ← (IDB)	Load the index register IY with the data on the IDB.
P11 = 1	Enable Flags	Enable the output of the CCR.
P14,13,12	cc = CC	Select the desired CCR flag from FIGURE 8.9
P15 = 0	Load Flags	Load the condition codes register with the Flags from the ALU

P17,16 = 0,1	Shift Left	Shift the data passing through the ALU shifter to the left one bit
P17,16 = 1,1	Shift Right	Shift the data passing through the ALU shifter to the right one bit
P24-P18	ALU = ADD DZ+1 ALU = SUB AX ALU = OR AZ ALU = EXNOR DZ	These bits define the ALU function, source, and CI input values. Functions are selected from bits P20,19,18, and include PAR, PAS, ADD, SUB, AND, OR, XOR and NOT. Sources are selected from bits P24,23,22, and include DZ, AZ, AB, AD, ZD, BZ, BA, and BD. The status of the CI input to the ALU is defined by bit P21.
P25 = 0	(A) ← (ALU)	Load accumulator A with the data on the ALU bus.
P26 = 0	(B) ← (ALU)	Load accumulator B with the data on the ALU bus.
P34-27		Branch address indicated as a hexadecimal $NN in the JPL instruction.
P36,35 = 0,0	[Z]	Set the next microinstruction address to the beginning of the fetch cycle.
P36,35 = 0,1	[μPC]	Set the next microinstruction address to follow the current one.
P36,35 = 1,0	[JPL $NN]	Set the next microinstruction address to the branch address field, P34-27.
P36,35 = 1,1	[MAP]	Set the next microinstruction address to the mapping ROM output value.

TABLE 8.2 RTL statement summary.

NAME: DATE: / MNEMONIC: OPCODE:		μP ROM ADDRESS		COMPUTER CONTROL UNIT										ARITHMETIC LOGIC UNIT									STATUS AND SHIFT							PROGRAM CONTROL UNIT										
				LOAD CTRL I1	LOAD CTRL I0	BRANCH ADDRESS A7	A6	A5	A4	A3	A2	A1	A0	LDB	LDA	SOURCE I2	SOURCE I1	SOURCE I0	CI	FCTN I2	FCTN I1	FCTN I0	R/L	SE	LD	FLAG SELECT A2	FLAG SELECT A1	FLAG SELECT A0	OE	IY	IX	VMA	R/W	SP	PC	IDB S2	IDB S1	IDB S0	DBB	IR
				36	35	34	33	32	31	30	29	28	27	26	25	24	23	22	21	20	19	18	17	16	15	14	13	12	11	10	09	08	07	06	05	04	03	02	01	00
T0	PC→IDB Path, Memory READ,	0	0	0	1											0	0	0	1	0	0	0										1	1			0	0	1		
	ALU = PAR D+1, [μPC]			μPC												PAR			C	D												RD				PC				
T1	(IR)←(EDB), (PC)←(PC)+1,	0	1	1	1																											1	1		0	0	0	1		0
	PC→IDB Path, Memory READ, [MAP]			MAP																												RD			P	PC				I

FIGURE 8.20 Fetch microprogram.

NAME:	DATE:	µP ROM ADDRESS	LOAD CTRL I1 I0 (36 35)	BRANCH ADDRESS A7–A0 (34–27)	LDB (26)	LDA (25)	SOURCE I2 I1 I0 (24 23 22)	CI (21)	FCTN I2 I1 I0 (20 19 18)	R/L (17)	SE (16)	LD (15)	FLAG SELECT A2 A1 A0 (14 13 12)	OE (11)	IY (10)	IX (09)	VMA (08)	R/W (07)	SP (06)	PC (05)	IDB S2 S1 S0 (04 03 02)	DBB (01)	IR (00)
MNEMONIC: CLAA	OPCODE: 4F																						
T0	ALU = AND DZ, (A)←(ALU),	0 2	0 0			0	0 0 0		1 0 0			0											
	Set Flags, [Z]		Z		LDA		DZ		AND			L											

FIGURE 8.21 CLAA instruction microprogram.

NAME:	DATE:	µP ROM ADDRESS	LOAD CTRL I1 I0 (36 35)	BRANCH ADDRESS A7–A0 (34–27)	LDB (26)	LDA (25)	SOURCE I2 I1 I0 (24 23 22)	CI (21)	FCTN I2 I1 I0 (20 19 18)	R/L (17)	SE (16)	LD (15)	FLAG SELECT A2 A1 A0 (14 13 12)	OE (11)	IY (10)	IX (09)	VMA (08)	R/W (07)	SP (06)	PC (05)	IDB S2 S1 S0 (04 03 02)	DBB (01)	IR (00)
MNEMONIC: LDAA #$NN	OPCODE: 86																						
T0	(DBB)←(EDB), PC→IDB Path,	0 8	0 1				0 0 0	1	0 0 0												0 0 1	0	
	ALU = PAR D+1, [µPC]		µPC				D	C	PAR												PC	D	
T1	(PC)←(PC)+1, DBB→IDB Path,	0 9	0 0			0	0 0 0	0	0 0 0											0	0 0 0		
	ALU = PAR D, (A)←(ALU), [Z]		Z		LDA		D		PAR											P	DBB		

FIGURE 8.22 LDAA #$NN microprogram worksheet.

NAME:	DATE:	µP ROM ADDRESS	LOAD CTRL I1 I0 (36 35)	BRANCH ADDRESS A7–A0 (34–27)	LDB (26)	LDA (25)	SOURCE I2 I1 I0 (24 23 22)	CI (21)	FCTN I2 I1 I0 (20 19 18)	R/L (17)	SE (16)	LD (15)	FLAG SELECT A2 A1 A0 (14 13 12)	OE (11)	IY (10)	IX (09)	VMA (08)	R/W (07)	SP (06)	PC (05)	IDB S2 S1 S0 (04 03 02)	DBB (01)	IR (00)
MNEMONIC: LDAA $NN	OPCODE: 96																						
T0	(DBB)←(EDB), PC→IDB Path,	1 0	0 1				0 0 0	1	0 0 0												0 0 1	0	
	ALU = PAR D+1, [µPC]		µPC				D	C	PAR												PC	D	
T1	(PC)←(PC)+1, DBB→IDB Path,	1 1	0 1														1	1		0	0 0 0	1	
	Memory Read, [µPC]		µPC														RD			P	DBB		
T2	(DBB)←(EDB), DBB→IDB Path,	1 2	0 0			0	0 0 0	0	0 0 0												0 0 0	0	
	ALU = PAR D, (A)←(ALU), [Z]		Z		LDA		D		PAR												DBB	D	

FIGURE 8.23 LDAA $NN microprogram worksheet.

NAME:	DATE:																													
MNEMONIC: LDAA $NN,X	OPCODE: A6	μP ROM ADDRESS	COMPUTER CONTROL UNIT LOAD CTRL I1 (36)	I0 (35)	BRANCH ADDRESS A7–A0 (34–27)	ARITHMETIC LOGIC UNIT LDB (26)	LDA (25)	SOURCE I2 (24)	I1 (23)	I0 (22)	CI (21)	FCTN I2 (20)	I1 (19)	I0 (18)	STATUS AND SHIFT R/L (17)	SE (16)	LD (15)	FLAG SELECT A2 A1 A0 (14–12)	OE (11)	PROGRAM CONTROL UNIT IY (10)	IX (09)	VMA (08)	R/W (07)	SP (06)	PC (05)	IDB S2 (04)	S1 (03)	S0 (02)	DBB (01)	IR (00)
T0	(DBB)←(EDB), PC→IDB Path.	1C	0	1				0	0	0	1	0	0	0												0	0	1	0	
	ALU = PAR D+1, [μPC]		μPC					D			C	PAR														PC			D	
T1	(PC)←(PC)+1, X+OPERAND→IDB Path.	1D	0	1																		1	1		0	0	1	1	1	
	Memory Read, [μPC]		μPC																			RD			P	X+0				
T2	(DBB)←(EDB), DBB→IDB Path.	1E	0	0			0	0	0	0	0	0	0	0												0	0	0	0	
	ALU = PAR D, (A)←(ALU), [Z]		Z			LDA		D				PAR														DBB			D	

FIGURE 8.24 LDAA $NN,X microprogram worksheet.

NAME:	DATE:																													
MNEMONIC: JMP $NN	OPCODE: 7E	μP ROM ADDRESS	COMPUTER CONTROL UNIT LOAD CTRL I1 (36)	I0 (35)	BRANCH ADDRESS A7–A0 (34–27)	ARITHMETIC LOGIC UNIT LDB (26)	LDA (25)	SOURCE I2 (24)	I1 (23)	I0 (22)	CI (21)	FCTN I2 (20)	I1 (19)	I0 (18)	STATUS AND SHIFT R/L (17)	SE (16)	LD (15)	FLAG SELECT A2 A1 A0 (14–12)	OE (11)	PROGRAM CONTROL UNIT IY (10)	IX (09)	VMA (08)	R/W (07)	SP (06)	PC (05)	IDB S2 (04)	S1 (03)	S0 (02)	DBB (01)	IR (00)
T0	(DBB)←(EDB), DBB→IDB Path.	30	0	1				0	0	0	0	0	0	0												0	0	0	0	
	ALU = PAR D, [μPC]		μPC					D				PAR														DBB			D	
T1	(PC)←(ALU), [Z]	31	0	0																					0					
			Z																						P					

FIGURE 8.25 JMP $NN, microprogram worksheet.

NAME:	DATE:	µP ROM ADDRESS	LOAD CTRL I1 I0 (36 35)	BRANCH ADDRESS A7 A6 A5 A4 A3 A2 A1 A0 (34–27)	LDB (26)	LDA (25)	SOURCE I2 I1 I0 (24 23 22)	CI (21)	FCTN I2 I1 I0 (20 19 18)	R/L (17)	SE (16)	LD (15)	FLAG SELECT A2 A1 A0 (14 13 12)	OE (11)	IY (10)	IX (09)	VMA (08)	R/W (07)	SP (06)	PC (05)	IDB S2 S1 S0 (04 03 02)	DBB (01)	IR (00)
MNEMONIC: BRA *+$NN	OPCODE: 20																						
T0	(DBB)←(EDB), DBB→IDB Path,	3 2	0 1				1 0 0	1	0 1 0												0 0 0	0	
	ALU ← ADD DP, [µPC]		µPC				DP	C	ADD												DBB	D	
T1	(PC)←(ALU), [Z]	3 3	0 0																	0			
			Z																	P			

*FIGURE 8.26 BRA *+NN, microprogram worksheet.*

NAME:	DATE:	µP ROM ADDRESS	LOAD CTRL I1 I0 (36 35)	BRANCH ADDRESS A7 A6 A5 A4 A3 A2 A1 A0 (34–27)	LDB (26)	LDA (25)	SOURCE I2 I1 I0 (24 23 22)	CI (21)	FCTN I2 I1 I0 (20 19 18)	R/L (17)	SE (16)	LD (15)	FLAG SELECT A2 A1 A0 (14 13 12)	OE (11)	IY (10)	IX (09)	VMA (08)	R/W (07)	SP (06)	PC (05)	IDB S2 S1 S0 (04 03 02)	DBB (01)	IR (00)
MNEMONIC: BEQ *+$NN	OPCODE: 27																						
T0	(DBB)←(EDB), PC→IDB Path,	3 4	0 1	0 0 1 1 0 1 1 0			0 0 0	1	0 0 0				1 0 0	1							0 0 1	0	
	ALU ← PAR D+1, cc=EQ, PL-T2, [µPC]		µPC	JPL $36 if Z=1			D	C	PAR				cc=EQ								PC	D	
T1	(PC)←(ALU), [Z]	3 5	0 0																	0			
			Z																	P			
T2	DBB→IDB Path, ALU ← ADD DP, [µPC]	3 6	0 1				1 0 0	1	0 1 0												0 0 0	0	
			µPC				DP	C	ADD												DBB	D	
T3	(PC)←(ALU), [Z]	3 7	0 0																	0			
			Z																	P			

*FIGURE 8.27 BEQ *+NN, microprogram worksheet.*

<table>
<tr><td colspan="2">NAME:</td><td>DATE:</td><td rowspan="3">µP
ROM
ADDRESS</td><td colspan="10">COMPUTER CONTROL UNIT</td><td colspan="9">ARITHMETIC LOGIC UNIT</td><td colspan="7">STATUS AND SHIFT</td><td colspan="11">PROGRAM CONTROL UNIT</td></tr>
<tr><td colspan="2" rowspan="2">MNEMONIC:</td><td rowspan="2">OPCODE:</td><td colspan="2">LOAD C'RL</td><td colspan="8">BRANCH ADDRESS</td><td rowspan="2">LDB</td><td rowspan="2">LDA</td><td colspan="3">SOURCE</td><td rowspan="2">CI</td><td colspan="3">FCTN</td><td rowspan="2">R/L</td><td rowspan="2">SE</td><td rowspan="2">LD</td><td colspan="3">FLAG SELECT</td><td rowspan="2">OE</td><td rowspan="2">IY</td><td rowspan="2">IX</td><td rowspan="2">VMA</td><td rowspan="2">R/W</td><td rowspan="2">SP</td><td rowspan="2">PC</td><td colspan="3">IDB</td><td rowspan="2">DBB</td><td rowspan="2">IR</td></tr>
<tr><td>I1</td><td>I0</td><td>A7</td><td>A6</td><td>A5</td><td>A4</td><td>A3</td><td>A2</td><td>A1</td><td>A0</td><td>I2</td><td>I1</td><td>I0</td><td>I2</td><td>I1</td><td>I0</td><td>A2</td><td>A1</td><td>A0</td><td>S2</td><td>S1</td><td>S0</td></tr>
<tr><td colspan="3"></td><td></td><td>36</td><td>35</td><td>34</td><td>33</td><td>32</td><td>31</td><td>30</td><td>29</td><td>28</td><td>27</td><td>26</td><td>25</td><td>24</td><td>23</td><td>22</td><td>21</td><td>20</td><td>19</td><td>18</td><td>17</td><td>16</td><td>15</td><td>14</td><td>13</td><td>12</td><td>11</td><td>10</td><td>09</td><td>08</td><td>07</td><td>06</td><td>05</td><td>04</td><td>03</td><td>02</td><td>01</td><td>00</td></tr>
<tr><td>TØ</td><td colspan="2"></td><td></td><td></td><td></td><td></td><td></td><td></td><td></td><td></td><td></td><td></td><td></td><td></td><td></td><td></td><td></td><td></td><td></td><td></td><td></td><td></td><td></td><td></td><td></td><td></td><td></td><td></td><td></td><td></td><td></td><td></td><td></td><td></td><td></td><td></td><td></td><td></td><td></td><td></td></tr>
</table>

Microprogramming Worksheet

NAME: DATE:	µP ROM ADDRESS	COMPUTER CONTROL UNIT										ARITHMETIC LOGIC UNIT									STATUS AND SHIFT							PROGRAM CONTROL UNIT										
MNEMONIC: OPCODE:		LOAD CTRL		BRANCH ADDRESS								LDB	LDA	SOURCE			CI	FCTN			R/L	SE	LD	FLAG SELECT			OE	IY	IX	VMA	R/W	SP	PC	IDB			DBB	IR
		I1	I0	A7	A6	A5	A4	A3	A2	A1	A0			I2	I1	I0		I2	I1	I0				A2	A1	A0								S2	S1	S0		
		36	35	34	33	32	31	30	29	28	27	26	25	24	23	22	21	20	19	18	17	16	15	14	13	12	11	10	09	08	07	06	05	04	03	02	01	00
T0																																						
T1																																						

Microprogramming Worksheet

NAME: DATE:

MNEMONIC: OPCODE:

	µP ROM	COMPUTER CONTROL UNIT										ARITHMETIC LOGIC UNIT									STATUS AND SHIFT							PROGRAM CONTROL UNIT										
		LOAD CTRL		BRANCH ADDRESS								LDB	LDA	SOURCE			CI	FCTN			R/L	SE	LD	FLAG SELECT			OE	IY	IX	VMA	R/W	SP	PC	IDB			DBB	IR
	ADDRESS	I1	I0	A7	A6	A5	A4	A3	A2	A1	A0			I2	I1	I0		I2	I1	I0				A2	A1	A0								S2	S1	S0		
		36	35	34	33	32	31	30	29	28	27	26	25	24	23	22	21	20	19	18	17	16	15	14	13	12	11	10	09	08	07	06	05	04	03	02	01	00
T0																																						
T1																																						
T2																																						

Microprogramming Worksheet

NAME: DATE:	µP	COMPUTER CONTROL UNIT										ARITHMETIC LOGIC UNIT									STATUS AND SHIFT							PROGRAM CONTROL UNIT										
MNEMONIC: OPCODE:	ROM	LOAD CTRL		BRANCH ADDRESS								LDB	LDA	SOURCE			CI	FCTN			R/L	SE	LD	FLAG SELECT			OE	IY	IX	VMA	R/W	SP	PC	IDB			DBB	IR
	ADDRESS	I1	I0	A7	A6	A5	A4	A3	A2	A1	A0			I2	I1	I0		I2	I1	I0				A2	A1	A0								S2	S1	S0		
		36	35	34	33	32	31	30	29	28	27	26	25	24	23	22	21	20	19	18	17	16	15	14	13	12	11	10	09	08	07	06	05	04	03	02	01	00
T0																																						
T1																																						
T2																																						
T3																																						

Microprogramming Worksheet

NAME: DATE:	µP ROM ADDRESS	COMPUTER CONTROL UNIT										ARITHMETIC LOGIC UNIT									STATUS AND SHIFT							PROGRAM CONTROL UNIT										
MNEMONIC: OPCODE:		LOAD CTRL		BRANCH ADDRESS								LDB	LDA	SOURCE			CI	FCTN			R/L	SE	LD	FLAG SELECT			OE	IY	IX	VMA	R/W	SP	PC	IDB			DBB	IR
		I1	I0	A7	A6	A5	A4	A3	A2	A1	A0			I2	I1	I0		I2	I1	I0				A2	A1	A0								S2	S1	S0		
		36	35	34	33	32	31	30	29	28	27	26	25	24	23	22	21	20	19	18	17	16	15	14	13	12	11	10	09	08	07	06	05	04	03	02	01	00
T0																																						
T1																																						
T2																																						
T3																																						
T4																																						

Microprogramming Worksheet

NAME: DATE:	µP ROM ADDRESS	COMPUTER CONTROL UNIT										ARITHMETIC LOGIC UNIT									STATUS AND SHIFT							PROGRAM CONTROL UNIT										
MNEMONIC: OPCODE:		LOAD CTRL		BRANCH ADDRESS								LDB	LDA	SOURCE			CI	FCTN			R/L	SE	LD	FLAG SELECT			OE	IY	IX	VMA	R/W	SP	PC	IDB			DBB	IR
		I1	I0	A7	A6	A5	A4	A3	A2	A1	A0			I2	I1	I0		I2	I1	I0				A2	A1	A0								S2	S1	S0		
		36	35	34	33	32	31	30	29	28	27	26	25	24	23	22	21	20	19	18	17	16	15	14	13	12	11	10	09	08	07	06	05	04	03	02	01	00
T0																																						
T1																																						
T2																																						
T3																																						
T4																																						
T5																																						

Microprogramming Worksheet

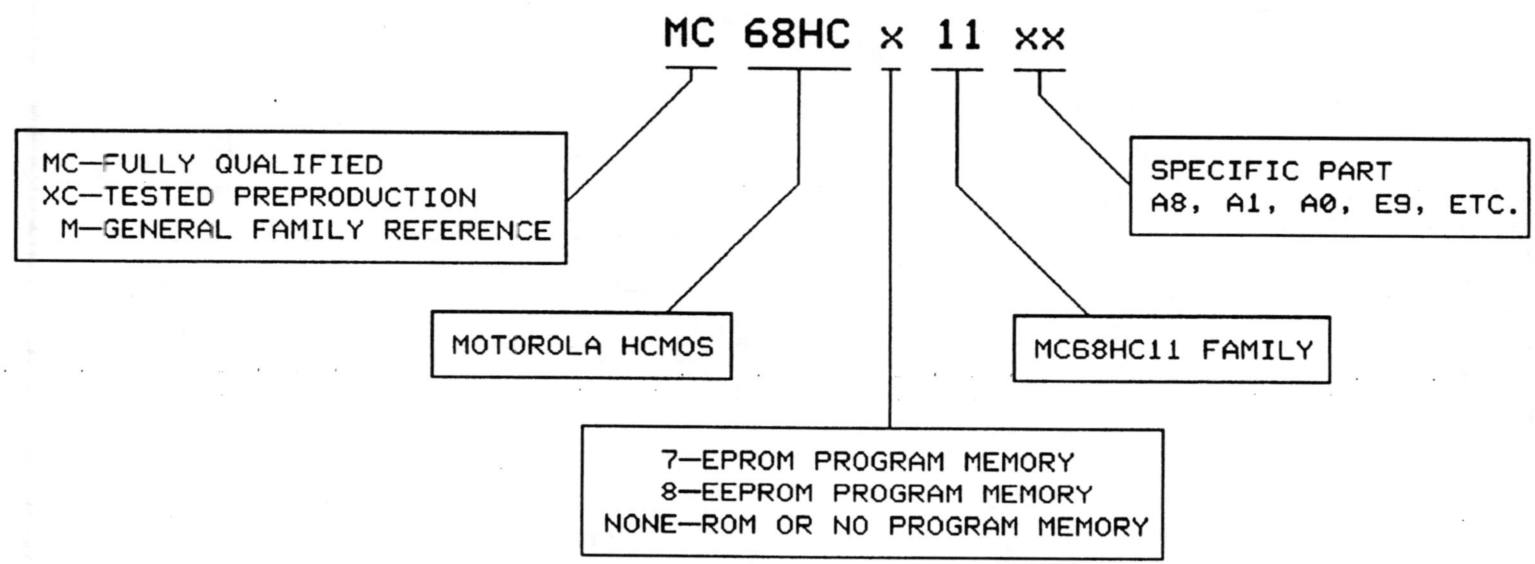

FIGURE 9.1 ***MC68HC11 family naming convention.***

Part Number	EPROM	ROM	EEPROM	RAM	CONFIG[2]	Comments
MC68HC11A8		**8K**	**512**	**256**	**$0F**	***Family built around this device***
MC68HC11A1			**512**	**256**	**$0D**	***'A8 with ROM disabled***
MC68HC11A0				**256**	**$0C**	***'A8 with ROM and EEPROM disabled***
MC68HC811A8			**8K+512**	**256**	**$0F**	***EEPROM Emulator for 'A8***
MC68HC11E9		**12K**	**512**	**512**	**$0F**	***Four input capture***
MC68HC11E1			**512**	**512**	**$0D**	***'E9 with ROM disabled***
MC68HC11E0				**512**	**$0C**	***'E9 with ROM and EEPROM disabled***
MC68HC811E2			**2K**[1]	**256**	**$FF**	***No ROM part for expanded systems***
MC68HC711E9	**12K**		**512**	**512**	**$0F**	***One-time programmable version of 'E9***
MC68HC11D3		**4K**		**192**	**N/A**	***Low-cost 40-pin version***
MC68HC711D9	**4K**			**192**	**N/A**	***One-time programmable version of 'D3***
MC68HC11F1			**512**[1]	**1K**	**$FF**	***High-performance non-multiplexed***
MC68HC11K4		**24K**	**640**	**768**	**$FF**	***>1 Meg memory space, PWM, CS***
MC68HC711K4	**24K**		**640**	**768**	**$FF**	***One-time programmable version of 'K4***
MC68HC11L6		**16K**	**512**	**512**	**$0F**	***'E9 with more ROM and more I/O***
MC68HC711L6	**16K**		**512**	**512**	**$0F**	***One-time programmable version of 'L4***

Notes: 1. The EEPROM is relocatable to the top of any 4K memory page. Relocation is done with the upper four bits of the control register.
2. CONFIG register values in this table reflect the value programmed prior to shipment from Motorola.

TABLE 9.1 MC68HC11 family devices.

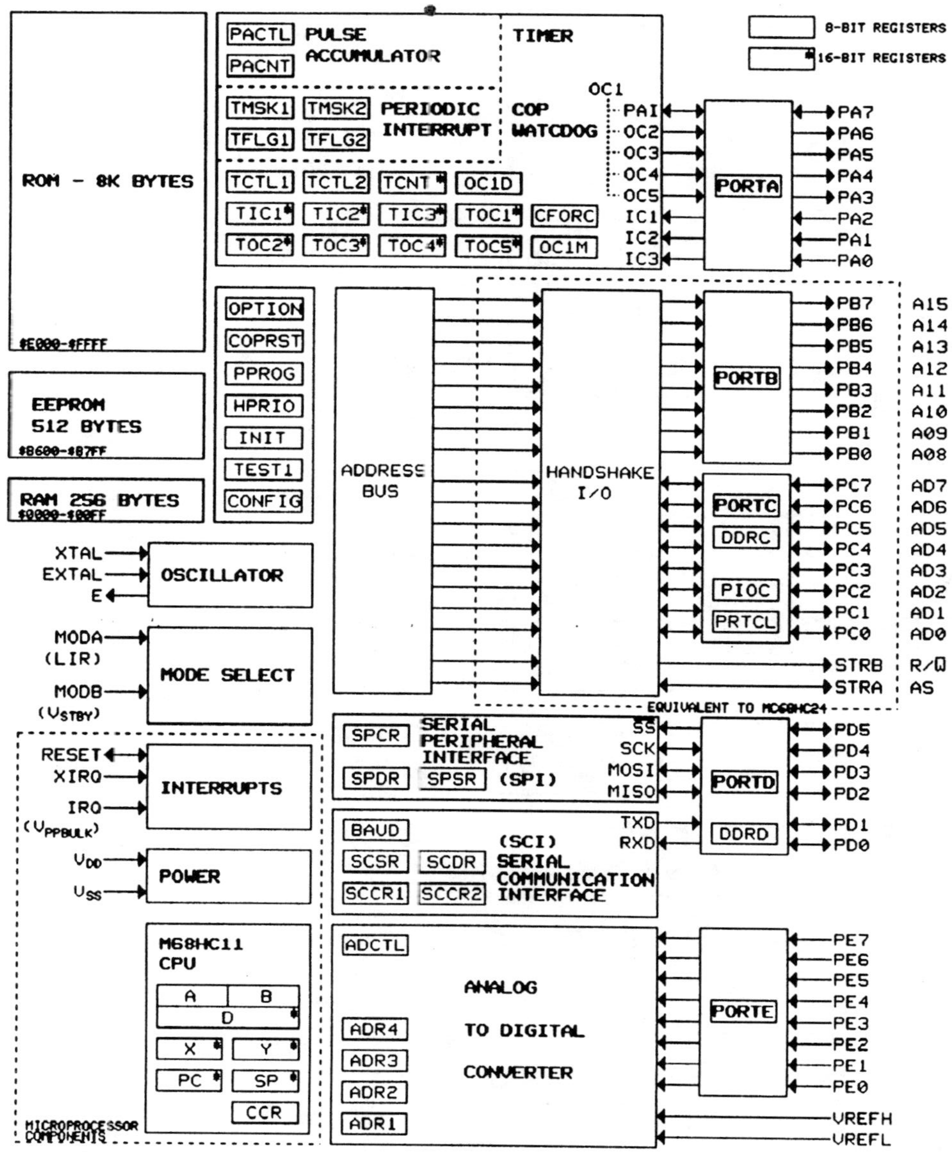

FIGURE 9.2

MC68HC11 block diagram.

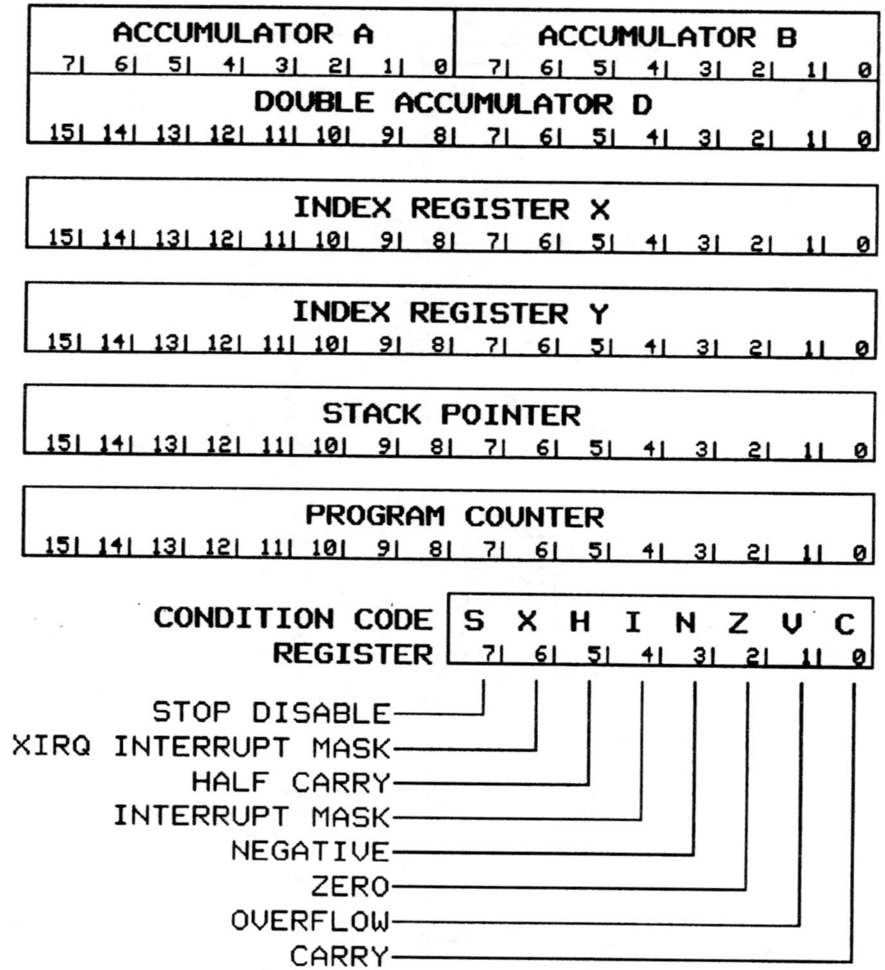

FIGURE 9.3 MC68HC11 programmer's model.

Inputs	*Mode*	*Control bits in HPRIO (latched at reset)*				
MODB	*MODA*	*Description*	*RBOOT*	*SMOD*	*MDA*	*IRV*
1	***0***	***Normal Single Chip***	***0***	***0***	***0***	***0***
1	***1***	***Normal Expanded***	***0***	***0***	***1***	***0***
0	***0***	***Special Bootstrap***	***1***	***1***	***0***	***1***
0	***1***	***Special Test***	***0***	***1***	***1***	***1***

TABLE 9.2 Hardware mode select summary.

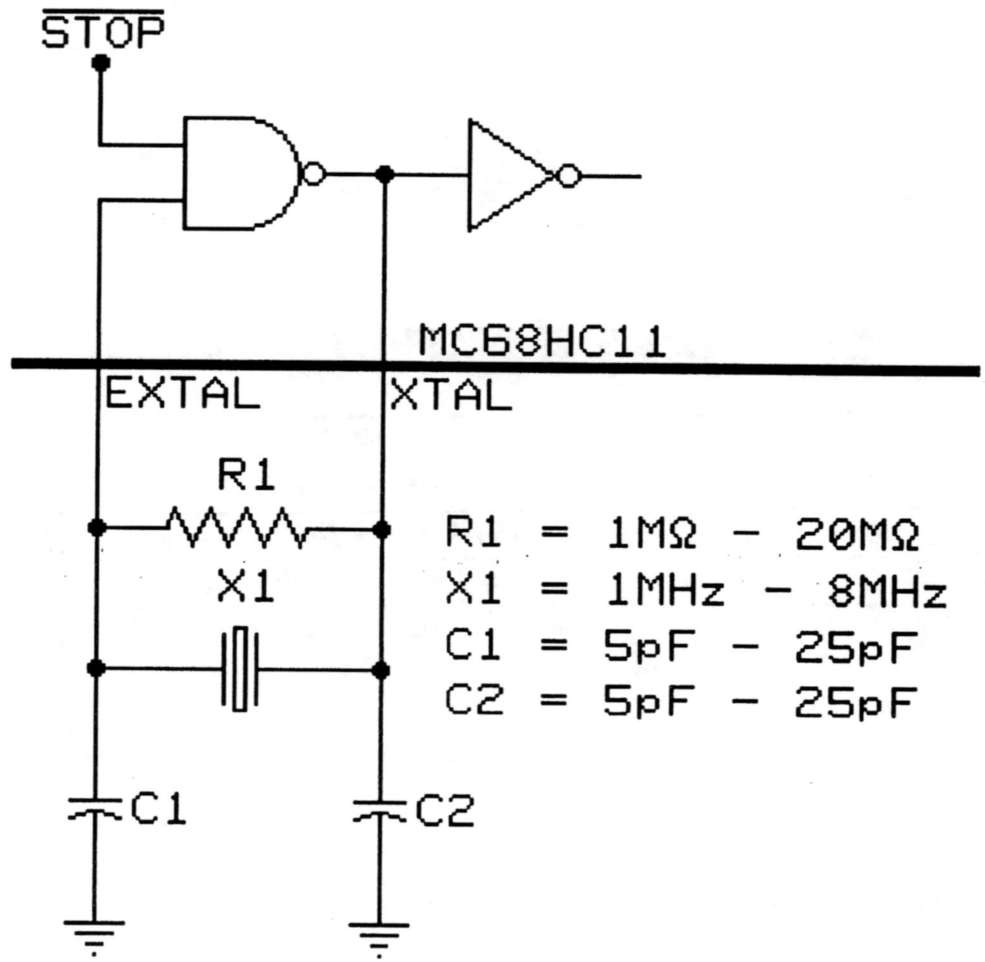

FIGURE 9.4 Crystal oscillator circuit.

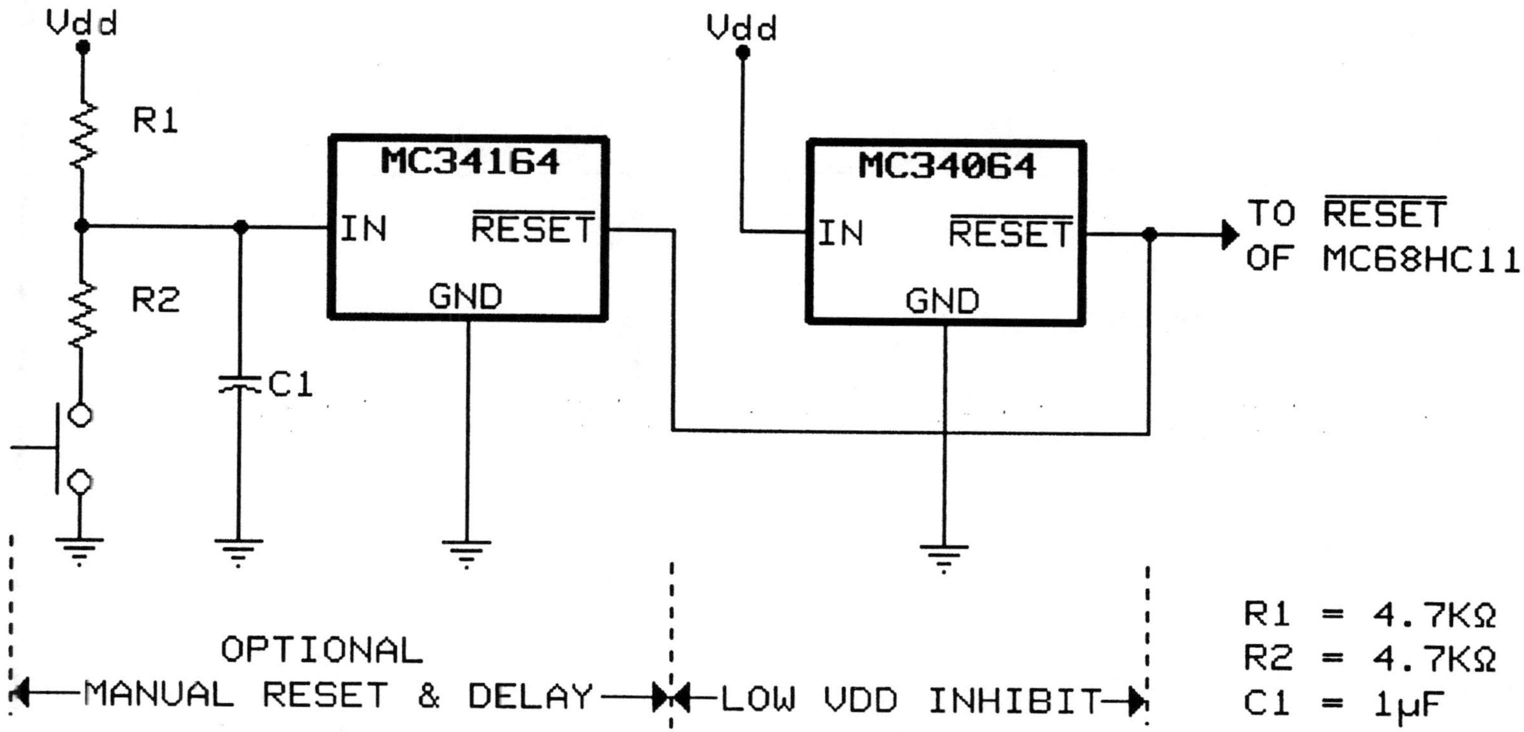

FIGURE 9.5 MC68HC11 reset circuitry.

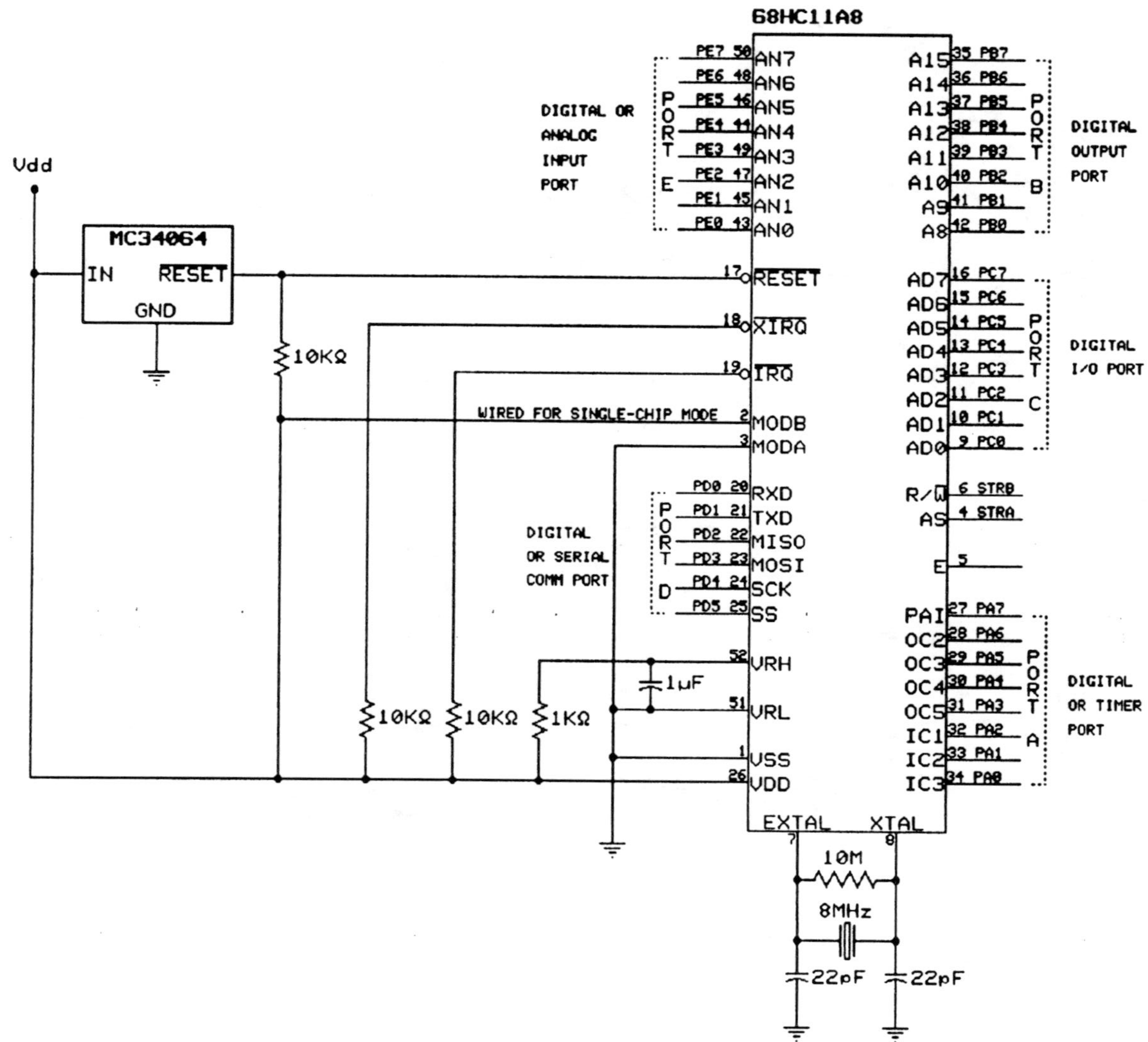

FIGURE 9.6 MC68HC11A8 microcontroller minimum single-chip configuration.

VECTOR ADDRESS	INTERRUPT SOURCE	CCR MASK	LOCAL MASK
$FFFE:FFFF	RESET	NONE	NONE
$FFFC:FFFD	COP CLOCK MONITOR FAIL (RESET)	NONE	CME
$FFFA:FFFB	COP FAILURE (RESET)	NONE	NOCOP
$FFF8:FFF9	ILLEGAL OPCODE TRAP	NONE	NONE
$FFF6:FFF7	SWI SOFTWARE INTERRUPT	NONE	NONE
$FFF4:FFF5	XIRQ PIN (PSEUDO NON-MASKABLE INTERRUPT)	X BIT	NONE
$FFF2:FFF3	IRQ (EXTERNAL PIN OR PARALLEL I/O)	I BIT	
$FFF0:FFF1	REAL TIME INTERRUPT	I BIT	RTII
$FFEE:FFEF	TIMER INPUT CAPTURE 1	I BIT	IC1I
$FFEC:FFED	TIMER INPUT CAPTURE 2	I BIT	IC2I
$FFEA:FFEB	TIMER INPUT CAPTURE 3	I BIT	IC3I
$FFE8:FFE9	TIMER OUTPUT CAPTURE 1	I BIT	OC1I
$FFE6:FFE7	TIMER OUTPUT CAPTURE 2	I BIT	OC2I
$FFE4:FFE5	TIMER OUTPUT CAPTURE 3	I BIT	OC3I
$FFE2:FFE3	TIMER OUTPUT CAPTURE 4	I BIT	OC4I
$FFE0:FFE1	TIMER OUTPUT CAPTURE 5	I BIT	OC5I
$FFDE:FFDF	TIMER OVERFLOW	I BIT	TOI
$FFDC:FFDD	PULSE ACCUMULATOR OVERFLOW	I BIT	PAOVI
$FFDA:FFDB	PULSE ACCUMULATOR INPUT EDGE	I BIT	PAII
$FFD8:FFD9	SPI SERIAL TRANSFER COMPLETE	I BIT	SPIE
$FFD6:FFD7	SCI SERIAL SYSTEM	I BIT	
$FFD4:FFD5 • • $FFC0:FFC1	RESERVED	—	—

INTERRUPT CAUSE	LOCAL MASK
RECEIVE DATA REGISTER FULL	RIE
RECEIVER OVERRUN	RIE
IDLE LINE DETECT	ILIE
TRANSMIT DATA REGISTER EMPTY	TIE
TRANSMIT COMPLETE	TCIE

INTERRUPT CAUSE	LOCAL MASK
EXTERNAL PIN	NONE
PARALLEL I/O HANDSHAKE	STAI

FIGURE 9.7 Interrupt vector ROM segment.

Bit 7	Bit 6	Bit 5	Bit 4	Bit 3	Bit 2	Bit 1	Bit 0	Address	Name	Description
BIT07	—	—	—	—	—	—	BIT00	$101F	TOC5	OUTPUT COMPARE 5 REGISTER
BIT15	—	—	—	—	—	—	BIT08	$101E		
BIT07	—	—	—	—	—	—	BIT00	$101D	TOC4	OUTPUT COMPARE 4 REGISTER
BIT15	—	—	—	—	—	—	BIT08	$101C		
BIT07	—	—	—	—	—	—	BIT00	$101B	TOC3	OUTPUT COMPARE 3 REGISTER
BIT15	—	—	—	—	—	—	BIT08	$101A		
BIT07	—	—	—	—	—	—	BIT00	$1019	TOC2	OUTPUT COMPARE 2 REGISTER
BIT15	—	—	—	—	—	—	BIT08	$1018		
BIT07	—	—	—	—	—	—	BIT00	$1017	TOC1	OUTPUT COMPARE 1 REGISTER
BIT15	—	—	—	—	—	—	BIT08	$1016		
BIT07	—	—	—	—	—	—	BIT00	$1015	TIC3	INPUT CAPTURE 3 REGISTER
BIT15	—	—	—	—	—	—	BIT08	$1014		
BIT07	—	—	—	—	—	—	BIT00	$1013	TIC2	INPUT CAPTURE 2 REGISTER
BIT15	—	—	—	—	—	—	BIT08	$1012		
BIT07	—	—	—	—	—	—	BIT00	$1011	TIC1	INPUT CAPTURE 1 REGISTER
BIT15	—	—	—	—	—	—	BIT08	$1010		
BIT07	—	—	—	—	—	—	BIT00	$100F	TCNT	TIMER COUNT REGISTER
BIT15	—	—	—	—	—	—	BIT08	$100E		
OC1D7	OC1D6	OC1D5	IC1D4	OC1D3				$100D	OC1D	OC1 ACTION DATA REGISTER
OC1M7	OC1M6	OC1M5	OC1M4	OC1M3				$100C	OC1M	OC1 ACTION MASK REGISTER
FOC1	FOC2	FOC3	FOC4	FOC5				$100B	CFORC	COMPARE FORCE REGISTER
BIT07	—	—	—	—	—	—	BIT00	$100A	PORTE	INPUT PORT E
		BIT05	—	—	—	—	BIT00	$1009	DDRD	DATA DIRECTION FOR PORT D
		BIT05	—	—	—	—	BIT00	$1008	PORTD	I/O PORT D
BIT07	—	—	—	—	—	—	BIT00	$1007	DDRC	DATA DIRECTION FOR PORT C
								$1006		RESERVED
BIT07	—	—	—	—	—	—	BIT00	$1005	PORTCL	ALTERNATE LATCHED PORT C
BIT07	—	—	—	—	—	—	BIT00	$1004	PORTB	OUTPUT PORT B
BIT07	—	—	—	—	—	—	BIT00	$1003	PORTC	I/O PORT C
STAF	STAI	CWOM	HNDS	OIN	PLS	EGA	INVB	$1002	PIOC	PARALLEL I/O CONTROL REGISTER
—	—	—	—	—	—	—	—	$1001		RESERVED
BIT07	—	—	—	—	—	—	BIT00	$1000	PORTA	I/O PORT A

FIGURE 9.8a MC68HC11 Register and Control Segment, $1000-$101F.

Bit 7	Bit 6	Bit 5	Bit 4	Bit 3	Bit 2	Bit 1	Bit 0	Address	Register	Description
				NOSEC	NOCOP	ROMON	EEON	$103F	CONFIG	COP, ROM, AND EEPROM ENABLES
TILOP		OOCR	CBYP	DISR	FCM	FCOP	TCON	$103E	TEST1	FACTORY TEST CONTROL REGISTER
RAM3	RAM2	RAM1	RAM0	RAM3	RAM2	RAM1	RAM0	$103D	INIT	RAM AND I/O MAPPING REGISTER
RBOOT	SMOD	MDA	IRV	PSEL3	PSEL2	PSEL1	PSEL0	$103C	HPRIO	HIGHEST PRIORITY I-BIT INT AND MISC
ODD	EVEN		BYTE	ROW	ERASE	EELAT	EEPGM	$103B	PPROG	EEPROM PROGRAMMING CONTROL REGISTER
BIT07	—	—	—	—	—	—	BIT00	$103A	COPRST	ARM/RESET COP TIMER CIRCUITRY
ADPU	CSEL	IRQE	DLY	CME		CR1	CR0	$1039	OPTION	SYSTEM CONFIGURATION OPTIONS
								$1038		RESERVED
								$1037		RESERVED
								$1036		RESERVED
								$1035		RESERVED
BIT07	—	—	—	—	—	—	BIT00	$1034	ADR4	A/D RESULT REGISTER 4
BIT07	—	—	—	—	—	—	BIT00	$1033	ADR3	A/D RESULT REGISTER 3
BIT07	—	—	—	—	—	—	BIT00	$1032	ADR2	A/D RESULT REGISTER 2
BIT07	—	—	—	—	—	—	BIT00	$1031	ADR1	A/D RESULT REGISTER 1
CCF		SCAN	MULT	CD	CC	CB	CA	$1030	ADCTL	A/D CONTROL REGISTER
BIT07	—	—	—	—	—	—	BIT00	$102F	SCDR	SCI DATA (READ RDR, WRITE TDR)
TDRE	TC	RDRF	IDLE	OR	NF	FE		$102E	SCSR	SCI STATUS REGISTER
TIE	TCIE	RIE	ILIE	TE	RE	RWU	SBK	$102D	SCCR2	SCU CONTROL REGISTER 2
R8	T8		M	WAKE				$102C	SCCR1	SCI CONTROL REGISTER 1
TCLR		SCP1	SCP0	RCKB	SCR2	SCR1	SCR0	$102B	BAUD	SCI BAUD RATE CONTROL
BIT07	—	—	—	—	—	—	BIT00	$102A	SPDR	SPI DATA REGISTER
SPIF	WCOL		MODF					$1029	SPSR	SPI STATUS REGISTER
SPIE	SPE	DWOM	MSTR	CPOL	CPHA	SPR1	SPR0	$1028	SPCR	SPI CONTROL REGISTER
BIT07	—	—	—	—	—	—	BIT00	$1027	PACNT	PULSE ACCUMULATOR COUNT REGISTER
DDRA7	PAEN	PAMOD	PEDGE			RTR1	RTR0	$1026	PACTL	PULSE ACCUMULATOR CONTROL REGISTER
TOF	RTIF	PAOVF	PAIF					$1025	TFLG2	TIMER INTERRUPT FLAG REGISTER 2
TOI	RTII	PAOVI	PAII			PR1	PR0	$1024	TMSK2	TIMER INTERRUPT MASK REGISTER 2
OC1F	OC2F	OC3F	OC4F	OC5F	IC1F	IC2F	IC3F	$1023	TFLG1	TIMER INTERRUPT FLAG REGISTER 1
OC1I	OC2I	OC3I	OC4I	OC5I	IC1I	IC2I	IC3I	$1022	TMSK1	TIMER INTERRUPT MASK REGISTER 1
		EDG1B	EDG1A	EDG2B	EDG2A	EDG3B	EDG3A	$1021	TCTL2	TIMER CONTROL REGISTER 2
OM2	OL2	OM3	OL3	OM4	OL4	OM5	OL5	$1020	TCTL1	TIMER CONTROL REGISTER 1

FIGURE 9.8b MC68HC11 Register and Control Segment, $1020-$103F.

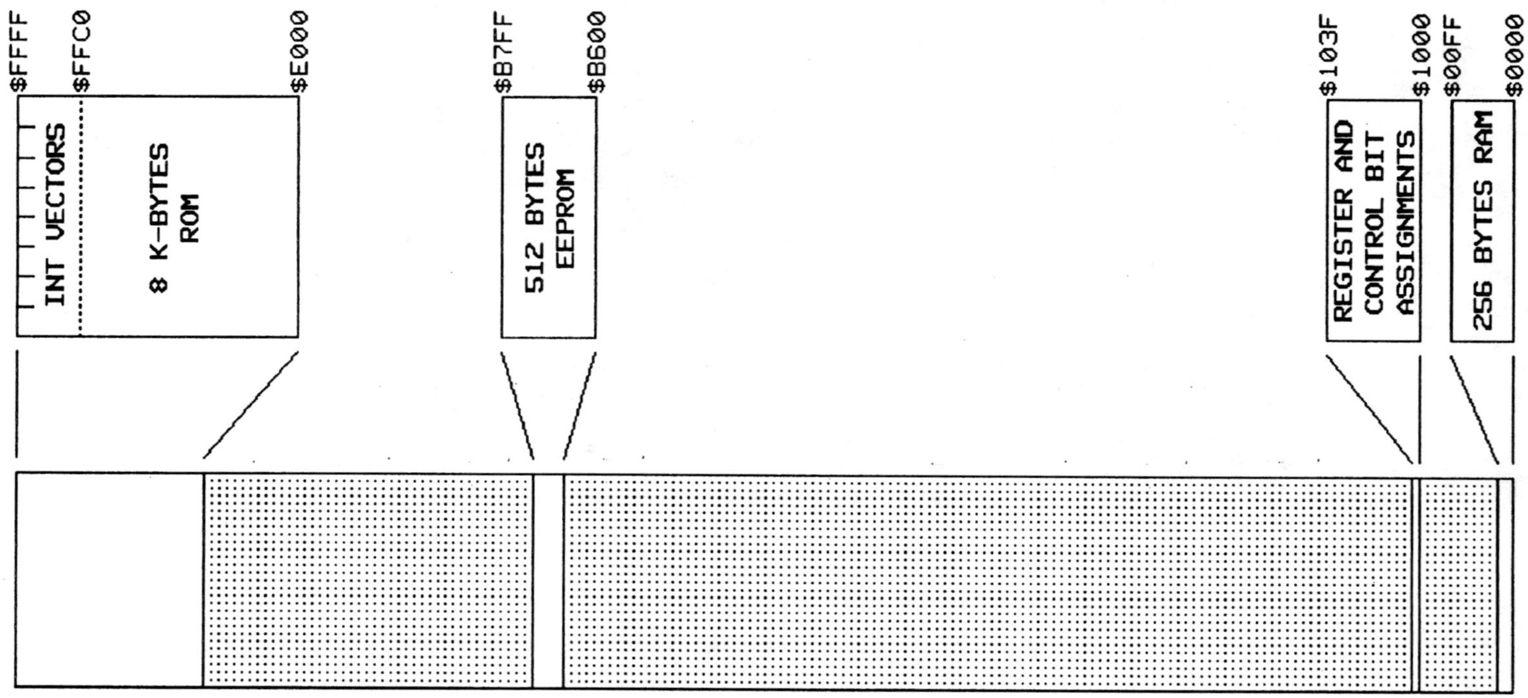

FIGURE 9.9 MC68HC11A8 Memory map.

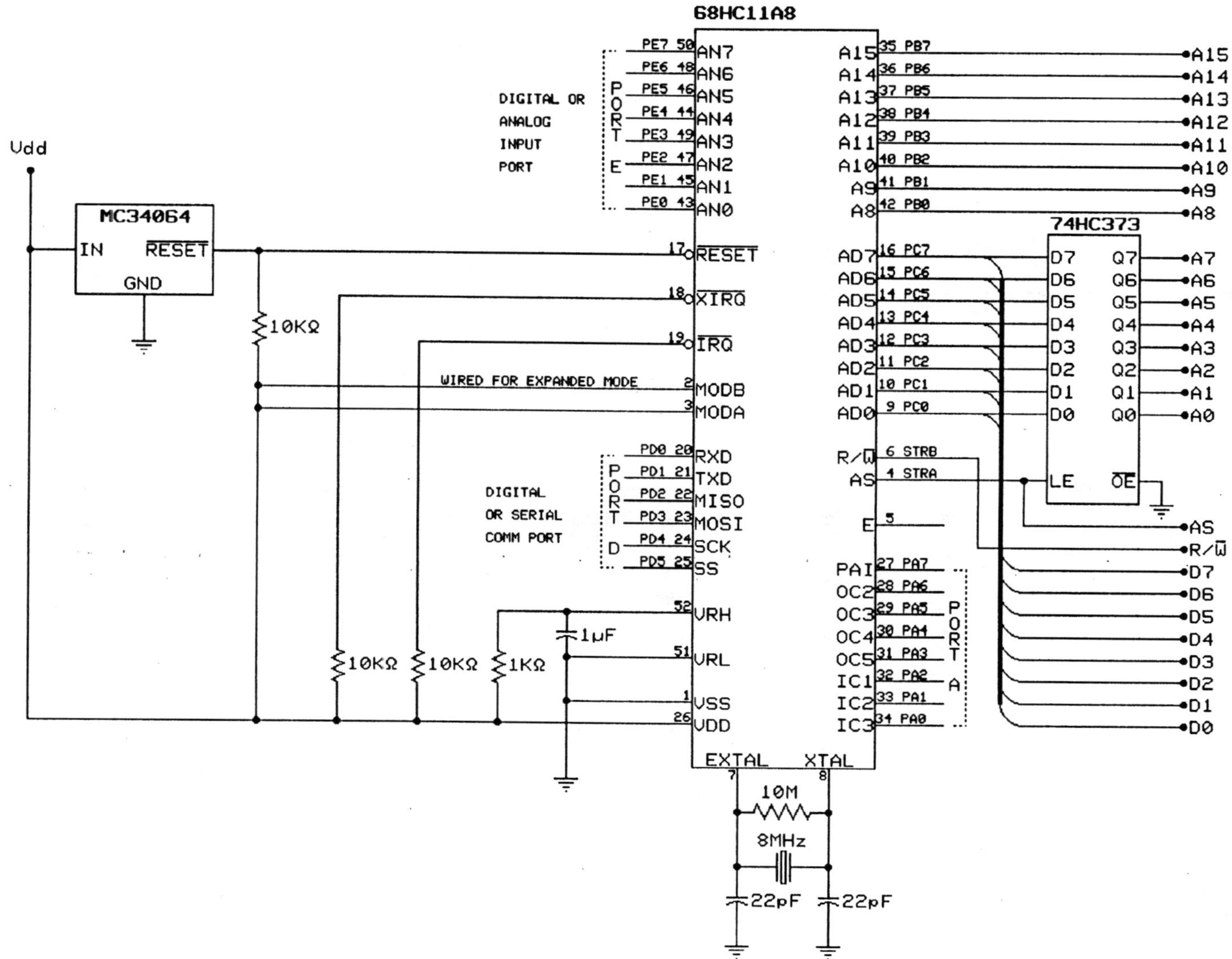

FIGURE 9.10 MC68HC11A8 microcontroller, minimum expanded mode configuration.

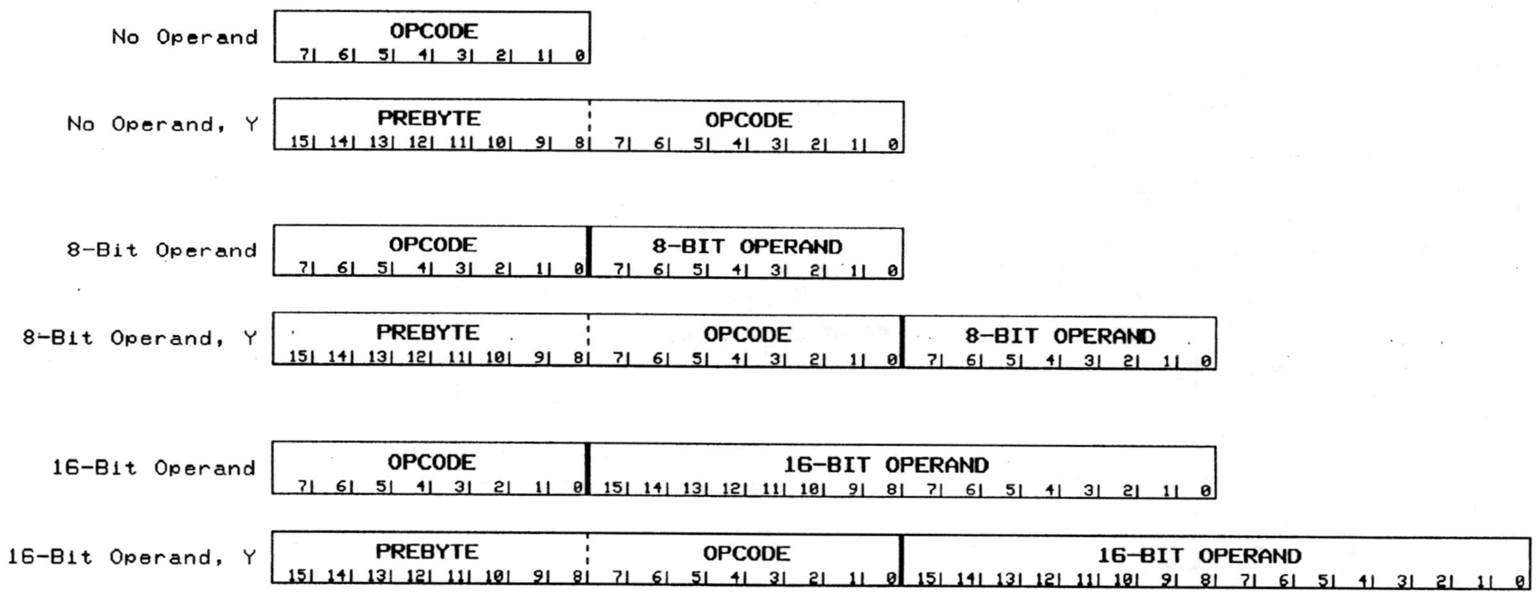

FIGURE 9.11 MC68HC11 Instruction Formats.

Addressing Mode	Description	Assembly Language	Effective Address
Inherent (or implied)	The opcode is self-explanatory. No operand is required.	No operand	None
Immediate	The operand is an eight or sixteen-bit data constant	#$NN #$NNNN	PC
Direct	The operand is an eight-bit address.	$NN	Operand
Extended	The operand is a sixteen-bit address.	$NNNN	Operand
Indexed	The address is calculated by adding the index register contents to the eight-bit operand.	$NN,X $NN,Y	Operand + X Operand + Y
Relative	The address is calculated by adding the program counter contents to the eight-bit operand.	*+$NN	Operand + PC

TABLE 9.3 MC68HC11 family addressing modes.

Instruction	Operation	T	M	D	E	X	Y	R	H	I	N	Z	V	C
ABA	Add Accumulators	*							*		*	*	*	*
ABX	Add B to X	*												
ABY	Add B to Y	*												
ADCA <ea>	Add with Carry to A		*	*	*	*	*		*		*	*	*	*
ADCB <ea>	Add with Carry to B		*	*	*	*	*		*		*	*	*	*
ADDA <ea>	Add Memory to A		*	*	*	*	*		*		*	*	*	*
ADDB <ea>	Add Memory to B		*	*	*	*	*		*		*	*	*	*
ADDD <ea>	Add 16-Bit Memory to D		*	*	*	*	*				*	*	*	*
ANDA <ea>	AND Memory to A		*	*	*	*	*				*	*	0	
ANDB	AND Memory to B		*	*	*	*	*				*	*	0	
ASL <ea>	Arithmetic Shift Left Memory				*	*	*				*	*	*	*
ASLA	Arithmetic Shift Left A	*									*	*	*	*
ASLB	Arithmetic Shift Left B	*									*	*	*	*
ASLD	Arithmetic Shift Left D	*									*	*	*	*
ASR	Arithmetic Shift Right Memory				*	*	*				*	*	*	*
ASRA	Arithmetic Shift Right A	*									*	*	*	*
ASRB	Arithmetic Shift Right B	*									*	*	*	*

Table 9.4a M68HC11 Instruction Set

Instruction	Operation	T	M	D	E	X	Y	R	H	I	N	Z	V	C
BCC <disp>	Branch if Carry Clear							*						
BCLR <ea>,m	Clear mask bits in memory			*		*	*				*	*	0	
BCS <disp>	Branch if Carry Set							*						
BEQ <disp>	Branch if Equal to Zero							*						
BGE <disp>	Branch if > or = to Zero							*						
BGT <disp>	Branch if > Zero							*						
BHI <disp>	Branch if Higher							*						
BHS <disp>	Branch if Higher or Same							*						
BITA <ea>	Bit Test A AND Memory		*	*	*	*	*				*	*	0	
BITB <ea>	Bit Test B AND Memory		*	*	*	*	*				*	*	0	
BLE <disp>	Branch if < or = to Zero							*						
BLO <disp>	Branch if Lower (Unsigned <)							*						
BLS <disp>	Branch if Lower or Same							*						
BLT <disp>	Branch if Less Than Zero							*						
BMI <disp>	Branch if Minus							*						
BNE <disp>	Branch if Not Equal to Zero							*						
BPL <disp>	Branch if Plus							*						

Table 9.4b M68HC11 Instruction Set

Instruction	Operation	T	M	D	E	X	Y	R	H	I	N	Z	V	C
BRA <disp>	Branch Always							*						
BRCLR <ea>,m	Branch if m Bit(s) Clear			*		*	*							
BRN <disp>	Branch Never							*						
BRSET <ea>,m	Branch if m Bit(s) Set							*						
BSET <ea>,m	Set Bits <ea> AND m			*		*	*				*	*	0	
BSR <disp>	Branch to Subroutine							*						
BVC	Branch if Overflow Clear							*						
BVS	Branch if Overflow Set							*						
CBA	Compare A to B (A-B)	*									*	*	*	*
CLC	Clear Carry Bit	*												0
CLI	Clear Interrupt Mask	*								0				
CLR <ea>	Clear Memory Byte				*	*	*				0	1	0	0
CLRA	Clear Accumulator A	*									0	1	0	0
CLRB	Clear Accumulator B	*									0	1	0	0
CLV	Clear Overflow Flag	*											0	
CMPA <ea>	Compare A to Memory (A-M)		*	*	*	*	*				*	*	*	*
CMPB <ea>	Compare B to Memory (B-M)		*	*	*	*	*				*	*	*	*

Table 9.4c M68HC11 Instruction Set

Instruction	Operation	T	M	D	E	X	Y	R	H	I	N	Z	V	C
COM <ea>	1's Complement Memory				*	*	*				*	*	0	*
COMA	1's Complement A	*									*	*	0	*
COMB	1's Complement B	*									*	*	0	*
CPD <ea>	Compare D to Memory (D-M)		*	*	*	*	*				*	*	*	*
CPX <ea>	Compare X to Memory (X-M)		*	*	*	*	*				*	*	*	*
CPY <ea>	Compare Y to Memory (Y-M)		*	*	*	*	*				*	*	*	*
DAA	Decimal Adjust A	*									*	*	*	*
DEC <ea>	Decrement Memory Byte				*	*	*				*	*	*	
DECA	Decrement Accumulator A	*									*	*	*	
DECB	Decrement Accumulator B	*									*	*	*	
DES	Decrement Stack Pointer	*												
DEX	Decrement Index Register X	*										*		
DEY	Decrement Index Register Y	*										*		
EORA <ea>	Exclusive OR Memory to A		*	*	*	*	*				*	*	0	
EORB <ea>	Exclusive OR Memory to B		*	*	*	*	*				*	*	0	
FDIV	Divide X=D/X, D=rem (D<X)	*										*	*	*
IDIV	Divide X=D/X, D=remainder	*										*	0	*

Table 9.4d M68HC11 Instruction Set

Instruction	Operation	T	M	D	E	X	Y	R	H	I	N	Z	V	C
INC <ea>	Increment Memory Byte				*	*	*				*	*	*	
INCA	Increment Accumulator A	*									*	*	*	
INCB	Increment Accumulator B	*									*	*	*	
INS	Increment Stack Pointer	*												
INX	Increment Index Register X	*										*		
INY	Increment Index Register Y	*										*		
JMP <ea>	Jump				*	*	*							
JSR <ea>	Jump to Subroutine			*	*	*	*							
LDAA <ea>	Load Accumulator A		*	*	*	*	*				*	*	0	
LDAB <ea>	Load Accumulator B		*	*	*	*	*				*	*	0	
LDD <ea>	Load Double Accumulator D		*	*	*	*	*				*	*	0	
LDS <ea>	Load Stack Pointer		*	*	*	*	*				*	*	0	
LDX <ea>	Load Index Register X		*	*	*	*	*				*	*	0	
LDY <ea>	Load Index Register Y		*	*	*	*	*				*	*	0	
LSL <ea>	Logical Shift Left Memory				*	*	*				*	*	*	*
LSLA	Logical Shift Left A	*									*	*	*	*
LSLB	Logical Shift Left B	*									*	*	*	*

Table 9.4e M68HC11 Instruction Set

Instruction	Operation	T	M	D	E	X	Y	R	H	I	N	Z	V	C
LSLD	Logical Shift Left D (A:B)	*									*	*	*	*
LSR <ea>	Logical Shift Right Memory				*	*	*				0	*	*	*
LSRA	Logical Shift Right A	*									0	*	*	*
LSRB	Logical Shift Right B	*									0	*	*	*
LSRD	Logical Shift Right D (A:B)	*									0	*	*	*
MUL	Multiply D = A x B	*												*
NEG <ea>	2's Complement Memory				*	*	*				*	*	*	*
NEGA	2's Complement A	*									*	*	*	*
NEGB	2's Complement B	*									*	*	*	*
NOP	No Operation (2 clock cycles)	*												
ORAA <ea>	OR Memory to Acc A		*	*	*	*	*				*	*	0	
ORAB <ea>	OR Memory to Acc B		*	*	*	*	*				*	*	0	
PSHA	Push A onto the Stack	*												
PSHB	Push B onto the Stack	*												
PSHX	Push X onto Stack, Lo First	*												
PSHY	Push Y onto Stack, Lo First	*												
PULA	Pull A from the Stack	*												

Table 9.4f M68HC11 Instruction Set

Instruction	Operation	T	M	D	E	X	Y	R	H	I	N	Z	V	C
PULB	Pull B from the Stack	*												
PULX	Pull X from Stack, Hi First	*												
PULY	Pull Y from Stack, Hi First	*												
ROL <ea>	Rotate Left (with Carry)				*	*	*				*	*	*	*
ROLA	Rotate Left Accumulator A	*									*	*	*	*
ROLB	Rotate Left Accumulator B	*									*	*	*	*
ROR <ea>	Rotate Right (with Carry)				*	*	*				*	*	*	*
RORA	Rotate Right Accumulator A	*									*	*	*	*
RORB	Rotate Right Accumulator B	*									*	*	*	*
RTI	Return from Interrupt	*									*	*	*	*
RTS	Return from Subroutine	*							*	*	*	*	*	*
SBA	Subtract B from A	*												
SBCA <ea>	A = A - <ea> - C-bit		*	*	*	*	*				*	*	*	*
SBCB <ea>	B = B - <ea> - C-bit		*	*	*	*	*				*	*	*	*
SEC	Set Carry Flag	*												1
SEI	Set Interrupt Mask	*								1				
SEV	Set Overflow Flag	*											1	

Table 9.4g M68HC11 Instruction Set

Instruction	Operation	T	M	D	E	X	Y	R	H	I	N	Z	V	C
STAA <ea>	Store Accumulator A to Mem			*	*	*	*				*	*	0	
STAB <ea>	Store Accumulator B to Mem			*	*	*	*				*	*	0	
STD <ea>	Store Accumulator D to Mem			*	*	*	*				*	*	0	
STOP	Stop Internal Clocks	*												
STS <ea>	Store Stack Pointer to Mem			*	*	*	*				*	*	0	
STX <ea>	Store Index Register X			*	*	*	*				*	*	0	
STY <ea>	Store Index Register Y			*	*	*	*				*	*	0	
SUBA <ea>	Subtract Memory from A		*	*	*	*	*				*	*	*	*
SUBB <ea>	Subtract Memory from B		*	*	*	*	*				*	*	*	*
SUBD <ea>	Subtract Memory from D		*	*	*	*	*				*	*	*	*
SWI	Software Interrupt (Call ISR)	*								1				
TAB	Transfer A to B	*									*	*	0	
TAP	Transfer A to CCR	*							*	*	*	*	*	*
TBA	Transfer B to A	*									*	*	0	
TPA	Transfer CCR to A	*												
TST <ea>	Test for Zero or Minus				*	*	*				*	*	0	0
TSTA	Test A for Zero or Minus	*									*	*	0	0

Table 9.4h M68HC11 Instruction Set

Instruction	Operation	T	M	D	E	X	Y	R	H	I	N	Z	V	C
TSTB	Test B for Zero or Minus	*									*	*	0	0
TSX	Transfer Stack Pointer to X	*												
TSY	Transfer Stack Pointer to Y	*												
TXS	Transfer X to Stack Pointer	*												
TYS	Transfer Y to Stack Pointer	*												
WAI	Wait for an Interrupt	*												
XGDX	Exchange D with X	*												
XGDY	Exchange D with Y	*												

Table 9.4i M68HC11 Instruction Set

Practical Example 9.1: Port A as Digital I/O

```
0000                 ; PORT A (DIGITAL I/O PORT)
0000                 ;
0000                         .EQU    PORTA,$00     ;Port A Data Register
0026                         .EQU    PACTL,$26     ;Pulse Acc. Control Register
1000                         .EQU    REGLIST,$1000
E100                         .ORG    $E100         ;Locate following code
E100                 PREX01  LDX     #REGLIST      ;X points to internal registers
E103                 ;
E103                 ;  1. LOAD B[7,2,1,0] FROM PORT A
E103                 ;
E103 1D2680                  BCLR    PACTL,X,$80 ;Clear DDRA for input
E106 E600                    LDAB    PORTA,X       ;Input data on PA7,2,1,0.
E108                 ;
E108                 ;  2. WRITE B[6,5,4,3] OUT PORT A  (PA7 = INPUT)
E108                 ;
E108 1D2680                  BCLR    PACTL,X,$80 ;Clear DDRA for input
E10B A600                    LDAA    PORTA,X       ;Get current Port A data
E10D 9487                    ANDA    b'10000111    ;Keep only Bits A7,2,1,0
E10F D478                    ANDB    b'01111000    ;Keep only Bits B6,5,4,3
E111 DA00                    ABA                   ;Merge A and B data
E113 A700                    STAA    PORTA,X       ;Output data on PA6,5,4,3.
E115                 ;
E115                 ;  3. WRITE B[7,6,5,4,3] OUT PORT A  (PA7 = OUTPUT)
E115                 ;
E115 1C2680                  BSET    PACTL,X,$80 ;Set DDRA for output
E118 E700                    STAB    PORTA,X       ;Output data on PA7,6,5,4,3.
E11A                         .END
```

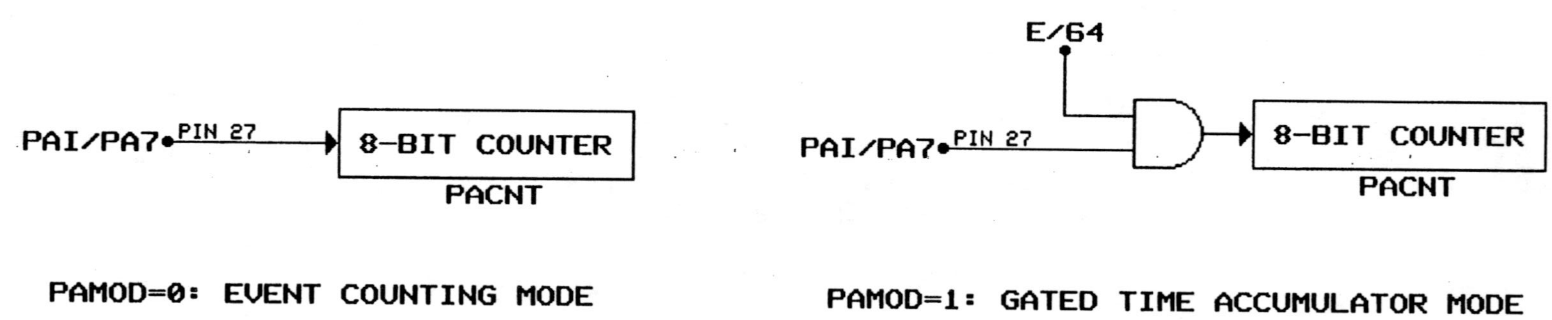

FIGURE 9.12 Pulse Accumulator Operating Modes.

Bit 7	Bit 6	Bit 5	Bit 4	Bit 3	Bit 2	Bit 1	Bit 0	Address	Register
BIT07	—	—	—	—	—	—	BIT00	$1027	PACNT
DDRA7	PAEN	PAMOD	PEDGE			RTR1	RTR0	$1026	PACTL
TOF	RTIF	PAOVF	PAIF					$1025	TFLG2
TOI	RTII	PAOVI	PAII			PR1	PR0	$1024	TMSK2

FIGURE 9.13 Pulse Accumulator Control and Status Registers

```
0000                ; PORT A (PULSE ACCUMULATOR)
0000                ;
0000                        .EQU    PORTA,$00       ;Port A Data Register
0024                        .EQU    TMSK2,$24       ;Timer Int Mask Register 2
0025                        .EQU    TFLG2,$25       ;Timer Int Flag Register 2
0026                        .EQU    PACTL,$26       ;Port Acc. Control Register
0027                        .EQU    PACNT,$27       ;Pulse Acc. Count Register
1000                        .EQU    REGLIST,$1000   ;List of internal registers.
E120                        .ORG    $E120           ;Locate following code
E120
E120                ; CAUSE AN INTERRUPT AFTER 100 EVENTS.
E120 CE1000         PREX02  LDX     #REGLIST        ;Point to list of registers.
E123 1D2680                 BCLR    PACTL,X,$80     ;Clear DDRA for input
E126 8664                   LDAA    #100            ;100 events on PA7
E128 40                     NEGA                    ;2's comp..for count down.
E129 A727                   STAA    PACNT,X
E12B 1D2630                 BCLR    PACTL,X,$30     ;PEDGE=0; Active-low edge
E12E                                                ;PAMOD=0; Ext Event Counting
E12E 1C2420                 BSET    TMSK2,X,$20     ;PAOVI=1; Enable overflow int
E131 1C2640                 BSET    PACTL,X,$40     ;PAEN=1; Enable pulse acc.
E134 39                     RTS
E135
E135                ; WAIT FOR 100 EVENTS
E135 1D2680         WT100   BCLR    PACTL,X,$80     ;Clear DDRA for input
E138 8664                   LDAA    #100            ;100 events on PA7
E13A 40                     NEGA                    ;2's comp. for count down.
E13B A727                   STAA    PACNT,X
E13D 1D2630                 BCLR    PACTL,X,$30     ;PEDGE=0; Active-low edge
E140                                                ;PAMOD=0; Ext Event Counting
E140 1C2420                 BCLR    TMSK2,X,$20     ;PAOVI=0; Disable overflw int
E143 1C2640                 BSET    PACTL,X,$40     ;PAEN=1; Enable pulse acc.
E146 1F2520FC WT100A        BRCLR   TFLG2,X,$20,WT100A  ;Repeat if PAOVF=0
E14A 1D2520                 BCLR    TFLG2,X,$20     ;Clear PAOVF
E14D 39                     RTS
E14E
E14E                ; COUNT EVENTS FOR 1 SECOND
E14E 1D2680         CT1SET  BCLR    PACTL,X,$80     ;Clear DDRA for input
E151 6F27                   CLR     PACNT,X         ;Clear event counter
E153 1D2630                 BCLR    PACTL,X,$30     ;PEDGE=0; Active-low edge
E156                                                ;PAMOD=0; Ext event counting
E156 1C2420                 BCLR    TMSK2,X,$20     ;PAOVI=0; Disable overflw int
E159 1C2640                 BSET    PACTL,X,$40     ;PAEN =1; Enable pulse acc.
E15C BD____                 JSR     SECOND          ;Wait for one second
E15F 1D2640                 BCLR    PACTL,X,$40     ;PAEN =0; Disable pulse acc.
E162 39                     RTS
```

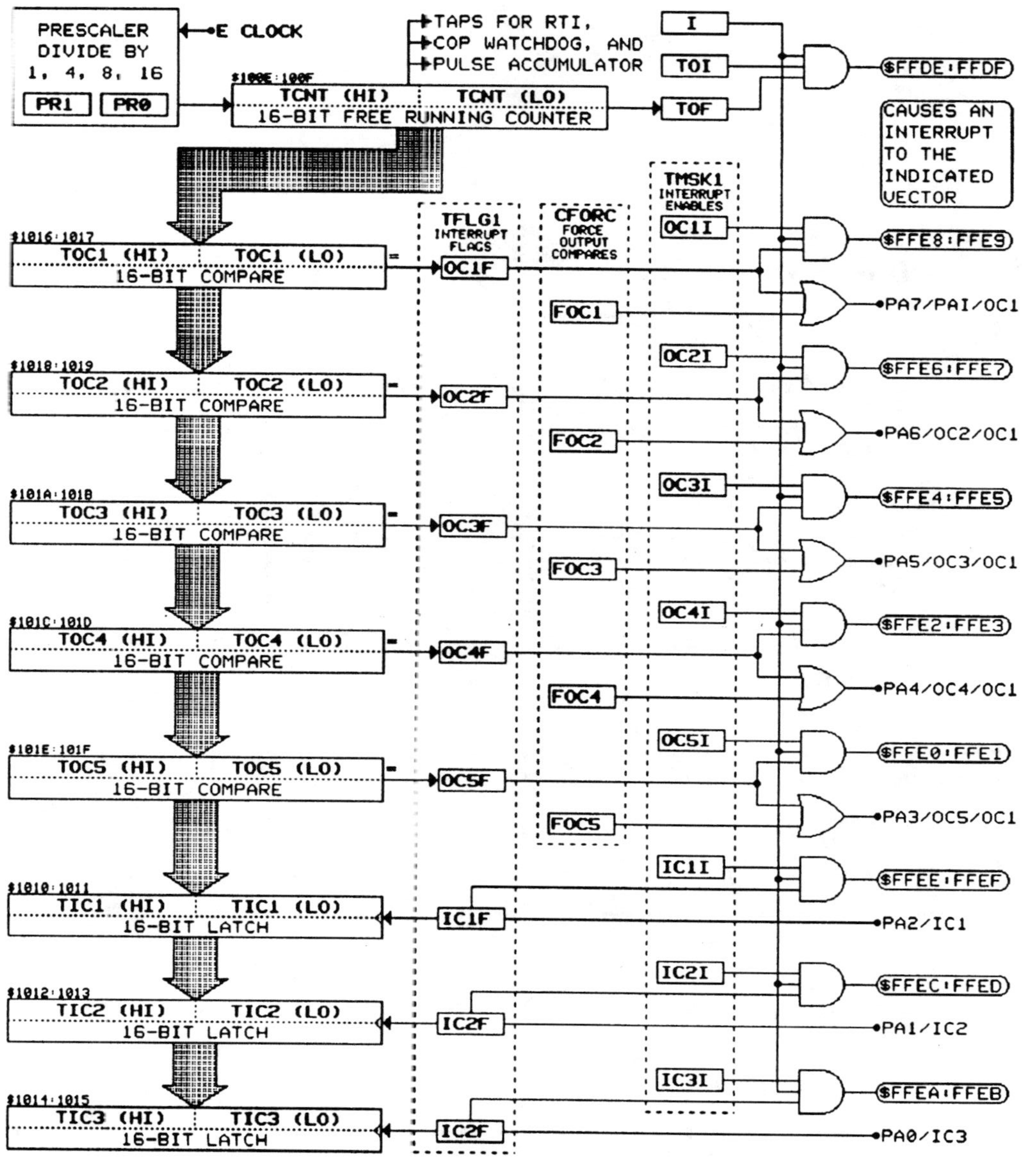

FIGURE 9.14
Main timer system block diagram.

PR1	PR0	Prescale Division	If E = 2 MHz, Count Rate = Overflow Rate =	If E = 2.1 MHz, Count Rate = Overflow Rate =
0	0	÷ 1	500 ns 32.77 ms	477 ns 31.25 ms
0	1	÷ 4	2 μs 131.1 ms	1.91 μs 125 ms
1	0	÷ 8	4 μs 262.1 ms	3.81 μs 250 ms
1	1	÷ 16	8 μs 524.3 ms	8 μs 500 ms

TABLE 9.5 Timer prescaler functions

DDRA7	PAEN	PAMOD	PEDGE			RTR1	RTR0	$1026	PACTL
TOF	RTIF	PAOVF	PAIF					$1025	TFLG2
TOI	RTII	PAOVI	PAII			PR1	PR0	$1024	TMSK2
OC1F	OC2F	OC3F	OC4F	OC5F	IC1F	IC2F	IC3F	$1023	TFLG1
OC1I	OC2I	OC3I	OC4I	OC5I	IC1I	IC2I	IC3I	$1022	TMSK1
		EDG1B	EDG1A	EDG2B	EDG2A	EDG3B	EDG3A	$1021	TCTL2
OM2	OL2	OM3	OL3	OM4	OL4	OM5	OL5	$1020	TCTL1
FOC1	FOC2	FOC3	FOC4	FOC5				$100D	OC1D
OC1M7	OC1M6	OC1M5	OC1M4	OC1M3				$100C	OC1M
OC1D7	OC1D6	OC1D5	IC1D4	OC1D3				$100B	CFORC

FIGURE 9.15 Main timer control and status registers.

RTR1	RTR0	$E \div 2^{13}$ Divided By	If E = 2 MHz, Interrupt Rate =	If E = 2.1 MHz, Interrupt Rate =
0	0	÷1	4.1 ms	3.91 ms
0	1	÷2	8.19 ms	7.81 ms
1	0	÷4	16.38 ms	15.62 ms
1	1	÷8	32.77 ms	31.25 ms

TABLE 9.6 Real-time interrupt rates as a function of RTR1:RTR0

EDGxB	*EDGxA*	*Configuration*
0	*0*	*Capture disabled*
0	*1*	*Capture on rising edges only*
1	*0*	*Capture on falling edges only*
1	*1*	*Capture on any edge (rising or falling)*

Table 9.7 Input capture edge polarity as a function of EDGxB:EDGxA.

```
0000                    ; PORT A (INPUT CAPTURE):  Measure pulse width on PA0 & PA1
0000                    ;
0013                            .EQU    TIC2,$13           ;Input Capture Register 2
0015                            .EQU    TIC3,$15           ;Input Capture Register 3
0021                            .EQU    TCTL2,$21          ;Timer Control Register 2
0022                            .EQU    TMSK1,$22          ;Timer Interrupt Flag Reg 1
0023                            .EQU    TFLG1,$23          ;Timer Interrupt Flag Reg 1
0024                            .EQU    TMSK2,$24          ;Timer Interrupt Mask Reg 2
0025                            .EQU    TFLG2,$25          ;Timer Interrupt Flag Reg 2
1000                            .EQU    REGLIST,$1000      ;Internal 68HC11 Registers
E200                            .ORG    $E200              ;Locate following code
E200                    ;
E200 CE1000     PREX03  LDX     #REGLIST           ;Point to list of registers
E203 86FF               LDAA    #$FF               ;Disable timer interrupts
E205 A722               STAA    TMSK1,X
E207 8601               LDAA    #$01               ;Set prescale to divide by 4
E209 A724               STAA    TMSK2,X            ; and disable TOI, RTI
E20B 8609               LDAA    #$09               ;Set PA0/IC3 rising edge
E20D A721               STAA    TCTL2,X            ;     PA1/IC2 falling edge
E20F 1E2302FC   PWCAPT1 BRSET   TFLG1,X,$02,PWCAPT1  ;Wait for input to fall
E213 1D2580             BCLR    TFLG2,X,$80          ;Wait for overflow flag
E216 1F2580FC   PWCAPT2 BRCLR   TFLG2,X,$80,PWCAPT2
E21A 4F                 CLRA                       ;Clear capture registers
E21B 5F                 CLRB                       ;
E21C ED15               STD     TIC3,X
E21E ED13               STD     TIC2,X
E220                    ;
E220                    ; Initialization done, input capture counters are counting
E220                    ; up.  TIC3 will stop first, then TIC2, so we will wait for
E220                    ; the interrupt flag on TIC2.
E220                    ;
E220 1F2302EB   PWCAPT3 BRCLR   TFLG1,X,$02,PWCAPT1  ;Wait for input to fall
E224                    ;
E224 EC13               LDD     TIC2,X             ;Calculate TIC2 - TIC3
E226 A315               SUBD    TIC3,X             ;Accumulator D contains time
E228                                               ; in 2 µs increments.
E228 39                 RTS
```

```
0000                   ; PORT A (REAL TIME INTERRUPT):  Monitor and control a temp
0000                   ;
0000                           .EQU    PORTA,$00         ;Port A Data Register
0026                           .EQU    PACTL,$26         ;Pulse Acc. Control Register
0024                           .EQU    TMSK2,$24         ;Timer Interrupt Mask Reg 2
1000                           .EQU    REGLIST,$1000     ;Address of register list
FFF0                           .ORG    $FFF0             ;Initialize RTI Vector
FFF0 F000                      .DW     TMPMON
E200                           .ORG    $E200             ;Prog to exercise the system
E200 CE1000        PREX04      LDX     #REGLIST          ;Point to list of registers
E203 1C2440                    BSET    TMSK2,X,$40       ;Disable RTI during setup
E206 8683                      LDAA    #$83              ;Set PA7 direction for input
E208 A726                      STAA    PACTL,X           ; and RTI rate to 8
E20A 1D2440                    BCLR    TMSK2,X,$40       ;Enable RTI
E20D 3E            MAIN1       WAI                       ;Wait for RTI
E20E 20FD                      BRA     MAIN1             ;Branch back for next int.
E210                   ;
F000                           .ORG    $F000             ;INTERRUPT SERVICE ROUTINE
F000 1F008003      TMPMON      BRCLR   PORTA,X,$80,TMPMON1  ;Exit if temp OK
F004 BD____                    JSR     ALARM             ;Call alarm if temp high
F007 3B            TMPMON1     RTI
```

BIT07	—	—	—	—	—	—	BIT00	$1004	PORTB
STAF	STAI	CWOM	HNDS	OIN	PLS	EGA	INVB	$1002	PIOC

FIGURE 9.16 Port B data and control registers.

Practical Example 9.5: PORT B as Digital Output

```
0000              ; PORT B (DIGITAL OUTPUT)
0000              ;
0004                     .EQU  PORTB,$04           ;Port B Data Register
1000                     .EQU  REGLIST,$1000       ;List of Internal Registers
E210                     .ORG  $E210
E210              ;
E210              ; The following segment writes all bits of A to Port B.
E210              ;
E210 A704         PREX05  STAA  PORTB,X              ;Port B = Accumulator A
E212              ;
E212              ; The following segment sets bit 2 of Port B to a logical 1
E212              ;
E212 1C0404              BSET  PORTB,X,$04         ;Port B[2] = 1
E215              ;
E215              ; The following segment copies the least significant nibble
E215              ; of Accumulator A to Port B without damaging the other four
E215              ; bits of Port B
                  ;
E215 E604                LDAB  PORTB,X             ;Get Port B data
E217 C4F0                ANDB  #$F0                ;Clear ls nibble of B data
E219 840F                ANDA  #$0F                ;Clear ms nibble of A data
E21B 1B                  ABA                       ;Merge the two together
E21C A704                STAA  PORTB,X             ;Write back to the port.
E21E                     .END
```

BIT07	—	—	—	—	—	—	BIT00	$1007	DDRC
BIT07	—	—	—	—	—	—	BIT00	$1005	PORTCL
BIT07	—	—	—	—	—	—	BIT00	$1003	PORTC
STAF	STAI	CWOM	HNDS	OIN	PLS	EGA	INVB	$1002	PIOC

FIGURE 9.17 Port C data and control registers.

Practical Example 9.6: Port C as Digital I/O

```
0000                ; PORT C (DIGITAL INPUT/OUTPUT)
0000                ;
0003                   .EQU  PORTC,$03         ;Port C Data Register
0007                   .EQU  DDRC,$07          ;Port C Data Direction Register
1000                   .EQU  REGLIST,$1000     ;List of Internal Registers
E500                   .ORG  $E500
E500                ;
E500                ; The following segment writes all bits of A to Port C.
E500                ;
E500 C6FF           PREX06  LDAB  #$FF         ;Configure Port C for output
E502 E707                   STAB  DDRC,X
E504 A703                   STAA  PORTC,X      ;Port C = Accumulator A
E506                ;
E506                ; The following segment sets bit 2 of Port C to a logical 1.
E506                ; It assumes that the data direction register has been
E506                ; initialized.
                    ;
E506 1C0304                 BSET  PORTC,X,$04  ;Port C[2] = 1
E509                ;
E509                ; The following segment copies the least significant nibble
E509                ; of Accumulator A to Port C without damaging the other four
E509                ; output bits
                    ;
E509 E603                  LDAB  PORTC,X       ;Get Port C data
E50B C4F0                  ANDB  #$F0          ;Clear least sig nibble of B data
E50D 840F                  ANDA  #$0F          ;Clear most sig nibble of A data
E50F 1B                    ABA                 ;Merge the two nibbles together
E510 A703                  STAA  PORTC,X       ;Write back to the port.
E512                       .END
```

BIT07	—	—	—	—	—	—	BIT00	$102F	SCDR
TDRE	TC	RDRF	IDLE	OR	NF	FE		$102E	SCSR
TIE	TCIE	RIE	ILIE	TE	RE	RWU	SBK	$102D	SCCR2
R8	T8		M	WAKE				$102C	SCCR1
TCLR		SCP1	SCP0	RCKB	SCR2	SCR1	SCR0	$102B	BAUD
BIT07	—	—	—	—	—	—	BIT00	$102A	SPDR
SPIF	WCOL		MODF					$1029	SPSR
SPIE	SPE	DWOM	MSTR	CPOL	CPHA	SPR1	SPR0	$1028	SPCR
		BIT05	—	—	—	—	BIT00	$1009	DDRD
		BIT05	—	—	—	—	BIT00	$1008	PORTD

FIGURE 9.18 Port D data, control, and status registers.

```
0000             ; PORT D (DIGITAL INPUT/OUTPUT)
0000             ;
0008                    .EQU  PORTD,$08         ;Port D Data Register
0009                    .EQU  DDRD,$09          ;Port D Data Direction Register
002C                    .EQU  SCCR1,$2C         ;SCI Control Register 1
0028                    .EQU  SPCR,$28          ;SPI Control Register
1000                    .EQU  REGLIST,$1000     ;List of Internal Registers
E520                    .ORG  $E520
E520             ;
E520             ; The following segment writes B[3,2,1] to Port D without
E520             ; damaging the other Port D data bits.  It then reads
E520             ; PORTD[5,4,0], writing them into B without destroying the
E520             ; remainder of the accumulator B bits.
                 ;
E520 CE1000 PREX08      LDX   #REGLIST          ;Point to register list.
E523 7F002C             CLR   SCCR1             ;Turn off SCI
E526 7F0028             CLR   SPCR              ;Turn off SPI
E529 8607               LDAA  #$07              ;Set Port D[3,2,1] as output
E52B 9709               STAA  DDRD
E52D              ;
E52D              ; Ready to write B[3,2,1] to PORTD[3,2,1]
E52D 9608               LDAA  PORTD             ;Get current Port D data
E52F 84F1               ANDA  #$F1              ;Keep the unchanged bits
E531 C407               ANDB  #$07              ;Keep the B bits to write
E533 1B                 ABA                     ;Merge them
E534 9708               STAA  PORTD             ;Write merged data to Port D
E536              ;
E536              ; Read PORTD[5,4,0] to B[5.4.0]
E536 9608               LDAA  PORTD             ;Get Port D data
E538 84F1               ANDA  #$F1              ;Isolate input Bits
E53A C407               ANDB  #$07              ;Clear corresponding bits in B
E53C 1B                 ABA                     ;Merge them
E53D 16                 TAB                     ;Write result to B
E53E                    .END
```

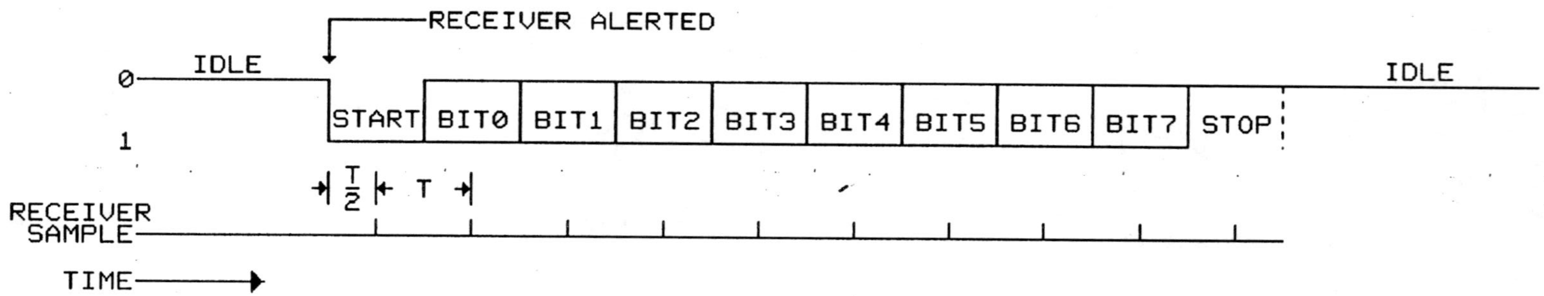

FIGURE 9.19 Asynchronous serial character format.

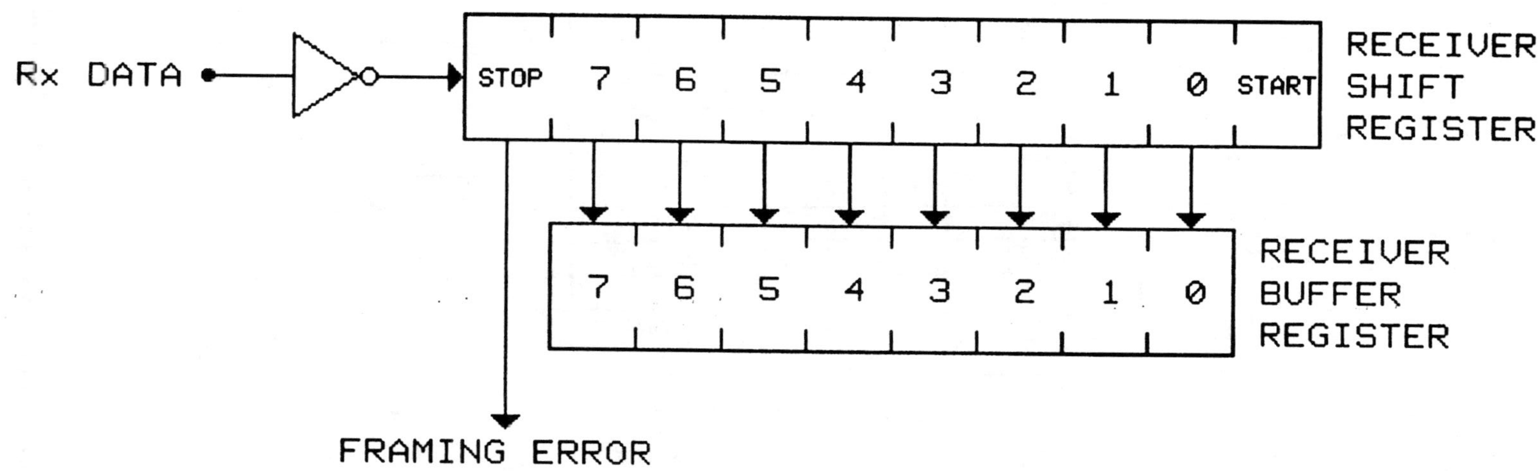

FIGURE 9.20 Asynchronous receiver buffer.

BAUD Register Contents					External Crystal Frequency				
Bit-5	*Bit-4*	*Bit-2*	*Bit-1*	*Bit-0*	2^{23} Hz	8MHz	4.9152 MHz	4MHz	3.6864 MHz
SCP1	*SCP0*	*SCR2*	*SCR1*	*SCR0*	Generated Baud Rate in Bits per Second				
0	0	0	0	0	131072	125000	76800	62500	57600
0	0	0	0	1	65536	62500	38400	31250	28800
0	0	0	1	0	32768	31250	19200	15625	14400
0	0	0	1	1	16384	1562.5	9600	7812.5	7200
0	0	1	0	0	8192	7812.5	4800	3906	3600
0	0	1	0	1	4096	3906	2400	195	1800
0	0	1	1	0	2048	1953	1200	977	900
0	1	0	0	0	43691	41666	25600	20833	19200
0	1	0	0	1	21845	20833	12800	10417	9600
0	1	0	1	0	10923	10417	6400	5208	4800
0	1	0	1	1	5461	5208	3200	2604	2400
0	1	1	0	0	2731	2604	1600	1302	1200
0	1	1	1	0	683	651	400	326	300
1	0	0	0	0	32768	31250	19200	15625	14400
1	0	0	0	1	16384	15625	9600	7812.5	7200
1	0	0	1	0	8192	7812.5	4800	3906	3600
1	0	0	1	1	4096	3906	2400	1953	1800
1	0	1	0	0	2048	1953	1200	977	900
1	0	1	1	0	512	488	300	244	225.5
1	0	1	1	1	256	244	150	122	112.5

TABLE 9.8 M68HC11 baud rates by crystal frequency.

Bit 7	Bit 6	Bit 5	Bit 4	Bit 3	Bit 2	Bit 1	Bit 0	Address	Register
BIT07	—	—	—	—	—	—	BIT00	$102F	SCDR (RDR on READ, TDR on WRITE)
TDRE	TC	RDRF	IDLE	OR	NF	FE		$102E	SCSR
TIE	TCIE	RIE	ILIE	TE	RE	RWU	SBK	$102D	SCCR2
R8	T8		M	WAKE				$102C	SCCR1
TCLR		SCP1	SCP0	RCKB	SCR2	SCR1	SCR0	$102B	BAUD
SPIE	SPE	DWOM	MSTR	CPOL	CPHA	SPR1	SPR0	$1028	SPCR
		SS	SCK	MOSI	MISO	TxD	RxD	$1008	PORTD

FIGURE 9.21 SCI control and data registers.

```
0000                    ; PORT D (ASYNCHRONOUS SERIAL TRANSMITTER)
0000                    ;
002B                            .EQU    BAUD, $2B         ;SCI Baud Rate Generator
002C                            .EQU    SCCR1,$2C         ;SCI Control Register 1
002D                            .EQU    SCCR2,$2D         ;SCI Control Register 2
002E                            .EQU    SCSR,$2E          ;SCI Status Register
002F                            .EQU    RDR,$2F           ;SCI Receive Data Register
002F                            .EQU    TDR,$2F           ;SCI Transmit Data Register
1000                            .EQU    REGLIST,$1000     ;List of Internal Registers
E540                            .ORG    $E540
E540                    ;
E540                    ; Configure the Port
E540 CE1000             PREX08  LDX     #REGLIST          ;Point to list of registers
E543 18CEE556                   LDY     #STRING           ;Point to string to send
E547 8623                       LDAA    #$23              ;Set to 2400 Baud
E549 972B                       STAA    BAUD,X
E54B 7F002C                     CLR     SCCR1,X           ;8-bit data
E54E 8608                       LDAA    #$08              ;Enable the Transmitter
E550 972D                       STAA    SCCR2,X
E552 BDE567                     JSR     SND_STR
E555 CF                         STOP
E556 54686973           STRING  .DB     "This is a string"
E55A 20697320
E55E 61207374
E562 72696E67
E566 04                         .DB     $04
E567                    ;
E567                    ; ************************************************************
E567                    ; * SUBROUTINE  S N D _ S T R (Y)                            *
E567                    ; *                                                          *
E567                    ; * This subroutine sends a string of ASCII characters out  *
E567                    ; * the asynchronous serial port.  Y points to the string,  *
E567                    ; * the string is teminated with a $04 character.           *
E567                    ; ************************************************************
E567 18A600             SND_STR LDAA    $00,Y             ;Get a string character
E56A 1808                       INY                       ;Increment string pointer
E56C 8104                       CMPA    #$04              ;End of the string?
E56E 2709                       BEQ     SNDSTR2           ;  If so, quit.
E570 7D002E             SNDSTR1 TST     SCSR,X            ;Is the transmitter ready?
E573 2AFB                       BPL     SNDSTR1           ;  If not, wait.
E575 972F                       STAA    TDR,X             ;Send character!
E577 20EE                       BRA     SND_STR           ;Go send another character
E579 39                 SNDSTR2 RTS
E57A
```

```
0000                ; PORT D (ASYNCHRONOUS SERIAL RECEIVER)
002B                        .EQU  BAUD, $2B         ;SCI Baud Rate Generator
002C                        .EQU  SCCR1,$2C         ;SCI Control Register 1
002D                        .EQU  SCCR2,$2D         ;SCI Control Register 2
002E                        .EQU  SCSR,$2E          ;SCI Status Register
002F                        .EQU  RDR,$2F           ;SCI Receive Data Register
002F                        .EQU  TDR,$2F           ;SCI Transmit Data Register
1000                        .EQU  REGLIST,$1000     ;List of Internal Registers
FFD6                        .ORG  $FFD6             ;SCI Interrupt Vector
FFD6 E5DC                   .DW   REC_CHR           ;   points to REC_CHR
E580                        .ORG  $E580
E580                ;
E580                ; Configure the Port
E580 CE1000         PREX09  LDX   #REGLIST          ;Point to list of registers
E583 18CEE59C               LDY   #STRING           ;Point to string buffer
E587 8632                   LDAA  #$32              ;Set to 2400 Baud
E589 972B                   STAA  BAUD,X
E58B 7F002C                 CLR   SCCR1,X           ;8-bit data
E58E 8624                   LDAA  #$24              ;Enable the Receiver and
E590 972D                   STAA  SCCR2,X           ;  receive interrupts.
E592 0E                     CLI                     ;Enable interrupts.
E593                        ;
E593                        ; Receive the String
E593                        ;
E593 3E             PREX09A WAI                     ;Wait for interrupts
E594 8104                   CMPA  #$04              ;End of string?
E596 2702                   BEQ   PREX09B           ;If so, quit
E598 20F9                   BRA   PREX09A
E59A 0F             PREX09B SEI                     ;Disable interrupts.
E59B CF                     STOP
E59C                STRING  .DS   $40               ;Reserve 64-byte buffer
E59C                ; ***********************************************************
E59C                ; * INTERRUPT SERVICE ROUTINE  R E C _ C H R (Y,A)          *
E59C                ; *                                                         *
E59C                ; * This interrupt service routine will receive an 8-bit    *
E59C                ; * character from the asynchronous serial port, and then   *
E59C                ; * write the character to memory pointed to by Y.          *
E59C                ; ***********************************************************
E59C 36             REC_CHR PSHA                    ;Save current accumulator A
E59D 962F                   LDAA  RDR,X             ;Get the received character
E59F 18A700                 STAA  $00,Y             ;Store into the buffer
E5A2 32                     PULA                    ;Restore previous A
E5A3 1808                   INY                     ;Increment the string pointer
E5A5 3B                     RTI                     ;Return from Interrupt
E5A6                        .END
```

Address	Vector Name
$00C4:00C6	SCI
$00C7:00C9	SPI
$00CA:00CC	Pulse Accumulator Input Edge
$00CD:00CF	Pulse Accumulator Overflow
$00D0:00D2	Timer Overflow
$00D3:00D5	Timer Output Capture 5
$00D6:00D8	Timer Output Capture 4
$00D9:00DB	Timer Output Capture 3
$00DC:00DE	Timer Output Capture 2
$00DF:00E1	Timer Output Capture 1

Address	Vector Name
$00E2:00E4	Timer Input Capture 3
$00E5:00E7	Timer Input Capture 2
$00E8:00EA	Timer Input Capture 1
$00EB:00ED	Real Time Interrupt
$00EE:00F0	IRQ
$00F1:00F3	XIRQ
$00F4:00F6	SWI
$00F7:00F9	Illegal Opcode
$00FA:00FC	COP Fail
$00FD:00FF	Clock Monitor Fail

TABLE 9.9 Bootstrap mode pseudo-vectors.

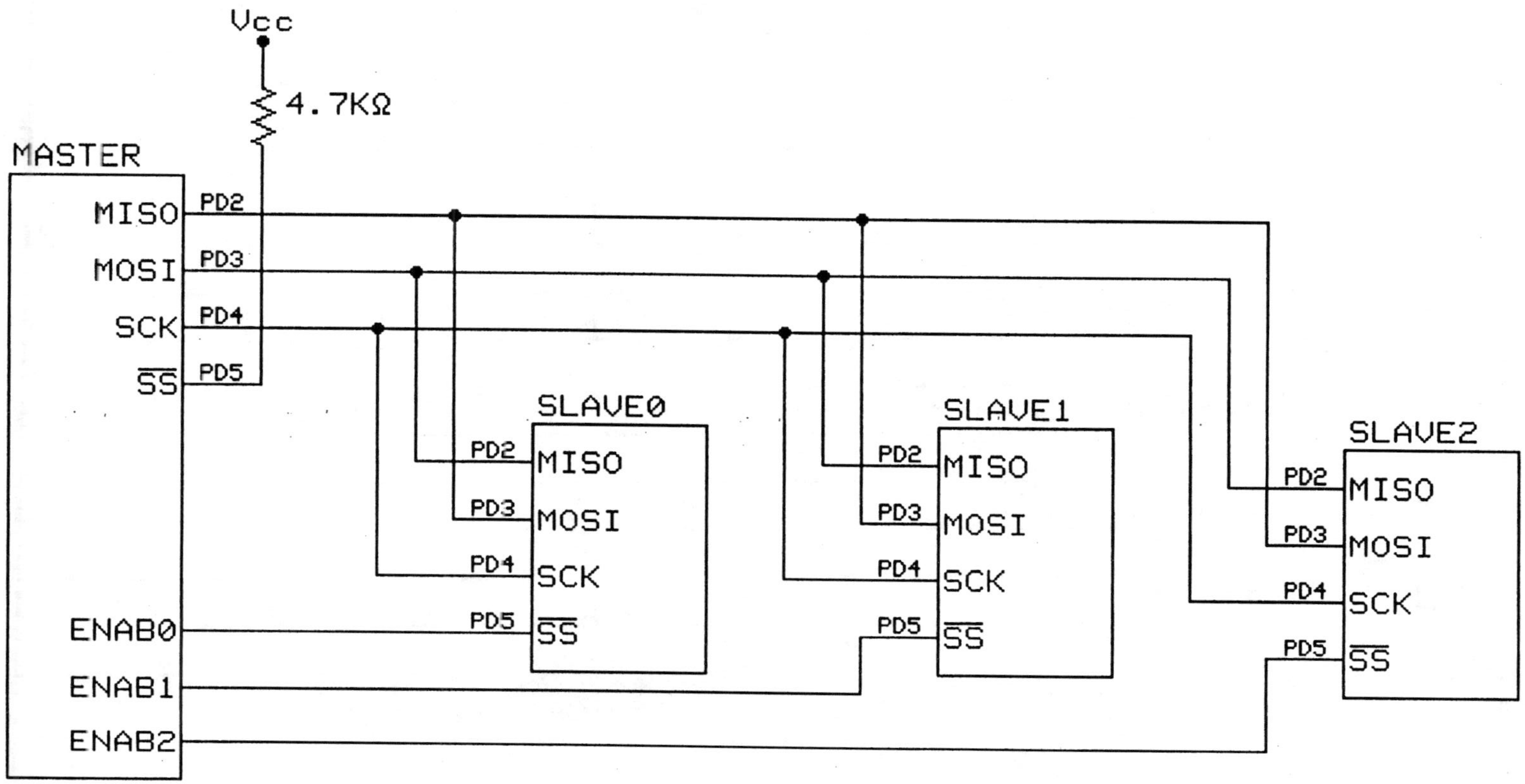

FIGURE 9.22 Synchronous interconnection of four MC68HC11 microcontrollers.

ADPU	CSEL	IRQE	DLY	CME		CR1	CR0	$1039	OPTION
BIT07	—	—	—	—	—	—	BIT00	$1034	ADR4
BIT07	—	—	—	—	—	—	BIT00	$1033	ADR3
BIT07	—	—	—	—	—	—	BIT00	$1032	ADR2
BIT07	—	—	—	—	—	—	BIT00	$1031	ADR1
CCF		SCAN	MULT	CD	CC	CB	CA	$1030	ADCTL
BIT07	—	—	—	—	—	—	BIT00	$100A	PORTE

FIGURE 9.23 A/D Converter Subsystem Status and Control Registers

MULT = 0					MULT = 1			
CC	CB	CA	Channel	Result Register	CD	CC	Channel	Result Register
0	0	0	PE0	ADR1	0	0	PE0	ADR1
0	0	1	PE1	ADR2			PE1	ADR2
0	1	0	PE2	ADR3			PE2	ADR3
0	1	1	PE3	ADR4			PE3	ADR4
1	0	0	PE4	ADR1	0	1	PE4	ADR1
1	0	1	PE5	ADR2			PE5	ADR2
1	1	0	PE6	ADR3			PE6	ADR3
1	1	1	PE7	ADR4			PE7	ADR4

TABLE 9.10 Result register assignments for A/D converter subsystem.

Practical Example 9.11: Port E Application.

```
0000                 ; PORT E (DIGITAL AND ANALOG INPUT)
0000                 ;
0039                          .EQU  OPTION,$39        ;Configuration Option Reg.
0034                          .EQU  ADR4,$34          ;A/D Result Register 4
0033                          .EQU  ADR3,$33          ;A/D Result Register 3
0032                          .EQU  ADR2,$32          ;A/D Result Register 2
0031                          .EQU  ADR1,$31          ;A/D Result Register 1
0030                          .EQU  ADCTL,$30         ;A/D Control Register
000A                          .EQU  PORTE,$0A         ;Port E Data Register
1000                          .EQU  REGLIST,$1000     ;List of Internal Registers
E5B0                          .ORG  $E5B0
E5B0 CE1000          PREX11   LDX   #REGLIST          ;Point to list of registers
E5B3                 ;
E5B3                 ; Power up the A/D converter subsystem
E5B3                 ;
E5B3 1C3980                   BSET  OPTION,X,$80      ;ADPU = 1, power up A/D
E5B6 1D3940                   BCLR  OPTION,X,$40      ;CSEL = 0, select E clock
E5B9                                                  ; if E < 750KHz set CSEL = 1
E5B9 8628                     LDAA  #40               ;Wait 100 microseconds.
E5BB 4A              PREX11A  DECA                    ; (Not usually necessary
E5BC 26FD                     BNE   PREX11A           ; because of time taken
E5BE                                                  ; for following code.)
E5BE                 ;
E5BE                 ; Read digital byte on Port E into accumulator A.
E5BE                 ;
E5BE A60A                     LDAA  PORTE,X           ;Get digital byte from E
E5C0                 ;
E5C0                 ; Get single reading from PE0 and place in accumulator A.
E5C0                 ;
E5C0 6F30                     CLR   ADCTL,X           ;SCAN=0, MULT=0, CD-CA = 0000
E5C2 1F3080FC        PREX11B  BRCLR ADCTL,X,$80,PREX11B  ;Wait for conversion
E5C6 A631                     LDAA  ADR1,X
E5C8                 ;
E5C8                 ; Start the A/D converter taking continuous readings on
E5C8                 ; PE7,6,5,4so that they can be obtained by the system at any
E5C8                 ; later time.
E5C8 9634                     LDAA  $34               ;SCAN=1, MULT=1, CC=1
E5CA A730                     STAA  ADCTL,X
E5CC 1F3080FC        PREX11C  BRCLR ADCTL,X,$80,PREX11C  ;Wait for first conv.
E5D0                 ;        ...
E5D0                          .END
```

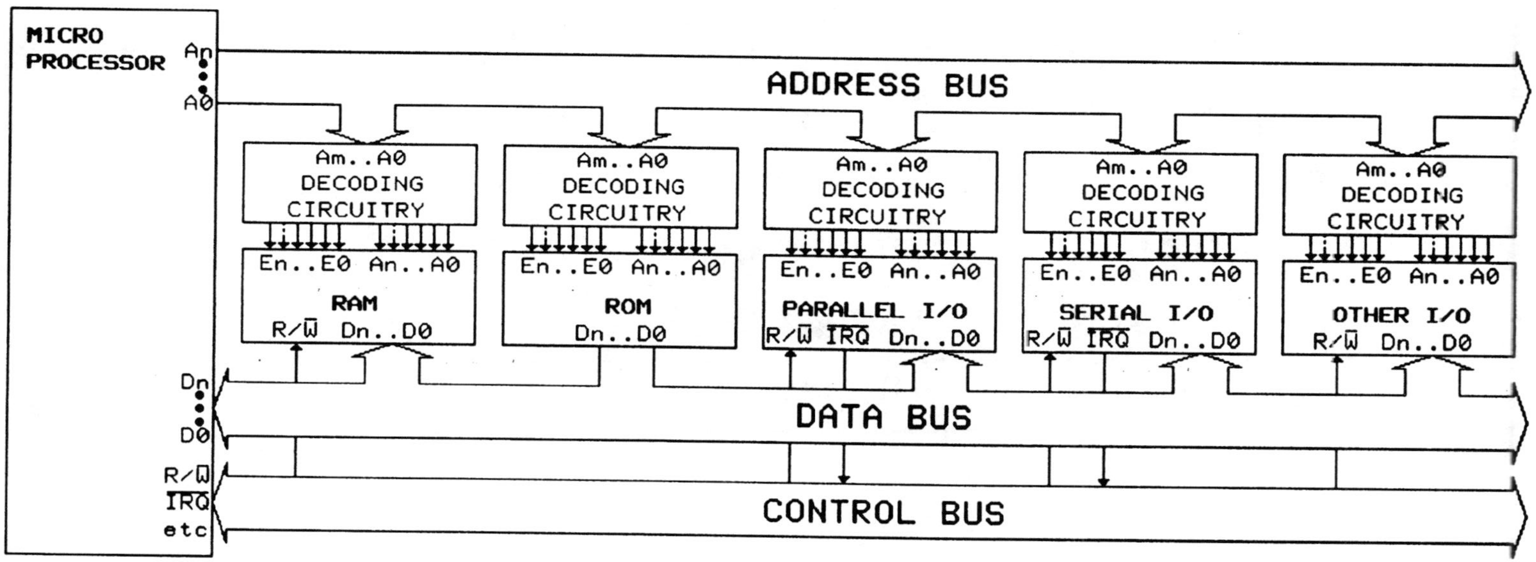

FIGURE 10.1 Microprocessor-based computer system

Device	Architecture	Chip-Select Labels
2K RAM	2048 x 8 bits	CS_0, CS_1, CS_2
4K ROM	4096 x 8 bits	OE_1, $!OE_2$
M6820 PIA (Parallel I/O)	4 x 8 bits	CS_0, $!CS_1$, $!CS_2$
M6850 ACIA (Serial I/O)	2 x 8 bits	CS_0, $!CS_1$, $!CS_2$
Octal Buffer	1 x 8 bits	G_0,$!G_1$

TABLE 10.1 Example devices.

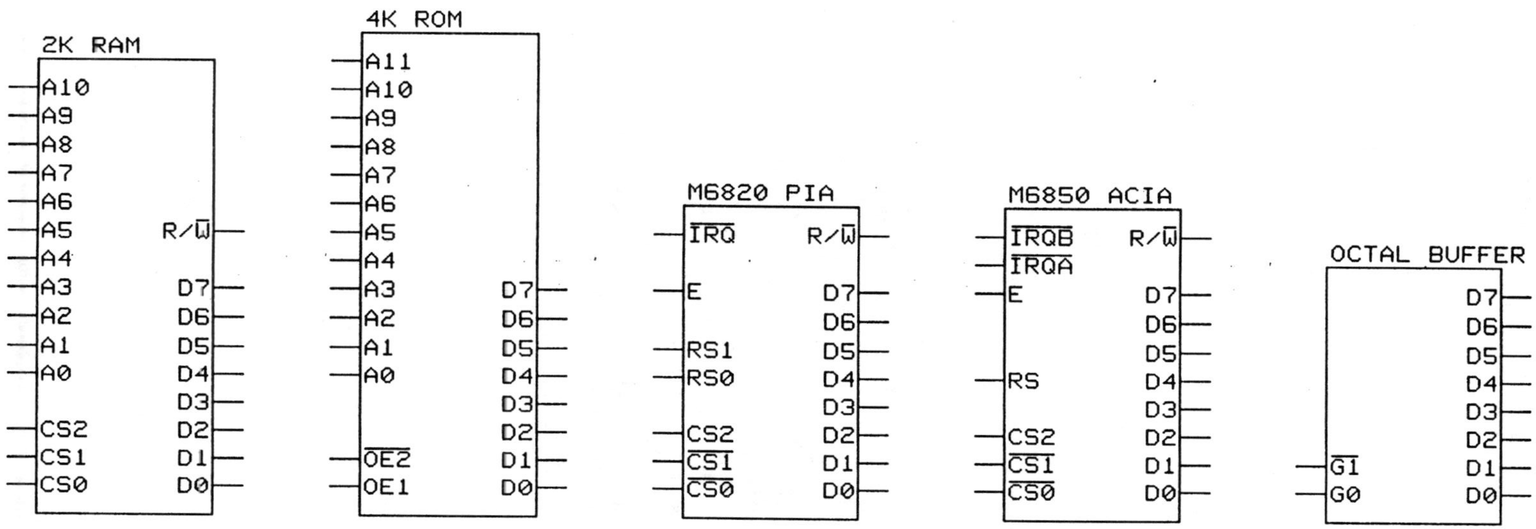

FIGURE 10.2 Example device outlines and processor connections.

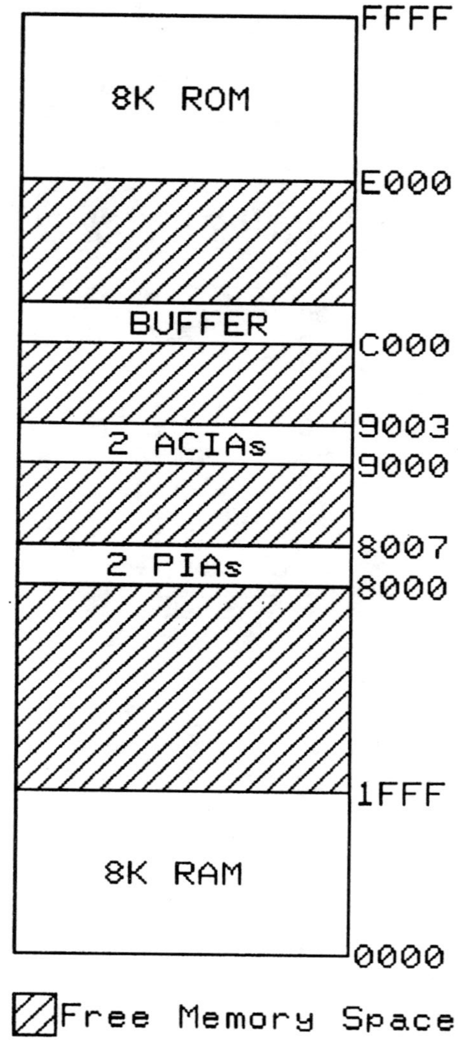

FIGURE 10.3   Example general memory map.

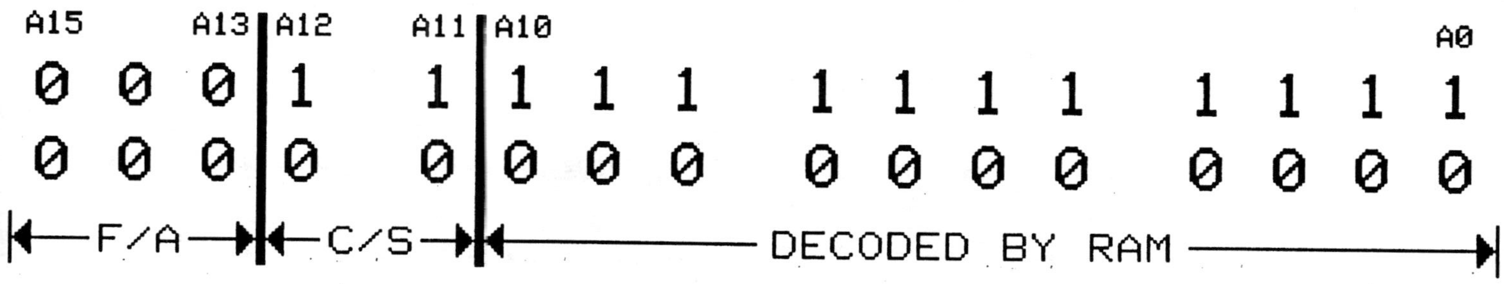

FIGURE 10.4 ***Binary memory map of the RAM block.***

$$Number\ of\ device\ address\ lines = \log_2(device\ address\ range)$$
$$Number\ of\ device\ address\ lines = \log_2(2048)$$
$$Number\ of\ device\ address\ lines = 11 \qquad Note: 2^{11} = 2048$$

EXPRESSION SET 10.1

$$Allocated\ chips\ per\ contiguous\ block = \frac{Block\ address\ range}{Chip\ address\ range}$$
$$Allocated\ chips\ per\ contiguous\ block = \frac{8K}{2K} = \frac{8192}{2048}$$
$$Allocated\ chips\ per\ contiguous\ block = 4$$

EXPRESSION SET 10.2

$$Number\ of\ C\,/\,S\ address\ lines = \log_2(Allocated\ chips)$$
$$Number\ of\ C\,/\,S\ address\ lines = \log_2(4)$$
$$Number\ of\ C\,/\,S\ address\ lines = 2 \qquad Note: 2^2 = 4$$

EXPRESSION SET 10.3

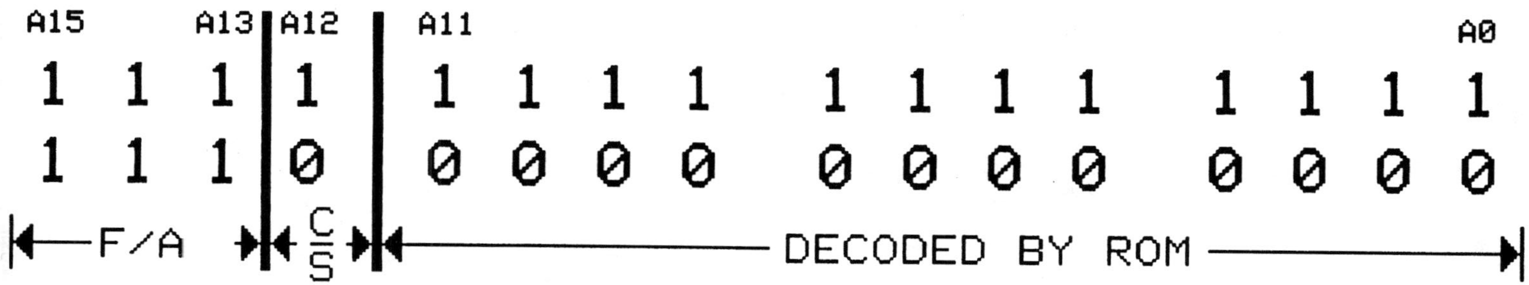

FIGURE 10.5 Binary memory map of the ROM block.

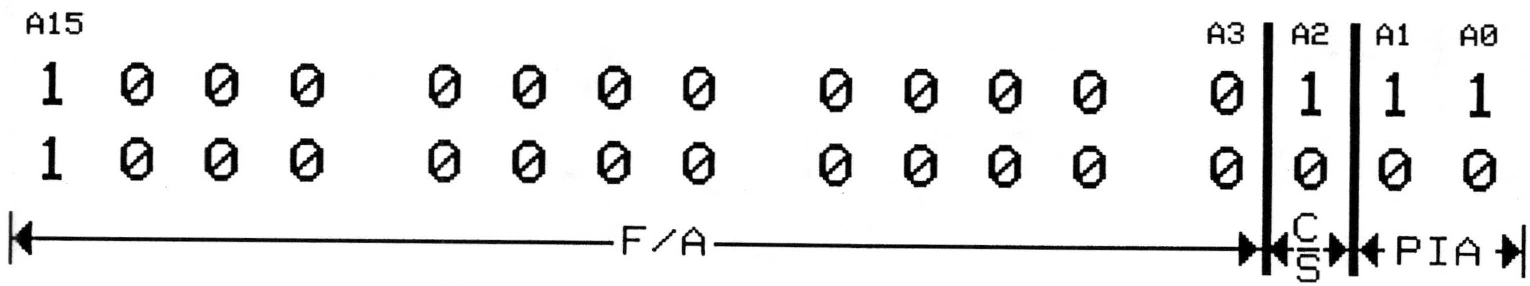

FIGURE 10.6 Binary memory map of the PIA block.

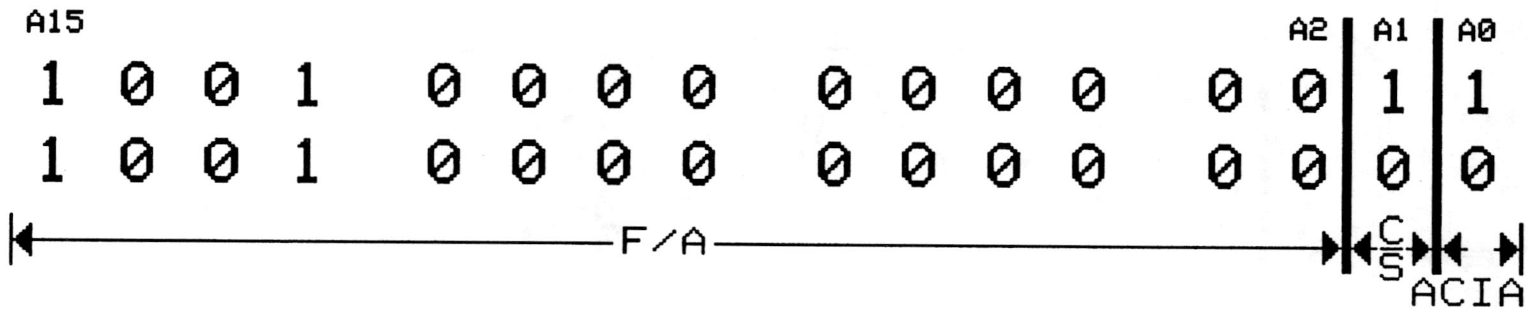

FIGURE 10.7 ***Binary memory map of the ACIA block.***

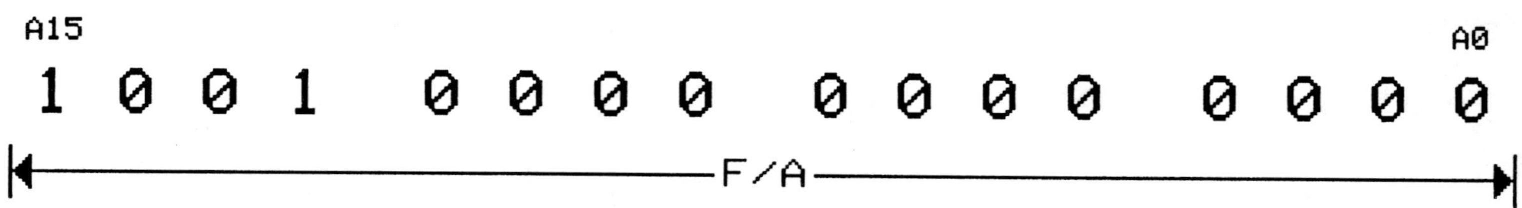

FIGURE 10.8 ***Binary memory map of the buffer block.***

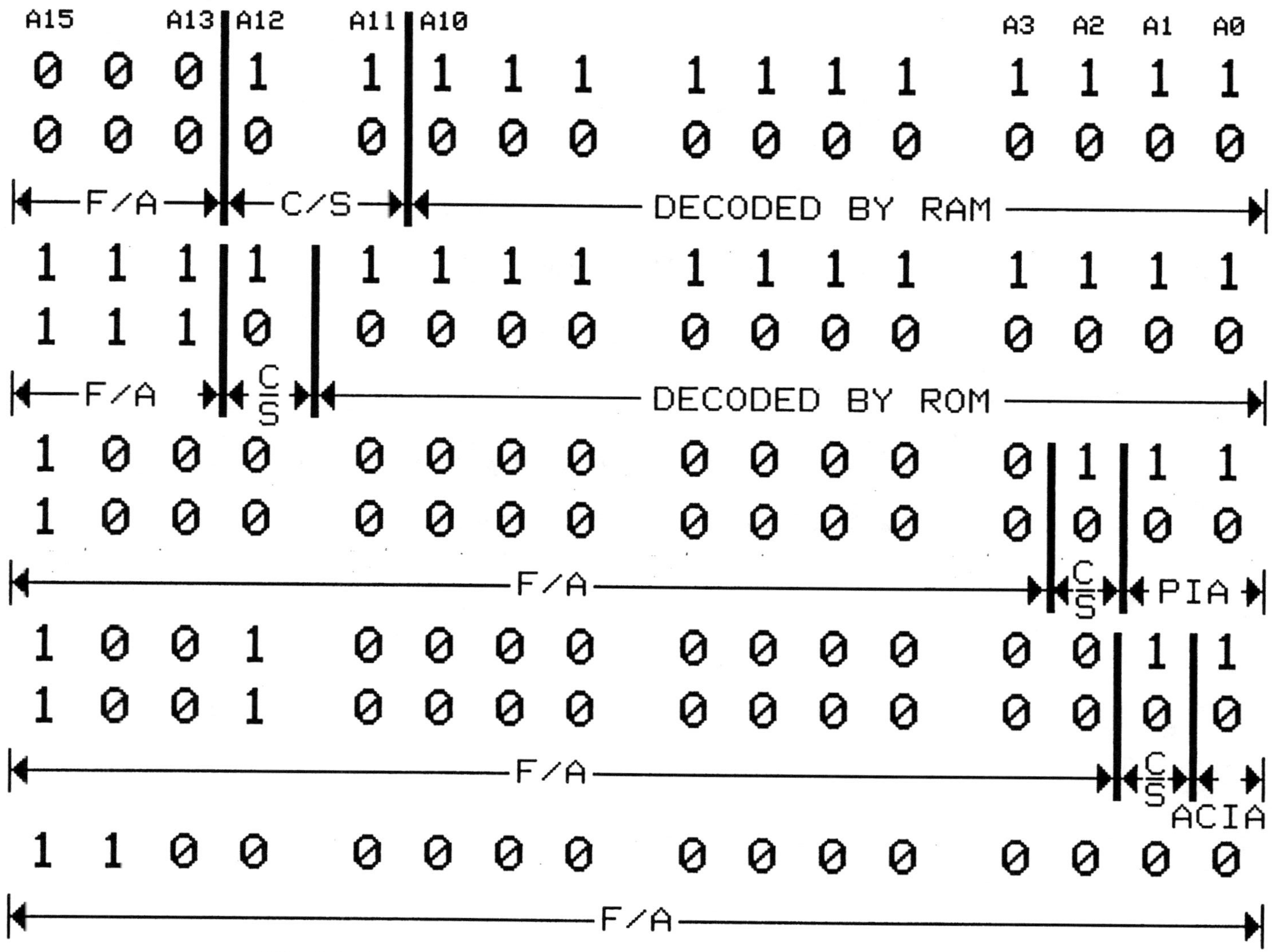

FIGURE 10.9 Summary of binary memory maps.

$$RAM\ Enable = \overline{A_{15}} \bullet \overline{A_{14}} \bullet \overline{A_{13}}$$

$$ROM\ Enable = A_{15} \bullet A_{14} \bullet A_{13}$$

$$PIA\ Enable = A_{15} \bullet \overline{A_{14}} \bullet \overline{A_{13}} \bullet \overline{A_{12}} \bullet \overline{A_{11}} \bullet \overline{A_{10}} \bullet \overline{A_{9}} \bullet \overline{A_{8}} \bullet \overline{A_{7}} \bullet \overline{A_{6}} \bullet \overline{A_{5}} \bullet \overline{A_{4}} \bullet \overline{A_{3}}$$

$$ACIA\ Enable = A_{15} \bullet \overline{A_{14}} \bullet \overline{A_{13}} \bullet A_{12} \bullet \overline{A_{11}} \bullet \overline{A_{10}} \bullet \overline{A_{9}} \bullet \overline{A_{8}} \bullet \overline{A_{7}} \bullet \overline{A_{6}} \bullet \overline{A_{5}} \bullet \overline{A_{4}} \bullet \overline{A_{3}} \bullet \overline{A_{2}}$$

$$Buffer\ Enable = A_{15} \bullet A_{14} \bullet \overline{A_{13}} \bullet \overline{A_{12}} \bullet \overline{A_{11}} \bullet \overline{A_{10}} \bullet \overline{A_{9}} \bullet \overline{A_{8}} \bullet \overline{A_{7}} \bullet \overline{A_{6}} \bullet \overline{A_{5}} \bullet \overline{A_{4}} \bullet \overline{A_{3}} \bullet \overline{A_{2}} \bullet \overline{A_{1}} \bullet \overline{A_{0}}$$

EXPRESSION SET 10.4.

DEVICE	A15	A14	A13	A12	A11	A10	A9	A8	A7	A6	A5	A4	A3	A2	A1	A0	MINIMUM LOGIC
RAM	(0)	0	0	0/1	0/1	0/1	0/1	0/1	0/1	0/1	0/1	0/1	0/1	0/1	0/1	0/1	$\overline{A15}$
ROM	1	1	(1)	0/1	0/1	0/1	0/1	0/1	0/1	0/1	0/1	0/1	0/1	0/1	0/1	0/1	$A13$
PIA	(1)	(0)	0	(0)	0	0	0	0	0	0	0	0	0	0/1	0/1	0/1	$A15\ \overline{A14}\ \overline{A12}$
ACIA	(1)	(0)	0	(1)	0	0	0	0	0	0	0	0	0	0	0/1	0/1	$A15\ \overline{A14}\ A12$
Buffer	1	(1)	(0)	0	0	0	0	0	0	0	0	0	0	0	0	0	$A14\ \overline{A13}$

FIGURE 10.10 *Free-allocation table.*

RAM	CHIP SELECT FIELD A12	A11	CS_0	CS_1	CS_2
0	0	0	$\overline{A11}$	$\overline{A10}$	$\overline{A15}$
1	0	1	$\overline{A11}$	A10	$\overline{A15}$
2	1	0	A11	$\overline{A10}$	$\overline{A15}$
3	1	1	A11	A10	$\overline{A15}$

① MAP ADDRESS LINES TO ASSERT CS LINES LEAVING ONE CS LINE OPEN

② CONNECT REMAINING CS LINE OF EACH DEVICE TO THE F/A LOGIC

FIGURE 10.11 Chip-select table for the RAM block.

ROM	A12	OE_0	$\overline{OE_1}$
0	0	A13	A12
1	1	A13	$\overline{A12}$

FIGURE 10.12 Chip-select table of the ROM block.

PIA	A2	CS_0	$\overline{CS_1}$	$\overline{CS_2}$
0	0	$\overline{A2}$	$\overline{A15 \bullet \overline{A14}}$	A12
1	1	A2	$\overline{A15 \bullet \overline{A14}}$	A12

FIGURE 10.13 ***Chip-select table for the PIA block.***

ACIA	A1	CS_0	$\overline{CS_1}$	$\overline{CS_2}$
0	0	$\overline{A1}$	$\overline{A15 \bullet \overline{A14}}$	$\overline{A12}$
1	1	A1	$\overline{A15 \bullet \overline{A14}}$	$\overline{A12}$

FIGURE 10.14 ***Chip select table of the ACIA block.***

BUFFER	G_0	$\overline{G_1}$
0	A14	A13

FIGURE 10.15 ***Chip-select table for the BUFFER block.***

```
/***************************************************************/
/* CUPL program to implement address bus decoding logic for */
/*                                                           */
/*    8K ROM    @ 'h'E000      ROM:   2Kx8, CS0,CS1,CS2      */
/*     2 PIAs   @ 'h'8000      PIA:    4x8, CS0,!CS1,!CS2    */
/*     2 ACIAs  @ 'h'9000      ACIA:   2x8, CS0,!CS1,!CS2    */
/*     1 Buffer @ 'h'C000      Buffer: 1x8, G0,!G1           */
/*    8K RAM    @ 'h'0000      RAM:   1Kx8, OE1,!OE2         */
/***************************************************************/

Name        busdcod1;
Revision    01;
Date        03/11/95;
Designer    J. Carter;
Company     University of North Carolina, Charlotte;
Device      P16L8;
/***************************************************************/
/*  ADDRESS BUS DECODE LOGIC                                  */
/***************************************************************/
/* Inputs from the Address Bus */
Pin  1 =  A1;
Pin  2 =  A2;
Pin  3 = A10;
Pin  4 = A11;
Pin  5 = A12;
Pin  6 = A14;
Pin  7 = A15;

/* Outputs to respective chip-select pins of memory & I/O */
Pin 12 =  !A1;             /* ACIA0:CS0             */
Pin 13 =  !A2;             /* PIA0:CS0              */
Pin 14 = !A10;             /* RAM0:CS1 RAM2:CS1 */
Pin 15 = !A11;             /* RAM0:CS0 RAM1:CS0 */
Pin 16 = !A12;             /* ROM1:OE1              */
Pin 17 = !(A15 & !A14); /* PIA0:CS1 PIA1:CS1 ACIA0:CS1 ACIA1:CS1 */
Pin 18 = !A15;             /* RAM0:CS0 RAM1:CS0 RAM2:CS0  RAM3:CS0 */
```

Device	Architecture	Chip-Select Labels
2K RAM	2048 x 8 bits	CS_0, CS_1, CS_2
4K ROM	4096 x 8 bits	OE_1, $!OE_2$
M6820 PIA (Parallel I/O)	4 x 8 bits	CS_0, $!CS_1$, $!CS_2$
M6850 ACIA (Serial I/O)	2 x 8 bits	CS_0, $!CS_1$, $!CS_2$
High-Speed RAM	256 x 8 bits	$!CS_0$, $!CS_1$

TABLE 10.2 Example devices.

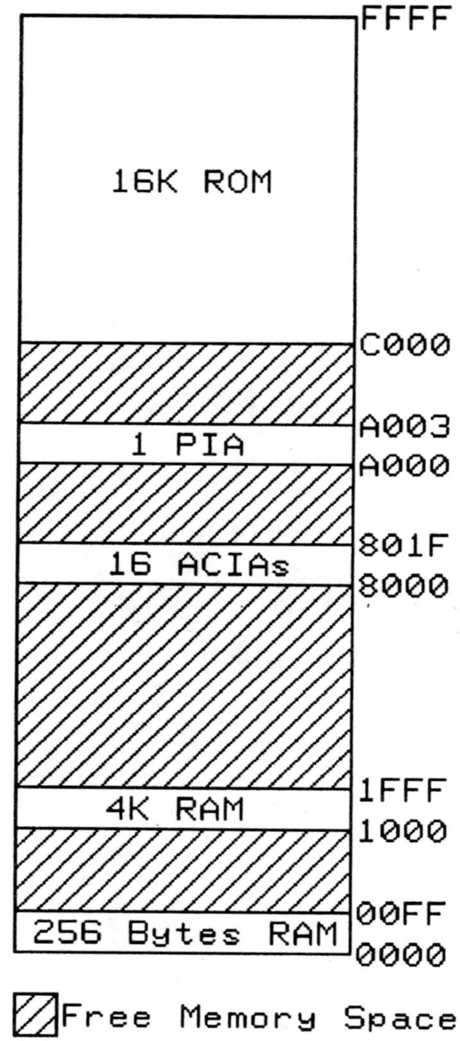

FIGURE 10.17 Practical Example 10.1, general memory map.

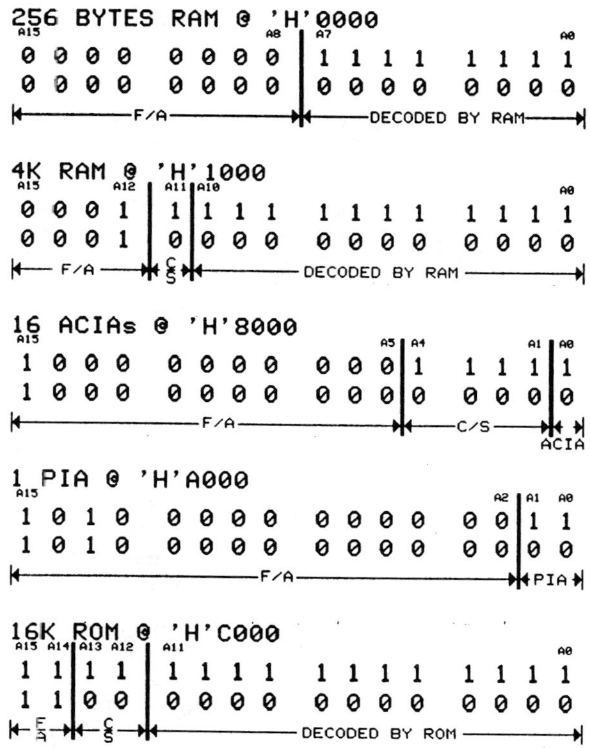

FIGURE 10.18 Practical Example 10.1, binary memory maps.

DEVICE	A15	A14	A13	A12	A11	A10	A9	A8	A7	A6	A5	A4	A3	A2	A1	A0	MINIMUM LOGIC
RAM0	(0)	0	0	(0)	0	0	0	0	0/1	0/1	0/1	0/1	0/1	0/1	0/1	0/1	$\overline{A15}$ $\overline{A12}$
RAM1	(0)	0	0	(1)	0/1	0/1	0/1	0/1	0/1	0/1	0/1	0/1	0/1	0/1	0/1	0/1	$\overline{A15}$ A12
ACIA	(1)	(0)	(0)	0	0	0	0	0	0	0	0	0/1	0/1	0/1	0/1	0/1	A15 $\overline{A14}$ $\overline{A13}$
PIA	1	(0)	(1)	0	0	0	0	0	0	0	0	0	0	0	0/1	0/1	$\overline{A14}$ A13
ROM	1	(1)	0/1	0/1	0/1	0/1	0/1	0/1	0/1	0/1	0/1	0/1	0/1	0/1	0/1	0/1	A14

FIGURE 10.19 Practical Example 10.1, free-allocation table.

ROM	C/S A13	C/S A12	OE_1	$\overline{OE_2}$
0	0	0	$\overline{A13}$	Q0
1	0	1	$\overline{A13}$	Q1
2	1	0	A13	Q0
3	1	1	A13	Q1

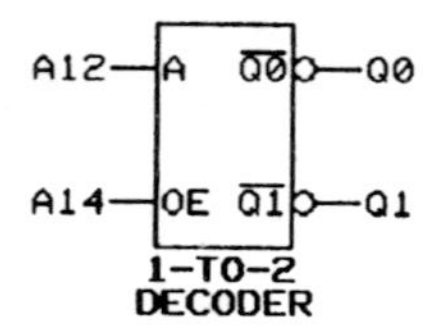

FIGURE 10.20 Practical Example 10.1, ROM chip-select table.

ACIA	CHIP SELECT A4	A3	A2	A1	CS_0	$\overline{CS_1}$	$\overline{CS_2}$
0	0	0	0	0	$\overline{A4}$	$\overline{A3}$	Q0
1	0	0	0	1	$\overline{A4}$	$\overline{A3}$	Q1
2	0	0	1	0	$\overline{A4}$	$\overline{A3}$	Q2
3	0	0	1	1	$\overline{A4}$	$\overline{A3}$	Q3
4	0	1	0	0	$\overline{A4}$	A3	Q0
5	0	1	0	1	$\overline{A4}$	A3	Q1
6	0	1	1	0	$\overline{A4}$	A3	Q2
7	0	1	1	1	$\overline{A4}$	A3	Q3
8	1	0	0	0	A4	$\overline{A3}$	Q0
9	1	0	0	1	A4	$\overline{A3}$	Q1
10	1	0	1	0	A4	$\overline{A3}$	Q2
11	1	0	1	1	A4	$\overline{A3}$	Q3
12	1	1	0	0	A4	A3	Q0
13	1	1	0	1	A4	A3	Q1
14	1	1	1	0	A4	A3	Q2
15	1	1	1	1	A4	A3	Q3

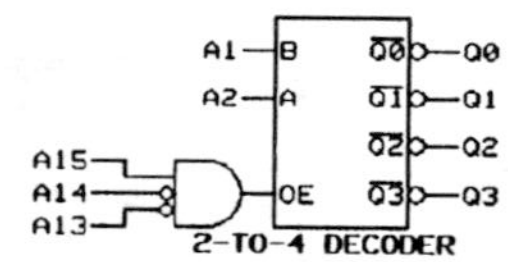

FIGURE 10.21 Practical Example 10.1, ACIA chip-select table.

256B RAM @'H'0000

RAM0	$\overline{CS_0}$	$\overline{CS_1}$
0	A15	A12

4K RAM @'H'1000

RAM1	C/S A11	CS_0	CS_1	CS_2
0	0	$\overline{A11}$	A12	$\overline{A15}$
1	1	A11	A12	$\overline{A15}$

PIA @ A000

PIA	CS_0	$\overline{CS_1}$	$\overline{CS_2}$
0	A13	A14	0

FIGURE 10.22 Practical Example 10.1, RAM and PIA chip-select tables.

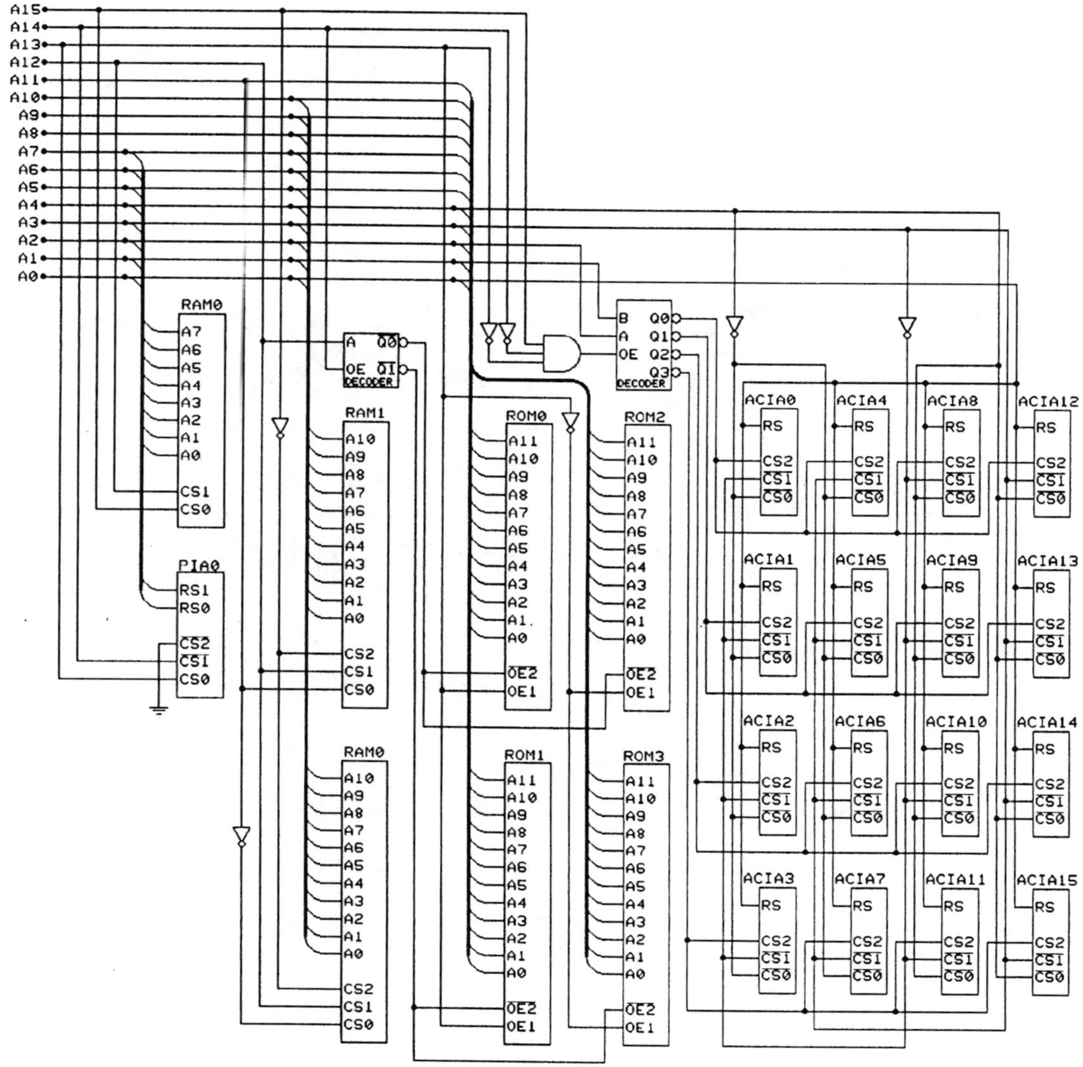

FIGURE 10.23 Practical Example 10.1, schematic diagram.

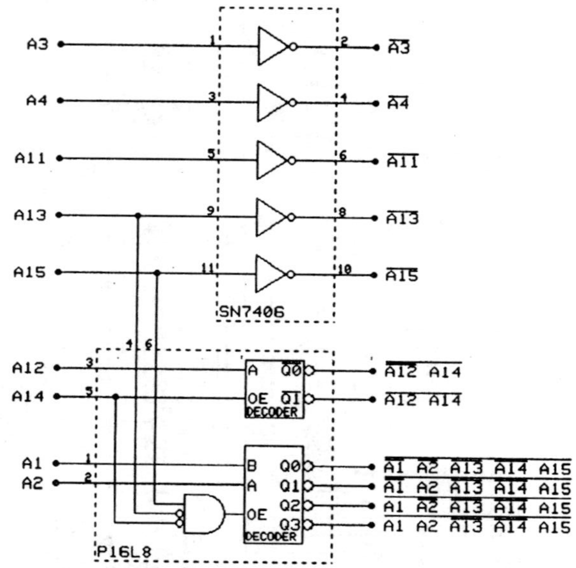

FIGURE 10.24 Practical Example 10.1, Decoding Circuitry

```
/*************************************************************/
/* CUPL program to implement address bus decoding logic for */
/*                                                          */
/*    256B RAM  @ 'h'0000      RAM:  256x8, !CS0,!CS1       */
/*    4K RAM    @ 'h'1000      RAM:   1Kx8, OE1,!OE2        */
/*    16 ACIAs  @ 'h'8000      ACIA:   2x8, CS0,!CS1,!CS2   */
/*     1 PIA    @ 'h'A000      PIA:    4x8, CS0,!CS1,!CS2   */
/*   16K ROM    @ 'h'C000      ROM:   2Kx8, CS0,CS1,CS2     */
/*************************************************************/

Name     prex-7-1;
Revision 01;
Date     03/13/95;
Designer J. Carter;
Company  University of North Carolina, Charlotte;
Device   P16L8;

/*************************************************************/
/*  ADDRESS BUS DECODE LOGIC                                 */
/*************************************************************/

/* Inputs from the Address Bus */

Pin  1 =  A1;
Pin  2 =  A2;
Pin  3 = A12;
Pin  4 = A13;
Pin  5 = A14;
Pin  6 = A15;
```

```
/* Outputs to respective chip-select pins of memory & I/O  */

Pin 12 = !A12&A14;                  /* ROM0:OE2 ROM2:OE2  */
Pin 13 =  A12&A14;                  /* ROM1:OE2 ROM3:OE2  */

Pin 14 = !A1&!A2&!A13&!A14&A15; /* CS2 of ACIA0,4,8,12    */
Pin 15 = !A1& A2&!A13&!A14&A15; /* CS2 of ACIA1,5,9,13    */
Pin 16 =  A1&!A2&!A13&!A14&A15; /* CS2 of ACIA2,6,10,14   */
Pin 17 =  A1& A2&!A13&!A14&A15; /* CS2 of ACIA3,7,11,15   */

/* The remaining decoding logic is implemented with an SN7406
   TTL inverter utilizing the following pin assignments:
   (Note that this entire program segment is a comment.)   */

/* Inputs from the Address Bus

Pin 1 = A3;
Pin 3 = A4;
Pin 5 = A11;
Pin 9 = A13;
Pin 11 = A15;                                              */

/* Output to respective chip-select pins of memory & I/O

Pin 2 = !A3;          CS1 of ACIA 0,1,2,3,8,9,10,11
Pin 4 = !A4;          CS0 of ACIA 0,1,2,3,4,5,6,7
Pin 6 = !A11;         CS0 of RAM0
Pin 8 = !A13;         CS0 of PIA0
Pin 10 = !A15;        CS2 of RAM0 and RAM1                 */
```

Device	Architecture	Chip-select labels
2K RAM	2048 x 8 bits	CS_0, CS_1, CS_2
4K RAM	4096 x 8 bits	!OE_1
High-speed RAM	256 x 8 bits	!CS_0, !CS_1
4K ROM	4096 x 8 bits	OE_1, !OE_2
16K ROM	16384 x 8 bits	!OE
M6820 PIA (Parallel I/O)	4 x 8 bits	CS_0, !CS_1, !CS_2
M6850 ACIA (Serial I/O)	2 x 8 bits	CS_0, !CS_1, !CS_2
74LS165 Parallel-load eight-bit shift register	1 x 8 bits	!LD
74LS240 Octal buffer	1 x 8 bits	! G_1, ! G_2

TABLE 10.3 Example devices.